ARTHUR W. JONES

University of Tennessee

Illustrations by
ALLAN D. JONES

INTRODUCTION TO PARASITOLOGY

 ADDISON-WESLEY PUBLISHING COMPANY

READING, MASSACHUSETTS · PALO ALTO · LONDON · DON MILLS, ONTARIO

This book is in the
Addison-Wesley Series in Life Science

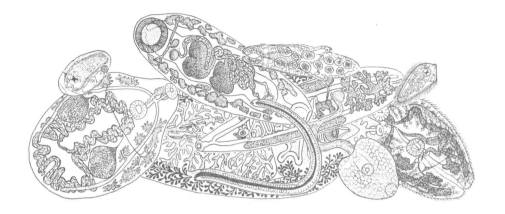

PREFACE

The importance of parasitology as a field of study cannot be acceptably estimated. Although, as Stoll pointed out in 1947, more than one-third of the people on earth probably harbor the large roundworm, *Ascaris*, and there are regions of the world where practically everyone has malaria, or hookworm, or filariasis, such statistics are almost meaningless. Numbers are not enough. *Ascaris* causes symptoms in only the more heavily parasitized victims, and while the nutritional erosion resulting from mild ascarid infections may, in a statistical sense, be significant, this extremely widespread condition is usually not thought of as "disease." Other parasites, much less common, may cause severe disorders affecting the circulation, the nervous system, or the reticulo-endothelial defenses of the body. "Mankind's most important disease," malaria, is on the decline, while schistosomiasis, a severe, disabling disorder, is increasing in incidence as it follows irrigation projects in underdeveloped countries. Urbanization and civilization favor the spread of the common pinworm, a medically unimportant parasite but an irritating challenge to epidemiologists. From the above brief mention of a few well-known human diseases, the importance of specific parasitisms is thus seen to be variable.

The importance of parasitism in general must be considered in its relationship to other problems which engage, or should engage, the energies and minds of man.

According to de Castro, three-fourths of the human population in 1958 suffered from malnutrition. According to Huxley (1959), population itself steals space, food, shelter, and clothing from hundreds of millions of men, women, and children, and by its unprecedented growth threatens to reduce the whole world to the condition of a miserably crowded slum. Although by treaty most nations have agreed to stop testing nuclear weapons in the atmosphere, still they develop "defensive" weapons, while their military scientists secretly plan for the chemical and biological devastation of the world. How important, among the problems of our time, are the parasites?

Parasitology is important enough to enlist the energies of thousands of hard-working biologists. Public health officers, especially in the newly independent nations of the tropics, realize that the prosperity and health of the people in their charge are seriously affected by parasite-caused diseases. Agricultural experts, whom we must depend upon to help increase farm production in the present desperate effort to feed the expanding populations, realize that parasitic worms and protozoa significantly reduce the productivity of man's agricultural animals. International agencies, such as WHO, have assigned a major part of their money and personnel to the eradication of malaria and to the study of such other parasitological diseases as trypanosomiasis, schistosomiasis, and hookworm. As Wright prophesied in 1951, the role of parasitology and parasitologists in today's world has become an important one.

It occurs to me, and no doubt to many of us, that we shall be both courageous and wise if we work for the health and happiness of mankind in the ways that are practical, even while the darkness thickens and immense dangers tempt us to despair. The desperate plight of humanity will not be relieved all at once, by some miracle. Perhaps the steady efforts of science and education will prevail against even that worst of enemies, man's inhumanity. These efforts, at any rate, will continue to yield measurable benefits in many fields, including the field of parasitology. I hope that this book will serve in its way to advance human welfare by assisting others to teach and learn about parasites and their effects on man.

A new textbook in parasitology should be neither written in haste nor accepted without caution. The existence of excellent contemporary texts, some of them tested by years of use and decades of revision, must make any author pause before presuming to add his book to the list and should cause any teacher of parasitology to deliberate carefully before choosing a new text for his course. Indeed, I waited to write this book until more than twenty years of teaching and research had given me confidence that I could contribute a text of value.

This textbook is intended for a short course in the junior or senior year of the biology curriculum. Such a course is widely offered. It appeals, or should appeal, to the premedical or preveterinary student, to the zoology or biology major or minor, and to the liberal arts student in search of upper division electives. It serves as a mere introduction to parasitology, however, which is a very large and important field of biology and cannot be presented in depth at the undergraduate level. One semester or two quarters can provide approximately enough time to introduce the principles of parasitology by examples and by laboratory work. The nonparasitologist can profit from such a course by learning some very interesting facts about human welfare on an international scale, by studying the versatility and evolutionary ingenuity of the animals which have become adapted to a parasitic way of life, and by observing the ecological and physiological interlocks by which the parasite-host relationship develops and is maintained. The future parasitologist can profit from such a course by seeing his professional field in the form of a survey which introduces him to the challenges of an important, rewarding, and rapidly expanding area of biology.

The teacher of an introductory course in parasitology should be able to use this book conveniently. It supplies basic data about some, but not all, of the parasites he may wish to discuss; thus he may rely on the book for the factual content of the course if he wishes, although his imagination and interest as a teacher may cause him to add material from his experience or the current literature. It indicates a number of relationships between parasitology and other subjects—geography, economics, history, sociology, anthropology, to name a few. Thus the teacher may, if so inclined, digress creatively when spurred on, perhaps, by questions from his students based on their reading of the text. It ventures to discuss some frontiers of research where parasitology interacts with immunology, pathology, and, on the theoretical level, population genetics and evolution.

Also, the book is written and illustrated for a nonmotivated reader, the student. Perhaps it is an unjust criticism of today's student to call him nonmotivated in terms of a subject in intermediate zoology, but it is the author's experience that even the most ambitious science major is attracted and aided by good writing and teaching, by enthusiasm at the teaching level, and by aesthetic considerations. The student who must take a certain number of courses to fulfill a major or minor requirement enters such a course as parasitology with some misgivings and little expectation of enjoyment. A readable, well-illustrated text can help the teacher lift such a student from dead weight to collaborator in the teaching-learning process.

To carry out the three aims expressed above (to tailor a book to a short course, to provide a useful teaching instrument, and to interest the student), I have selected certain things above others for inclusion. Major health problems caused by parasites are included because of their importance and interest, and because, obviously, more is known about such subjects than about other parts of parasitology. For the same reasons, veterinary parasitology receives more attention than most writers give it. Economic, political, cultural, and historical aspects of parasitology are discussed in appropriate places; the prevalence and incidence of many parasites of man and his animals are greatly affected by economic levels, by government action, by hygiene and food, and, notably, by the past and present behavior of individuals and nations. The principles of parasitology are included as part of the discussions of specific problems and are then treated in special chapters near the end of the book. This arrangement has value because students learn by observing rather than by being told, and their interest in particular problems like malaria will help them acquire the facts from which rather difficult concepts like premunition and racial resistance can be derived.

Some critics may object to the omission of important material from this text. There is a large and growing literature on the parasites of wild animals. Also, the ecology of animal parasites is a very interesting and significant part of modern research, continuing a tradition of inquiry that began long ago with the earliest attempts to work out life cycles. The user of the text may wish to introduce such materials, but he will find that in a short course he will have to omit other, perhaps equally important, matters. I have decided to leave the more esoteric aspects of parasitological literature to graduate courses and seminars, as I do in my own program of graduate teaching.

The book is organized in a superficially orthodox way. The parasites are arranged according to their places in the animal kingdom, beginning with the Protozoa. Each major group is discussed in an introductory chapter which is followed by a series of chapters on specific diseases. These chapters, however, are not repetitions of some formula such as "organism, life cycle, pathogenesis, treatment, and control." Each chapter has one or more important principles to illustrate. Thus Chapter 2 shows how a complex and ancient disease, malaria, has yielded, at last, to a worldwide program of eradication based upon over 200 years of research and study by an international team of malariologists. The discussion of two relatively unimportant worm parasites, *Clonorchis* and *Paragonimus,* show how certain unusual cultural practices cause men to become infected with parasites that are common in other animals; the concept of zoonosis emphasized here is also found in other chapters, while the practices of eating raw fish and raw crabs, sources of infection with the two parasites, respectively, is related later in the book to behavioral factors in the transmission of such familiar conditions as trichinosis and pinworms. Schistosomiasis, caused by blood flukes, is presented as a pressing problem, with treatment and control still requiring much research. Here, too, connections with agriculture, population expansion, and hydroengineering projects are emphasized. After a number of chapters serving the double function of giving information about parasites and introducing valuable principles and concepts, the book ends with a section devoted to discussion of the principles themselves. Host-parasite relationships are considered as a set of definitions, special cases of the ecological interdependence of living things. Immunity as a form of host response is approached by way of a general summary of modern immunological theory, and the special features of immunity involving animal parasites—premunition, negative phase immunity, and hypersensitivity—are given attention. Pathogenesis is treated as attack by parasites, and various injuries to tissue and physiological health are listed. Examples from the first part of the text are used freely in these chapters, which were anticipated in many brief discussions of particular parasites. A chapter on the evolution of parasitism follows; collected in one place are various ideas on the mutability and adaptivity of parasites, the development of drug-resistance, the occurrence of geographical and host limited varieties, and the peculiar population structures which should make parasites unusually interesting to students of genetics and population dynamics. Finally, a chapter on world public health puts in perspective the knowledge conveyed by the whole text and suggests the direction which parasitological studies may take in the future. The organization, in short, while it makes the book appear on the surface to be a catalogue of facts, actually serves to introduce to the student basic ideas in parasitology.

The illustrations are carefully designed to suit the overall purposes just described In most cases the figures supply descriptive material not written out. Frequently ideas have been best expressed by drawings or graphs. The figures throughout are intended to have an impact of their own; they are designed for teaching. Because of the brevity of this book, the figures must substitute for many pages of description, and the user is warned that the illustrations are important sources of information.

The teacher who uses this book may, of course, emphasize what he desires, and he has at his command wonderful resources for expansion in the fine books of Baer, Belding, Chandler and Read, Cheng, Faust and Russell, Rogers, Smyth, Swellengrebel, and others.

The information now available about parasites is vast, full of details, and, not surprisingly, much more extensive in some fields than in others. The best known parasites are those of man and his domestic animals. But even knowledge of these parasites has startling gaps, which causes serious students to wonder. Relatively little is known about the physiology of parasites. The host's reactions, immunological and other, are being studied now for the first time by a sizable number of workers. The effect of drugs is only beginning to be examined in more than an empirical way, although understanding the specific pathways by which a parasiticidal drug interferes with a parasite's metabolism seems essential to the effective development of chemical therapeutics. Epidemiological and ecological factors have long been widely recognized as important, through much research on life cycles, vectors, and the population levels of parasites. But broader ecological aspects—such as the economic cost of various parasitisms, the practical balance that must be drawn between the harm caused by a parasite and the expense or trouble required to eradicate it, and the probable effects of education, urbanization, industrialization, migration, and population growth upon the extent and importance of various parasitisms— have been little studied and are little understood. A textbook writer is tempted to avoid these difficulties by loading his book with data, thus concealing the gaps in knowledge which are the most stimulating part of his subject. But many of the parasitological data, in my opinion, are redundant. I have tried in this book to use relatively few examples to illustrate only major problems and principles, and I have not avoided mention of the ignorance which parasitologists bravely share with other scientists.

I should like to acknowledge my dependence on the books mentioned above as well as on various individuals. I have used several of the above books as texts in courses in general or advanced parasitology, and have consulted them freely for data and examples to use in lectures and to include here. Their authors have my gratitude. But two other groups of people have my especial thanks—the working parasitologists who built up the vast body of knowledge on which public health and animal disease specialists can act, and from which textbooks are derived, and a second group, the students, my own and others', who through youthful curiosity and the candor of their uncluttered minds keep all fields of study alive and growing, and who become the scientists and citizens of tomorrow. My personal thanks are due to the editorial and production staffs of Addison-Wesley Publishing Company for their expert guidance; to the special consultants who helped remove errors from my work, but who in no way are to blame for any errors that may persist; to a small group of graduate students who respectfully but with considerable enthusiasm criticized in depth most of the chapters in early versions; to the artist, my brother, who succeeded so well in perceiving and portraying the aesthetic qualities of a rather unaesthetic subject; and to my wife, who was very patient and understanding.

Knoxville, Tennessee   A.W.J.
November 1966

CONTENTS

ix

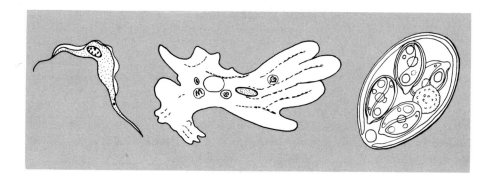

CHAPTER 1

PROTOZOA

Taxonomy, the internationally accepted system of arranging and naming the animals and plants known to science, is much more than the complex and forbidding science as it appears to the novice. The Latinized Greek names applied to such familiar groups as horses or oak trees seem an insult to common sense and common speech, and the published discussions of ways of arranging organisms in categories variously designated by terms such as phylum, class, order, family, etc., appear (improperly, of course) to be directly descended from the more futile kinds of medieval scholastic disputation. Nevertheless, the serious student of biology would be lost without taxonomy. He would be lost in a jungle of "common names" quite unintelligible to foreigners like himself, that is, to any but the native speakers of the language to which the name is "common." He would also be lost in a thicket of unsorted types, unable to predict the habit or guess the morphology of any unfamiliar animal or plant on the basis of its obvious relationship to known forms. In other words, without a taxonomic system universal in tongue and regular in arrangement, vast confusion would reign in place of the order under which biologists today study and discuss the complex world of living things.

Therefore it is proper to start each descriptive part of this book with a brief discussion of the taxonomy of the animal parasites belonging to that section. Before discussing the Protozoa, however, we shall attempt to clarify some general facts about naming and arranging organisms.

In 1758, Linnaeus published a list of animals which he named and organized in a special way. Each animal was given two names: a general name, which indicated that the animal belonged to a larger group, and a specific name, which, in Linnaeus' opinion, applied only to animals so like each other that their group could not be further subdivided. These two parts of the name thus designated a species; the scientific name of a species was defined as a binomial, and the system was based on such a binomial nomenclature. While it was to be expected that two hundred years of study and discovery would reveal flaws in such a simple method of naming, the basic idea promoted by Linnaeus has been retained. His use of Latin and Greek (dead languages, hence unchanging, and recognized in all nations of the western world as languages of classical scholarship) has made the binomial international in meaning; the Linnean names form a substantial base for the second aspect of taxonomy, that is, arrangement or classification.

Classification is a part of everyday thinking and expression. It consists simply of putting similar things together in groups. Groups, in turn, may be put together. Scientific classification uses carefully selected categories to arrange the species defined first by Linnaeus and added to in vast numbers by specialists ever since. This organization of details into categories permits shortcuts in description, prevents the loss of information through faulty "filing," and makes one intelligible whole out of the great mass of biological information now in existence. Incidentally, the taxonomic system often mirrors the underlying regularity which Darwin and the evolutionists discern in nature, i.e., the fact that organisms are usually most like each other when they are most closely related by descent.

We should remind ourselves that the systems of names and categories which we call classification are not identical with the actual animals and plants thus organized and represented. As semanticists point out, the reality beneath the structure of words is objects. The objects of taxonomy are such things as populations of animals, arborium specimens of plants, mammal collections, fossils, observed behavior such as migrations or fluctuations in populations, and even such fundamentals as the distribution and transmission of genetic codes. The system of classification is continually molded and modified by the discoveries of men working with actual organisms. Thus, while taxonomy overcomes the language barriers between peoples, and supports by its orderly structure the great body of biological research, taxonomy itself depends on growing and changing data for its validity and meaning. Such classification is a useful, but abstract and approximate, description of nature.

An example of a taxonomic system, the Protozoa, will introduce some important parasites of man and his domestic animals.

PHYLUM PROTOZOA

The Protozoa may be considered a major division of the animal kingdom, that is, a subkingdom along with the subkingdom Metazoa. However, since they are readily distinguishable from all other groups of animals, the Protozoa are usually put on the same level as the other distinct groups, which are called phyla (*s.* phylum). The phylum Protozoa consists of the single-celled animals, so-called by analogy to the well-known many-celled animals, but is perhaps just as properly called noncellular or acellular because their organization is based on the interaction of subcellular units rather than of tissues or organs composed of many cells. Early microscopists were correct in their naive recognition of liver, heart, and stomach in the tiny bodies of infusorians; the subcellular structures which they saw were in a sense analogous to the organs of metazoa, and the bodies of many kinds of protozoa are complex to a high degree, although they are without cells or tissues. The phylum Protozoa, like other phyla, is divided into classes. In this case, there are four such classes, three of which are important parasitologically.

CLASS MASTIGOPHORA

The class Mastigophora consists of flagellates. (See Fig. 1–1.) All of these are equipped, at some stage in their life cycle, with whiplike locomotor structures, the number and arrangement of which are used in classifying these organisms. Many of the flagellates are classified as plants; indeed the class itself is in a sense a borderline between the single-celled plants and the acellular animals. Like many other classes, the class Mastigophora is divided into subclasses. The subclass Phytomastigina contains the plantlike flagellates. These usually have pigmented bodies called chromoplasts or chromatophores in which photosynthesis occurs. Like other flagellates, they move by one or more flagella. Some of the Phytomastigina form colonies and thus suggest an evolutionary trend toward many-celled organisms. Botanists consider the Phytomastigina to be algae, within the plant kingdom. Here we see an example of a borderline which the excellent system of classification has not succeeded in either sharpening or erasing; this fact reemphasizes the artificial nature of taxonomy but should not be used to discredit taxonomists or to discourage further attempts to improve the system. The subclass Zoomastigina contains the animallike flagellates. These animals, which are similar by the fact that they all lack chromatophores, are greatly diversified in structure, life cycle, and habitat. The subclass is divided, like other classes or subclasses, into several orders, major groups of basically similar organisms.

Order 1, Rhizomastigida, is a group of flagellates which, while having flagella, move as well by amoeboid pseudopodial action. Order 2, Protomastigida, contains simple flagellates with one or two flagella and with plastic but not amoeboid bodies. It includes the extremely important family Trypanosomidae, with its disease-causing parasites in the genera *Trypanosoma*, *Schizotrypanum*, and *Leishmania*. Order 3, Polymastigida, includes uninucleate and binucleate species as well as some with many nuclei. It is quite common to find from five to eight flagella; all must have at least three. This order includes the interesting family Hexamitidae, to which belongs the genus *Giardia*, parasitic in man and other mammals. Order 4, Trichomonadida, includes either uninucleate or multinucleate (but not binucleate) flagellates with an internal rodlike axostyle and an internally interconnected flagellar complex, bearing 3 to 6 flagella. A prominent parabasal body (internal organ lying

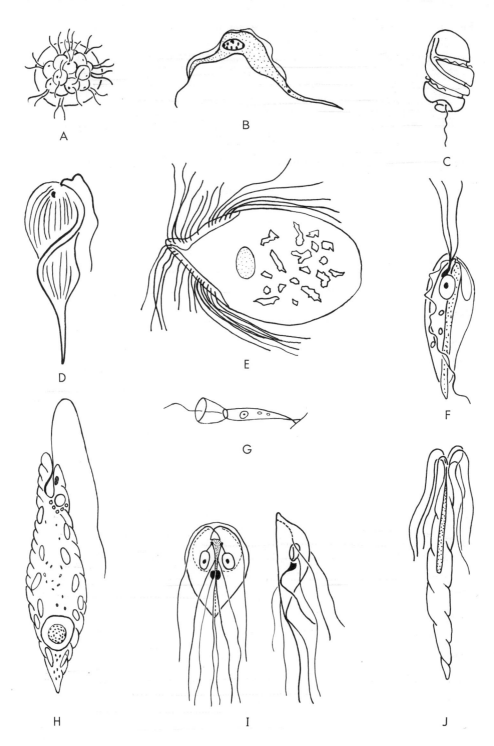

FIG. 1–1. Representative members of the class Mastigophora. [After Hall, 1953.] (A) *Pandorina* (order Phytomonadida, family Volvocidae), ×180. (B) *Trypanosoma* (order Protomastigida, family Trypanosomidae), ×1700. (C) A dinoflagellate (order Dinoflagellida), ×370. (D) *Phacus* (order Euglenida), ×420. (E) *Trichonympha* (order Hypermastigida), ×335. (F) *Tritrichomonas* (order Trichomonadida), ×2100. (G) A choanoflagellate (order Protomastigida, family Codosigidae), ×370. (H) *Euglena* (order Euglenida), ×370. (I) *Giardia lamblia* (order Polymastigida), ×1780. (J) *Streblomastix* (order Polymastigida), ×980. (A), (D), and (H), like many flagellates of the Phytomonadina and related groups, have chloroplasts.

beside the basal granules of the flagella) is present. The family Trichomonadidae contains several parasites of man and domestic animals, including the venereal pathogen, *Trichomonas vaginalis*. Order 5, Hypermastigida, consists of uninucleate flagellates with many flagella. These are intestinal parasites of wood-eating insects (termites and wood-roaches) as well as of the ancient and omnivorous cockroach. Order 6, Opalinida, is still considered by many as a subclass of the Ciliata. It includes multinucleate forms covered by rows of short flagella (or long cilia); these are intestinal parasites of Amphibia. All of the above orders of flagellates, as the examples show, are divided into families, which in turn contain genera. Within each genus are groups of species, the real base upon which the abstract higher categories have been erected.

CLASS SARCODINA

The class Sarcodina includes the complex planktonic protozoa belonging to the orders Foraminiferida and Radiolarida, the free-living and parasitic naked amoebas of the order Amoebida, and the amoebas with shells or tests of the order Testacida. Two other small groups in the class are the Helioflagellida, amoebas with flagella (cf. Rhizomastigida above), and the Heliozoida, which are spherically symmetrical and planktonic in fresh water and have pseudopods supported by axial rods radiating from skeletal elements in the outer layer of cytoplasm. Our concern in this book is with the relatively few members of the Sarcodina which are medically significant. (See Fig. 1–2.)

In the family Endamoebidae of order Amoebida is found the genus *Entamoeba* (Casagrande and Barbagallo, 1895), a genus which contains three of the amoebas parasitic in man. The name of this genus has been the subject of a controversy which illustrates some features of taxonomy not brought out elsewhere in this book. The name *Endamoeba* was used by an American, Leidy, in 1879, for a species of amoeba found by Bütschli, a German, in the cockroach. Later the genus *Entamoeba* was proposed by the Italians Casagrande and Barbagallo, in 1895, for the two

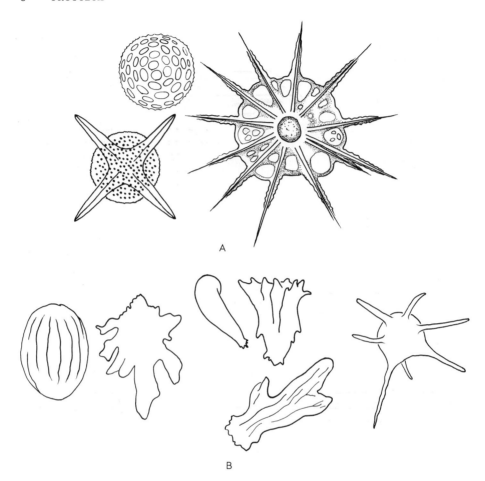

FIG. 1–2. Representative members of the class Sarcodina. [After Hall, 1953.] (A) Several members of the subclass Actinopoda, including two fossil radiolarians (left) and a living *Actinophrys sol*, ×260. (B) Subclass Rhizopoda, order Amoebida. Several free-living amoebas of characteristic forms. (C) *Amoeba proteus*, ×85. (D) Cyst of *Entamoeba histolytica*, ×2600. (E) Subclass Rhizopoda, order Foraminifera; (1) living foraminiferan, *Allogromia*, ×33, and (2) six fossils. (F) Order Testacida, *Difflugia pyriformis*, ×160.

species of amoeba, *histolytica* and *coli*, found in the human intestine. Since Leidy's name *Endamoeba* had priority, the International Commission on Zoological Nomenclature ruled, in 1928, that the name *Entamoeba* was a synonym of *Endamoeba* and should not be used. But in 1954, the Commission reversed its ruling on the grounds that the genus of amoebas found in cockroaches is different from that found in man and that therefore both names are valid. Hence it is proper to use *Endamoeba*

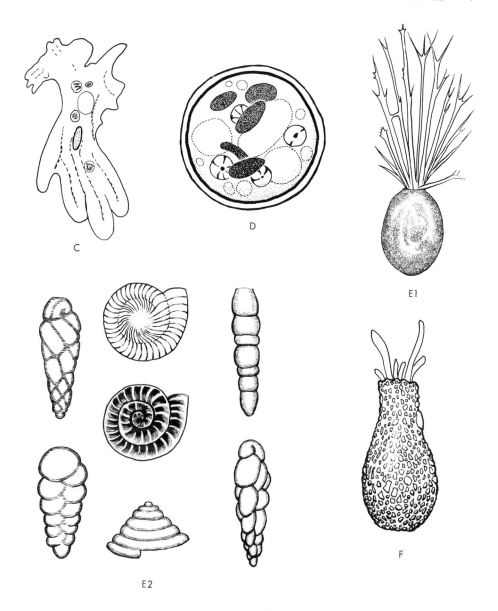

for the species *Endamoeba blattae* (Bütschli, 1878; Leidy, 1879) and to use *Entamoeba* for the species *E. histolytica* (Schaudinn, 1903), *E. coli* (Grassi, 1879; Casagrande and Barbagallo, 1895), and *E. gingivalis* (Gros, 1849; Brumpt, 1913), all parasitic in man. The reader can see that the naming of a species may be a slow process, involving scientists of many lands, and that disagreements over taxonomic matters require, at times, settlement by an international judiciary.

CLASS CILIATA

The class Ciliata contains those protozoa commonly called ciliates, relatively complex organisms moving by means of short, rodlike cilia. Such cilia beat in a coordinated manner for active swimming and are modified into various structures, such as membranes for feeding or leglike tufts for "walking" over a substrate. Ciliates have two kinds of nuclei: a macronucleus, which interacts with cytoplasmic systems in the control of metabolism and growth, and a micronucleus, which functions in sex and reproduction. (The analogy to somatic and germinal tissues in higher organisms is obvious.) As parasites, ciliates are relatively unimportant, but they hold great interest for biologists. Much has been learned from ciliates about sex, about cytoplasmic inheritance, and about nutrition, growth, and differentiation. They continue to be studied enthusiastically in many laboratories. The class is divided into two subclasses and fifteen orders, distinguishable from each other by patterns of cilia, body form, and various other features. The only ciliate parasite of man is *Balantidium coli*,* which belongs to the order Trichostomatida. *Paramecium*, the famous genus which has been the subject of genetic and other studies, belongs in the order Hymenostomatida. *Tetrahymena* is another genus of hymenostomates which has yielded scientific information; this has been in the field of nutrition. The order Entodiniomorpha contains the remarkable stomach ciliates of cattle, other ruminants, and horses; these ciliates supply significant amounts of protein in the diets of these animals. Finally, there is the order Suctorida, considered by some as a separate class of protozoa but probably related to the orders of ciliates characterized by uniform body ciliation. However, it is only in the buds or immature forms of Suctorida that this ciliary pattern is evident; mature Suctorida are sessile, living attached to other animals or objects and lacking cilia. They have tentacles with which they seize and devour other protozoa. A text on parasitology has space for only a brief discussion of ciliates, yet the importance of the ciliates is far greater than these few words might imply.

CLASS SPOROZOA

The Sporozoa are a class of protozoans which form spores, the disseminated phase of the life cycles of these parasites. (See Fig. 1–3.) All the Sporozoa are parasitic; they inhabit members of all other animal groups, including other protozoa. Probably the class is not a natural group; all of its members are highly adapted to parasitism, and they may have lost structures (flagella, pseudopods, etc.) which,

---

* As a ciliate parasitic in swine and occasionally man, this parasite lives in the colon of pigs, invading the mucosa and multiplying. It forms cysts, which, when ingested by another host, start a new infection. Balantidiasis is mild (a form of colitis) in swine, but may be severe in man. Its chief importance, therefore, is in its danger to man. Sanitary precautions should be observed around pigs, for the latter very commonly harbor *Balantidium* (up to 75% incidence).

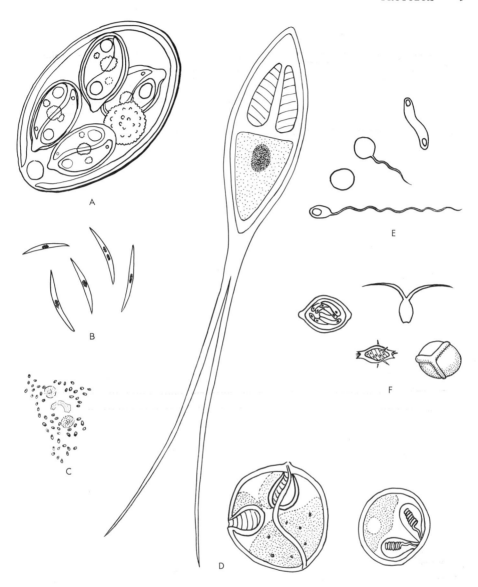

FIG. 1–3. Spores and sporozoites of several orders of Sporozoa. [After Hall, 1953.]
(A) Oocyst of *Eimeria* (order Coccidia) containing four sporocysts and a residual body.
Each sporocyst contains two sporozoites. (B) Sporozoites of *Plasmodium* (order
Haemosporidia) from a mosquito's salivary gland. (C) Sporozoites of *Theileria*
(order Haemosporidia) in the hemocoele of a tick. (D) Several of the complex spores
of members of the order Cnidosporidia. (E) Spores, two with extruded polar filaments,
of microsporidians. (F) Spores of several species of the order Gregarinida. All are
drawn to the same scale of magnification at about 1800×.

if present, would place certain sporozoans in the Mastigophora or Sarcodina. Some Sporozoa are perhaps related to the Fungi, a great group of primitive plants; these sporozoans would not be animals at all. Spores or sporozoites are produced by all sporozoans. Spores are tiny cells surrounded by tough membranes; they are usually capable of dispersion by wind or water and of long survival in adversity. Sporozoites are similar nucleated bits of protoplasm, but they do not have spore membranes; sporozoites require a host in which to survive and are usually injected by one host into another. The mature forms (as well as the spores) are highly varied in size, structure, habitat, and behavior. The class Sporozoa consists of three subclasses, the Telosporidea, the Cnidosporidea, and the Acnidosporidea. Thus, in the Telosporidea, simple spores are formed, or sporozoites may be produced without spore walls. In the Cnidosporidea each spore contains one or more polar capsules, within which are extrusible coiled polar filaments. Such spores may also be complex, being formed from several cells. The small subclass Acnidosporidea contains two groups of organisms, the spores of which differ from those of the Cnidosporidea in being without polar capsules; the life cycles of Acnidosporidea suggest no relationship to the Telosporidea. Some important telosporideans are the malaria organisms and the coccidians. The next chapter is devoted to malaria; and coccidiosis, an important disease of poultry and livestock, will be discussed in a succeeding chapter. Gregarines, also members of the Telosporidea, are common parasites of invertebrates; often these sporozoans are very large in their growing, or trophic, stages. Cnidosporideans cause severe diseases of bees and silkworms. The Acnidosporidea include *Sarcocystis,* which produces harmless cysts in the voluntary muscles of man yet kills sheep and other animals through a toxin.

The phylum Protozoa has been sketched above as a group of greatly varied animals, united for convenience by a taxonomic system; its complexity has been only intimated. A few important or interesting parasites were mentioned. The next five chapters concern the parasites which inflict upon man and his domestic animals severe and expensive disease.

SUGGESTED READINGS

CORLISS, J. O., *The Ciliated Protozoa.* Pergamon, New York, 1961.

HALL, R. P., *Protozoology.* Prentice-Hall, New York, 1953.

KUDO, R. R., *Protozoology,* 4th ed. Thomas, Springfield, Ill., 1954.

MANWELL, R. D., *Introduction to Protozoology.* St. Martin's, New York, 1961.

CHAPTER 2

MALARIA

Malaria is a disease which affects hundreds of millions of people directly, as it weakens, kills, and impoverishes the inhabitants of many of the nations of the earth. Malaria affects the remaining population of the earth indirectly, for the malaria-caused loss in labor and efficiency is a waste reflected in the cost of goods everywhere, and the misery of the malarious peoples is a threat, akin to the threat of contagious poverty and spreading revolution, to the relatively healthy and prosperous peoples of this small world. Yet the immense cost and the dangerous threat of malaria have aroused the leaders of men in many nations to join together in combating this disease of mankind; and the result of this united effort is the forseeable eradication of what may still be called, for a few more years at least, "man's most important disease."

Malaria, a disease of the blood, necessarily affects the total health of the body. Symptoms are chills, fever, weakness, and emaciation. The outcome may be recovery, death, or a period of variable duration after which relapse occurs. Different forms of malaria exist as a result of the many species of the parasite; people react in various ways to the same form of malaria, because of their general state of health, the treatment they can be given, or the hereditary resistance or susceptibility they may possess. A generalized description of the disease follows.

THE CYCLE

The parasites causing human malaria are all members of the class Sporozoa, order Haemosporidia, family Plasmodiidae, and genus *Plasmodium*, the latter so named because of the resemblance of its growing form to the shapeless amoeba, a protoplasmic mass or "plasmodium." Transmission is by anopheline mosquitoes. Development in the mosquito and human host follows a definite pattern in the four species of *Plasmodium* in man: *Plasmodium vivax*, *P. malariae*, *P. ovale*, and *P. falciparum*. Minor differences in form and behavior distinguish these species (Fig. 2–1).

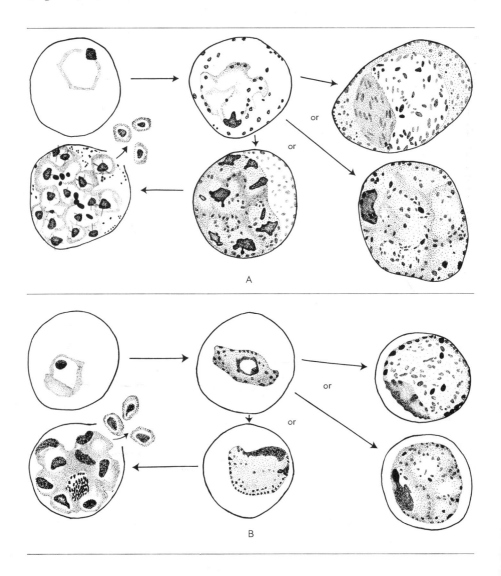

The life cycle (Fig. 2–2) of *Plasmodium vivax* may be used as an example. Sexual forms, called gametes, unite in the gut of the mosquito. The zygote, somewhat elongated or wormlike, passes through the intestinal wall of the host and comes to lie against the surface of the mosquito's stomach. There, surrounded by a double membrane, the zygote grows, its nucleus divides many times, and after 10–20 days (depending on temperature) it becomes a sporocyst containing several thousand sporozoites (threadlike bodies serving the same function—transmission to another host—as the spores of certain other protozoa or of fungi). When the sporocyst breaks open, sporozoites pass into the mosquito's blood and then enter

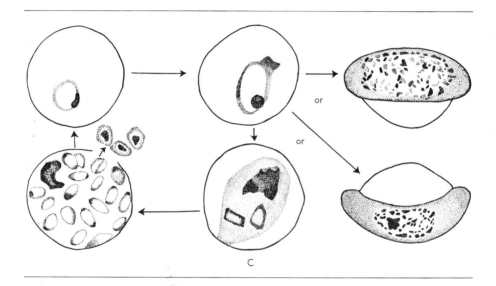

FIG. 2–1. The three common species of *Plasmodium*, as seen in blood smears. (A) *Plasmodium vivax*, showing at left the schizogonic cycle in red blood cells (ring form, trophozoite or early schizont, late schizont, and segmenter, with merozoites, 16 in number; these later escape from the ruptured host cell). At right are the two kinds of gametocytes, microgametocyte (upper) with rather diffuse nuclear material, and macrogametocyte (lower) with compact nuclear material. Schüffner's dots occur in the host cytoplasm, pigment granules in the cytoplasm of the parasite. (B) *Plasmodium malariae*, showing schizogony at left, with relatively dense, compact trophozoites and a low number (eight) of merozoites. Schüffner's dots are not present. At right are the micro- and macrogametocytes. (C) *Plasmodium falciparum*. The only forms seen in peripheral blood are the ring stages or youngest trophozoites (upper left) and the gametocytes (right). Schizogony occurs in cells trapped in capillaries of bone marrow, spleen, etc. [*Note: Plasmodium ovale*, the fourth human malaria (not illustrated), resembles *P. vivax* closely; the main difference is the distortion of infected erythrocytes caused by the former.]

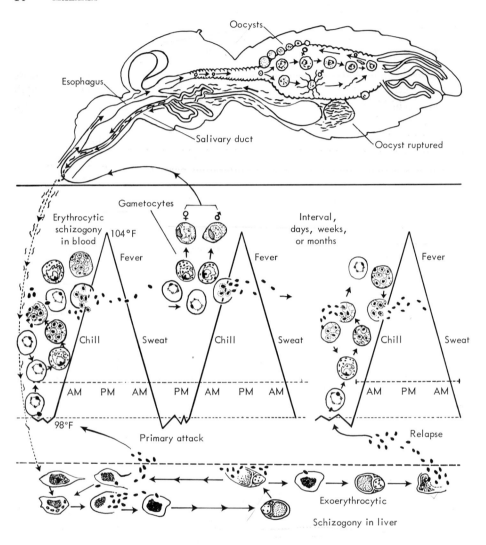

FIG. 2–2.   Life cycle of *Plasmodium vivax*. Gametocytes are taken with blood into the mosquito's stomach, where fertilization occurs. Oocysts form in the wall of the stomach and liberate sporozoites which invade the salivary gland. When injected into a human host, the sporozoites enter liver cells, where several cycles of schizogony occur, resulting after some days or weeks in a primary attack of malaria. Synchronized at 48-hour intervals, the attack is characterized by schizogonic cycles in the blood. The cycles include chills, fever, sweating, and eventual recovery. After an interval varying from days to months, relapse may occur because of the continued presence of exoerythrocytic stages in the liver. Relapse is characterized by the same symptoms and blood forms as were seen in the primary attack.

the cells of the salivary gland. Mixed with the anticoagulant secretions of the gland, sporozoites may then be injected, as the mosquito bites, into the bloodstream of a human host.

After injection by the mosquito, the parasites seem to disappear. For many years after the other stages in the life cycle were known, the cause of this so-called prepatent (hidden) phase of the life cycle had to be imagined. It was suggested that, as in avian malaria, the human malaria parasite had an exoerythrocytic (outside the red blood cells) stage, during which it multiplied in some tissue of the body, and after which it could be demonstrated in the blood. In 1948, however, Shortt and others actually found masses of proliferating parasites in cells in the liver of monkeys experimentally infected with very heavy initial doses of *Plasmodium cynomolgi*, an agent of monkey malaria. Afterward, exoerythrocytic stages were demonstrated in human infections with *Plasmodium falciparum, P. vivax, P. malariae,* and *P. ovale.* The prepatent period is therefore a result of a stage of multiplication outside the blood. In *vivax* malaria, the duration of prepatency is from 10 to 18 days.

The parasite multiplies by schizogony (multiple fission) in its human host, whether in the fixed tissues of various organs or in the traveling cells of the blood. Blood phases are well known; schizogony here follows a specific pattern. Merozoites (fragments) of the exoerythrocytic schizogony presumably are carried in the blood until they penetrate the surface of red blood cells. Then they grow to small "ring forms" (Fig. 2–1) which enlarge, becoming irregular in shape and increasing in both cytoplasmic and nuclear portions. After a host cell is nearly filled by the parasite and has become somewhat swollen, with characteristic stainable dots or granules (Schüffner's dots), the parasite undergoes schizogony, its nucleus dividing by mitosis four times to form a 12 to 18 nucleate schizont. The cytoplasm of the schizont furrows, dividing the cell into 12–18 merozoites (the "segmenter" stage). Then the host cell membrane ruptures, and merozoites are thrown into the blood plasma, together with metabolic wastes, pigment from the damaged host hemoglobin, and various other substances, some presumably toxic. The time between entry of the merozoite and liberation of its 12–18 descendants is, in *vivax* malaria, 48 hours. (In other malarias—*falciparum* and *ovale* malaria—the schizogonic cycle requires the same time or longer. *Plasmodium malariae* requires 72 hours.)

While the schizogonic cycle is continuing, certain merozoites, after invading red blood cells, develop into sexual forms. These are the male and female gametocytes necessary for completion of the life cycle of *Plasmodium*. When ingested by a suitable mosquito, these forms become mature in a way similar to that of spermatozoa; the female becomes an ovum or macrogamete, and the male divides to form four to six threadlike microgametes. As described above, a zygote, resulting from the union of a macrogamete with one microgamete, becomes a sporocyst, in which sporozoites develop, to invade the mosquito's salivary gland. After the time (10 days to 3 weeks) required for these occurrences, the parasites may infect a new human host.

SYMPTOMS

At first, the symptoms shown by the patient may be a slight, irregular fever, a headache, and general malaise. At this time, few parasites can be found in the blood, diagnosis being effective only by means of "thick films," which pile up the blood cells on a slide. After three or four days, however, the growth-schizogony cycles of all the parasites become synchronized, and the characteristic chills and fever of malaria appear. While the merozoites are being released from a billion cells at the same time (and very few 16-fold increases, or cell generations, are needed to infect vast numbers of host cells), the human body suffers such symptoms as lowered blood pressure, increased pulse and respiration, weakness, and feelings of intense cold, although the body temperature may be actually hypernormal and rising at the time. Such a chill is followed by fever during which the body temperature may rise dangerously. After 8 to 16 hours, the fever subsides and is followed by profuse sweating and exhaustion. Attacks recur with about equal intensity every 48 hours (or third day, hence the name "tertian" malaria); after one or two weeks, the attacks diminish in severity until the patient feels well.

The disease has not run its course, however. Exoerythrocytic stages remaining in the liver may again release merozoites, and relapses will occur with high probability but low predictability after three or four weeks or even a year. Some authorities distinguish between "recrudescences," "relapses," and "recurrences" in terms of an increasing time interval between the first and repeated acute stages. The variability in time of relapse among species and even among intraspecific strains of *Plasmodium*, combined with the likelihood of new or superimposed infection, makes such distinctions not particularly useful. Some forms of malaria become chronic diseases, weakening their victims over months and years.

PATHOLOGY

The pathology of malaria is of course related to the symptoms and damage caused by the parasite in its human host. Thus the release of toxic substances by the bursting segmenters may trigger a response in the temperature-regulating mechanisms of the body. Possibly the subjective chill is evidence of damage to the nervous center involved. The fever has its own effect, similar to the effects of other fevers. The collapse and sweating following a paroxysm may be due in part to physical exhaustion from the abnormally high metabolic rate that accompanies fever and in part to toxic products of such metabolism. Anemia naturally results from the destruction of blood cells, and splenomegaly (enlarged spleen), so characteristic in malarious regions even of individuals in presumably good health, is probably due to hypertrophy of lymphoid tissue of the spleen during the active phase of removal of malarial pigment and cellular debris while a malaria attack is going on. (This fact is used in surveys of malaria; the splenic index of a population tells the level of malaria which the people are experiencing.) These partly hypothetical suggestions actually show that many details concerning the way in which the parasite causes disease still remain to be discovered.

Pathological details differ with the species of *Plasmodium* and even with the particular strain infecting a victim. Individuality, condition, and even genotype of the host probably play a part, as does immunity.

Plasmodium vivax produces the pathological effects described above. The rare *P. ovale* is similar to *P. vivax,* although the red blood cells infected with *ovale* become oval and slightly irregular instead of swollen as in *vivax* infections. *P. ovale* is somewhat milder in its effect than *P. vivax. P. malariae* (which causes quartan malaria) sometimes stipples the red cells with fine granules (Ziemann's dots); this malaria is milder than the tertian malaria of *vivax* or even *ovale,* but may relapse after a longer period of time. *P. malariae* infections often (in up to 50% of the cases) lead to kidney disease. Perhaps the slower growth rate of the *malariae* parasite (only 6 to 12 merozoites) or the long 72-hour interval between paroxysms, accounts for both the mild nature and the unusually long duration, which, if very prolonged, leads to kidney malfunction. *Plasmodium falciparum* causes the severe disease, malignant tertian malaria. The infected host cells change in surface characteristics, so that they clump together, clogging capillaries. Consequently only the earliest blood forms—ring stages—are found in the circulating blood. Red blood cells are often doubly infected; such a condition is rarely if ever observed in other malarias. Also present in the blood are the gametocytes, characteristically crescent- or sausage-shaped. Depending on which organ is affected, blockage of the blood supply may produce gastrointestinal symptoms, heart failure, lung damage with pneumonia, or brain disorder simulating stroke. A disease called blackwater fever, in which the urine is colored dark by hemoglobin, is never unaccompanied by *falciparum* malaria. Unlike the other three malarias, however, *falciparum* malaria does not relapse after 6 to 8 months.

AVIAN MALARIA

Although there are members of the malaria group affecting animals other than man, the zoomalarias are relatively unimportant. *Plasmodium* species infect various birds; the canary has been widely used in laboratory studies aimed at the screening of drugs against human malaria. *Plasmodium gallinaceum* is highly pathogenic to chickens. Avian malarias are transmitted by culicine mosquitoes, as Ross was the first to demonstrate. A species of the genus *Haemoproteus* causes pigeon malaria. Hippoboscid flies (bloodsucking "louse-flies" which parasitize various species of birds and mammals) are the requisite intermediate hosts in which fertilization and sporogony occur, as in the case of *Plasmodium* in the mosquito. Pigeons may suffer no discernible effects, or they may have acute disease, resulting in death. Control of pigeon malaria depends on eradication of hippoboscids from the lofts. Other species of *Haemoproteus* have been reported from quail, turkeys, etc. The genus *Leucocytozoon* includes species parasitic in ducks. Black flies (*Simulium* spp.) serve as obligate intermediate hosts. The parasite causes acute and rapidly fatal disease in ducklings, but causes somewhat milder disease, characterized by weakness, loss of alertness, and occasionally death, in older birds. Adults, appearing normal except for unusual thinness, may be carriers. Losses from duck malaria may be very great in some outbreaks of the disease. A similar disease, caused by another species of *Leucocytozoon*, is seen in turkeys.

IMMUNITY

Immunity to malaria organisms takes a variety of forms. Individual recovery from attacks of malaria may be attributed to a special kind of immune response. The effectiveness of this response seems to depend upon phagocytosis of infected blood cells by spleen and liver macrophages; thus active cases of malaria show liver and spleen enlargement. The stimulation of liver and spleen macrophages both to phagocytose and to increase in number is due specifically to the presence of malaria organisms in the blood. This kind of immunity is temporary, for it is lost when the infection is lost. A special name, "premunition," has been given this response. In malaria as in other parasite-caused diseases premunition is an important concept.

Although premunition occurs in the presence of blood forms, it does not seem to be effective in the presence of the exoerythrocytic stages alone. The long duration of *vivax* and especially *malariae* malaria suggests that whatever response the body makes against parasites in the liver tissue is weak. Perhaps these malarias run a natural course, becoming cured when the reservoir of exoerythrocytic forms is exhausted. Certainly the differences in duration and in permanence of recovery in the various malarias seem hard to reconcile with the idea of a simple antiexoerythrocytic defense mechanism.

When the disease has been cured, immune response ceases; it may be elicited by reinfection with the same strain of malaria but in most cases will not occur upon infection with a different strain or species, against which a new immunity must be established during the course of a new attack of malaria. Individual immunity to malaria is thus seen to be highly specific.

Racial immunity or inherited resistance to malaria also exists. Thus in the United States, Negroes show a high resistance to clinically effective infection with the strains of *Plasmodium* ordinarily used in fever-therapy of syphilis. Such simple examples of racial immunity are rare. The resistance of resident populations in highly endemic areas is mixed (being partly inherent and partly due to almost universal early infection and continuous reinfection), so that the supposedly "racially" resistant individuals are also exhibiting premunition. In an attempt to study racial immunity, acquired immunity thus presents a problem and must be distinguished.

The origin of racial immunity is undoubtedly evolutionary and is the effect of long continued selection of inherently resistant individuals, resulting in the relative increase of "malaria-resistance genes" in the populations or races concerned. (One genetic basis for resistance to malaria happens to be well known. The reader is referred to Chapter 29, Evolution, in which the "sickling trait" and its relation to malaria are discussed.) In other words, malaria-susceptible individuals have failed to produce as many offspring as resistant individuals, because of early death or chronic illness. Only in areas of high incidence over a long period of time could such selective forces be effective; that is, populations of regions such as the Mediterranean lands, where malaria became severe in historic times, would not be expected to possess inherited resistance, whereas the descendants of prehistoric inhabitants of highly malarious regions such as West Africa should have had time to develop some racial immunity.

Occasionally very severe epidemics occur, with high (up to 30%) mortality. These epidemics are nearly always in subtropical or temperate regions, where seasonal changes (temperature or humidity) may reduce the incidence of infection to a low point. A seasonal return to conditions favorable for malaria then permits the rapid spread of malaria among a population only slightly or not at all immune and may have explosive and tragic effect. Similar epidemics may, however, occur in tropical climates, where little seasonal change exists; in such cases, the epidemics are due to the introduction of new or exotic strains of *Plasmodium*, against which no immunity could have been built up. "Fulminant malaria," as the above epidemics have been called (for their lightninglike, explosive quality), thus reinforces the idea that immunity to malaria is temporary and lasts only so long as some parasites remain in the blood.

As explained elsewhere (Chapter 27), theories about immunity to pathogens, to foreign substances, and to grafted or injected tissues are still being formulated. To the theoretical knowledge of immunity the facts about malaria may contribute. Premunition, a sort of feedback by which the infection, as long as it exists, controls or limits itself, may be significant in many kinds of disease. It seems a peculiarly appropriate adjustment in the animal host-parasite relationship, since it permits the host to live with a low-grade infection, the acute form of which would be fatal, and yet does not render the host unsuitable for future parasitic guests by rendering the host permanently immune. If, as seems probable from observations on malaria, there is some degree of permanent immunity, this immunity is so specific that different strains of the same parasite species may reinfect the host. Thus, in the evolutionary sense, premunition encourages variability among parasites, and at the same time it protects and preserves the host for future guests.

TREATMENT

Treatment of malaria, in a broad sense, includes measures to protect and to cure populations, since malaria is so serious and widespread that it is actually, as several authors have pointed out, a disease of nations, indeed of mankind. But treatment of a condition of man is so complex that it must be discussed later, along with a description of the cost and incidence of malaria. Treatment, in the narrow sense, means the use of drugs to prevent infection and to suppress or eliminate the parasite in clinical cases.

The first effective drug to be used against malaria was obtained from the bark of the cinchona tree of South America. Quinine and related substances extracted from this plant have been used since 1640 to destroy the blood forms of *vivax* and other malarias. Only recently (in the 1920's and during World War II) have new drugs been synthesized or discovered. Some of these are relatively nontoxic and, unlike quinine, may be effective against exoerythrocytic stages. None of these drugs is ideal. Atabrine was of great military value to the U.S. and its allies when the Japanese occupied the cinchona plantations of Indonesia during World War II; it effectively prevented the symptoms of malaria from appearing in troops operating in highly malarious regions, although it discolored the skin somewhat and failed to destroy exoerythrocytic stages or to prevent actual infections. During the twenties, plasmochin was tried successfully against gametocytes in the blood; it therefore served to prevent transmission. But plasmochin was rather toxic. The most effective drugs now in clinical use are chloroquine, amodiaquin, chlorquanide, primaquine, and daraprim. Several of these drugs are highly effective in suppressing symptoms. Either chloroquine or amodiaquin can quickly relieve clinical attacks of malaria by destroying the multiplying blood forms. Since these drugs are ineffective against exoerythrocytic stages, complete cure usually requires the administration of the rather slowly acting drug primaquine in conjunction with one of the clinically effective drugs. Drug-resistant strains of malaria are known and may be expected to become more numerous. Therefore it is important that research be continued on the chemotherapy of malaria, so that new drugs may be made available.

CONTROL

Even ideal drugs will not cure malaria in its aspect of a disease of mankind. The parasites in the liver and blood are only part of a complicated group of factors, events, conditions, even attitudes of mind that combine to produce and exaggerate the difficulties. Control or eradication depends on knowledge of these factors and their intelligent manipulation.

For many years, ever since Ross and Grassi demonstrated the mosquito transmission of malaria, it had seemed obvious to many that malaria control equalled mosquito control.

Various species of *Anopheles* or closely related genera, and no others, transmit human malaria. Studies of mosquitoes by medical entomologists have shown that different species exhibit different behavior patterns. Some mosquitoes live and die near human habitations, and "prefer" human blood to any other kind. Others, equally capable of transmitting malaria, are "wild" species, living far from houses or huts and seldom biting man. Some anophelines utilize small containers of water, such as tree-holes, the bases of the leaves of pineapplelike plants, or water-filled tin cans in refuse heaps. Others lay their eggs in permanent bodies of water, along the margins. The life span of female mosquitoes is an important factor in seasonal outbreaks and in the planned duration of control programs. Natural enemies of the mosquito—insectivorous vertebrates, predatory insects, even diseases of mosquitoes caused by microorganisms—have been and are being investigated. Mosquito-destroying chemicals, the famous DDT and more recently Dieldrin,

Chlordane, and others, have been tested and used in mosquito control programs. The new problems created by man-made reservoirs such as the Tennessee Valley Authority system in the United States, whereby thousands of miles of inland shorelines furnish new breeding places for mosquitoes, can be solved very effectively by controlled water level fluctuations. After the eggs have been laid and the larvae have hatched, a sudden drop in water level exposes these to death by drying. Specific habitats, such as small drainage ditches, household water tanks, water-holding refuse heaps, etc., can be either drained, covered, or sprayed, as their nature may require. But control of "wild" mosquitoes, which breed in jungles or swamp areas, will always tax the ingenuity of experts. However, in spite of great difficulties, not all of which are biological, the combined intelligence, skill, and knowledge of epidemiologists, ecologists, entomologists, and chemists have effectively controlled malaria-transmitting mosquitoes in many parts of the world. Chandler (1955) cited Soper's opinion that 90% of the world's malaria could be wiped out by 1960, indicating his belief that Soper's estimate was a reasonable hope.

This hope, based as it was on mosquito control largely by insecticides, has had to be abandoned. Actually, between 1950 and 1955, the World Health Organization's experts were giving up the concept of control (which assumed the survival of some malaria, at the level, perhaps, of typhoid fever in developed parts of the world today) in favor of a more daring concept, eradication. The reason for this shift in strategy was the alarming fact of biological variation.

Mosquitoes, like all other forms of life, have some genetic instability. In any population of mosquitoes subjected to extensive spraying with insecticides, some individuals will be found to be more resistant than others. Such survivors produce a new generation, which can inherit whatever genes for drug-resistance their parents possessed. It was found necessary in Greece to use alternately DDT and Dieldrin in order to control species of *Anopheles* only a few years after successful control with DDT had been initiated. It was seen that the widespread development, by selection, of insecticide-resistant mosquitoes could be a worldwide disaster; at worst, control could lead to uncontrollable epidemics, and at best it would result in a see-saw battle between mutating mosquitoes and inventive pharmacologists. The alternative to control is, of course, eradication.

ERADICATION

The eradication of malaria will probably be brought about in time. The difficulties —technical obstacles, costs, social and political obstructions—are tremendous. But the cost of malaria (to be discussed later), higher than that of any other disease, is much greater than the expense of eradication. In order for the great effort in malaria eradication to be successful, it is necessary that all responsible people understand the meaning of the eradication of malaria.

Russell (1958) pointed out that 76 countries were then planning, were carrying out, or had completed the eradication of malaria within their borders. Since that time more programs have started and many people have been freed from the threat

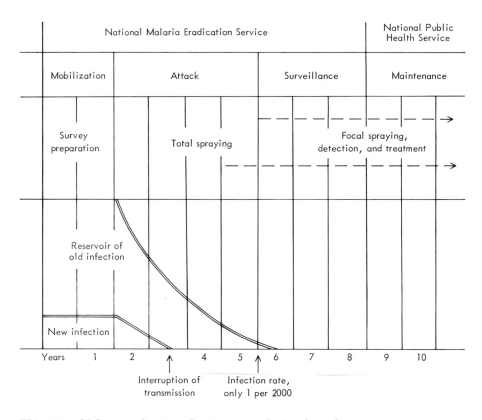

FIG. 2–3. Malaria eradication. During a period of perhaps five years, successive steps in an eradication program have stopped transmission, reduced existing infections to an extremely low level, and eventually resulted in a malaria-free nation where ordinary public health agencies can recognize, segregate, and treat new cases of malaria. [After Russell, 1958, and Alvarado and Bruce-Chwatt, 1962.]

of malaria. Although recent estimates (see Alvarado and Bruce-Chwatt, 1962) indicated that perhaps one sixth of the 1400 million people living in malarious lands were still not under any plan for eradication, the success of eradication programs should influence the governments of all malarious countries to take part in this gigantic effort.

Eradication, in Russell's words, includes four phases or stages: preparation, attack, consolidation, and surveillance (see Fig. 2–3).

Preparation. Although it actually began perhaps with the first speculations about the nature and cause of malaria and continues today in all malaria research programs, preparation means, in the practical sense, obtaining special knowledge about the particular area within which an eradication program is to begin. An epidemiological team, ideally consisting of a public health diagnostician, an entomologist, and an engineer, spends the necessary time investigating malaria in the assigned district. Blood smears or splenic indexes give a picture of incidence. By examining mosquitoes for sporozoites, the important vectors are identified, and the pattern of biting and resting behavior is sketched. The breeding places of the principal vectors are mapped, and the housing and other construction is studied for ease and effectiveness of potential spraying operations. Maps are drawn if necessary. Consideration is given to obtaining whatever local labor and cooperation may be needed. After such essential information is at hand, planning can proceed. The exact plan for each area and district will be unique; all sorts of curious factors—social, political, religious, economic—must be considered, in addition to the general facts about the type and incidence of malaria, vector behavior, and so on. The planners must consider costs, sources of funds, recruitment of trained and untrained personnel, and the estimated duration of the attack phase. Because of the above variants, it is impossible to describe a "generalized plan" for attack. Fortunately, plans have been worked out in many countries, and there are experts who can plan attacks, given the information obtained by the epidemiologists, with reasonable chances of success. But the program, in every stage, depends greatly on nonexperts, that is, the political leaders, local governments, and the people themselves.

Attack. As Russell points out, the aim of the attack is to empty the parasite reservoir. The resting places of mosquitoes must be sprayed with residual insecticide, which, upon contact with the resting mosquito's feet, causes death. A complete spraying program ensures that nearly all infected mosquitoes (which must rest while engorged and heavy with blood) will be killed before they have a chance to transmit the disease. The spraying program may involve teams of locally obtained workers, who methodically spray the walls of houses, huts, storerooms, and latrines, using hand-pumped sprays or pressure tanks to spread the insecticide. The wages of the workers and the cost of the insecticide are sizable portions of the eradication budget. The mosquito part of the parasite reservoir cannot, however, be emptied unless the source, human infection, is also treated. Thus skilled personnel equipped with antimalarial drugs of known effectiveness attempt to treat all known cases, until the human sources of malaria parasites have been cured. With this concentrated and energetic attack completed, the next step in malaria eradication can begin.

Consolidation consists of the finding of residues or pockets of malaria missed during the attack. With the elimination of malaria from these residues, the whole district under the eradication program is prepared for the fourth and most important part of the plan, surveillance.

Surveillance is essential for success. The attack, with subsequent consolidation, has usually lasted not less than three years. It has resulted in the virtual disappearance of malaria from the area. But unless adequate surveillance follows, the attack will have failed. For with the imperfection of present drugs and the inherent uncertainties of any plan of eradication, only vigilance for a considerable period can prevent the reemergence of malaria in the recently cleared region.

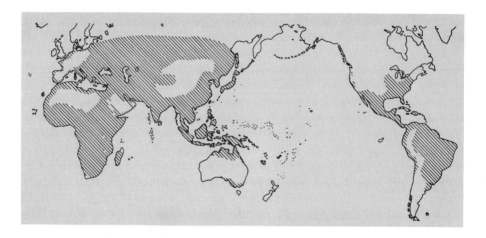

FIG. 2–4. The extent of malaria in the world before control and eradication measures were widely instituted. [After Alvarado and Bruce-Chwatt, 1962.]

Neither the anopheline mosquitoes nor the potential human hosts have disappeared; it must be made certain that the parasite reservoir is indeed empty.

This certainty comes in time. Since most individual malaria infections disappear within three years, this time may be considered a reasonable minimum for surveillance. If after three years (during which each reported case of malaria has been quickly and successfully treated and exposed persons have been carefully protected from infection) no locally acquired malaria can be found, then the attack can be declared successful.

Unfortunately, however, it is actually during surveillance that a malaria eradication program is most likely to fail. The apparent victory breeds carelessness. Compared with the cost of an active attack, surveillance is inexpensive. But to the taxpayer, the chief of a government, or the political leaders of a country, the cost of surveillance may seem excessive, since the results are not so dramatically evident as the effects of the first stages of the campaign. Paradoxically, it has been harder to obtain the relatively small appropriations for making sure of the success of eradication programs than it has been to obtain the initial large funds. This problem of maintaining support for a program that is nearly complete remains to be solved.

Once the attack has been consolidated and three years of surveillance have proven the success of the plan, then ordinary public health measures, that is, maintenance, can be relied upon. Every new case of malaria—almost invariably imported—if diagnosed and reported to the public health authorities, may be promptly treated, quarantined, and disposed of in the manner of any communicable disease. Malaria thus becomes only one of many "rare diseases," against which public health agencies know how to fight successfully. In the United States, once a highly malarious country, 24 cases of malaria were confirmed in 1957, of which only eight were contracted in the United States. In later years imported cases continue to be recognized in similarly low numbers. Other examples of the successful eradication of malaria and the relegation of malaria to the category of rare disease now exist throughout the world (see Fig. 2–4).

HISTORY AND IMPORTANCE

Malaria as a disease has been known for at least 2500 years. Unmistakable references to it occur in ancient Greek writings, the regularly intermittent fever, then as now, being a distinguishing feature. Hippocrates recognized the three types of malaria by the spacing of their paroxysms, and is credited with observing the relationship between malaria and swamps or stagnant water. Certainly, from early times, regions have been shunned as unhealthful because of malaria. And the events of history, battles, migrations, the rise and fall of civilizations, have often been affected by malaria, although records of "fevers," "plagues," etc., have seldom been detailed enough for positive identification of a particular disease. No demographic data regarding malaria in the remote past are available, although the report cited by Boyd (1930) that Empedocles of Agrigento, through drainage measures, freed a Sicilian city from an endemic fever, probably malaria, is evidence of an effect of malaria upon a population within the ancient world.

In recent times both the geographic extent and the medical importance of malaria have come to be recognized, if not by the general public, at least by malariologists and other health specialists. In 1940, perhaps the time of highest incidence of malaria, this disease was common in all temperate, subtropical, and tropical regions of the world, with a few oceanic islands excepted (see Fig. 2–4). The southeastern part of the United States, especially the coastal plain and the Mississippi Valley, was malarious; deaths, in some regions 5 per 100,000 people per year, occurred from New York State to Florida and Texas. In some parts of Mexico and Guatemala, an annual death rate of more than 1,000 per 100,000 was estimated to have existed over a ten-year period from 1928 to 1938. Similar heavy death rates have been reported in other parts of the world. Epidemics, which have occurred at times almost everywhere within the malaria belt of the globe, sometimes produce fantastically high mortality, such as that in Amritsar, mentioned by Russell (1958), where the death rate of 200 per 1,000 people during two months in 1908 completely disrupted life and labor in that Indian city of 160,000. From official reports and estimates, it has been concluded that in the 1940's there were 350,000,000 cases of malaria throughout the world, of which 1% or about 3,000,000 people died annually.

That such a disease is important must be evident. Its importance can be realized by utilizing data from varied sources—medical, economic, sociological.

Some data on certain diseases in the United States (Shepard, 1956) may be useful. In that year there were 11 million sufferers with arthritis, 10 million each with heart disease and mental illness, one million with tuberculosis, and less than one million and a half with cancer, cerebral palsy, muscular dystrophy, or poliomyelitis. Out of 160,000,000 people, about 34,000,000 suffered from one of the above eight diseases. For research and alleviation of these diseases, U.S. citizens contributed $140,000,000 annually. Although malaria by that time was virtually nonexistent in the United States, its estimated incidence in the rest of the world was 250,000,000 out of perhaps 2,500,000,000 people, or about 10%. In 1958, the total budget for the study and control of malaria, from all sources, approached $100,000,000 (Russell, 1958). The discrepancies in the expenditures are striking— ten times as much per heart victim as per malaria patient—and the similarities in numbers of victims are equally striking, about the same world percentage of malaria victims in 1958 as the U.S. percentage of heart and arthritis combined. (The dramatic reduction in malaria, which by 1962 was estimated to have a world incidence of 100,000,000 people, merely reduces, but does not wipe out, the above contrast.)

That the expenditures for study and relief ·of the various diseases are so disproportionate to the incidence and severity of the conditions is not surprising. Even in the U.S., as Shepard (1956) showed, almost half the money spent for research in the diseases mentioned was used for research in polio, the sufferers from which numbered less than 0.3% of the victims of the eight diseases mentioned. One third of the total sufferers, those with arthritis and rheumatism, benefited from less than 2% of the total expenditures. It is obvious that the public view of the importance of various diseases is influenced by such factors as publicity and organized campaigns rather than the actual incidence and damaging effects of the diseases themselves.

Although knowledge of the cost and incidence of malaria is available, it has not effectively been put before the public eye. Such true statements as "Malaria disables more people than any other disease" (Alvarado and Bruce-Chwatt, 1962) mean little, unless comparisons with other diseases can be made and costs can be estimated. The above data on U.S. incidence of arthritis, heart disease, and mental illness give some perspective to the 3% figure for the incidence of world malaria. Malaria exists as a threat to over 1.2 billion people throughout the world. The world incidence figures arise from the malarious countries, and the incidence rate where malaria exists is not the above 3% (of the world population) but is closer to 10% (of the population of the malarious nations). Of course, many regions, such as parts of Africa, India, and Central America, have vastly higher rates of malaria; some areas approach 100%. Malaria is one of the great diseases, ranking in incidence with arthritis, heart disease, and mental illness, and probably exceeding them in overall destructiveness. But how much does malaria cost? If that question can be answered, the expense of malaria eradication efforts will or will not appear justified in cold economic terms.

COST OF MALARIA

Malaria is usually a chronic, debilitating, periodically disabling disease. In situations where it is prevalent, the number of man-hours lost from productive labor multiplied by the number of malaria sufferers gives a figure that can be charged as loss in the manufacture of goods, in the production of crops, or in the earning of a "gross national product." On the basis of partial estimates, this figure comes to about $2,000,000,000 annually.* Of the world's 250,000,000 cases of malaria in 1958, India accounted for 100,000,000 cases, with an annual loss amounting to $500,000,000; the above world cost is an extrapolation from the Indian figures. Russell (1958) made it clear that while the cost of malaria is borne directly by the malarious peoples, it is shared by the malaria-free nations. He estimated the annual cost of imports to the U.S. from malaria-ridden lands to be $300,000,000 more than it would be if the imports had been produced without the burden of malaria.

Perhaps the above estimates of the cost of malaria are biased toward bigness, since they are made by malariologists. But the malariologists have omitted in their estimates a factor economists call the "multiplier effect."

It is commonplace in economic theory that economic expansion requires a surplus of production over consumption. This surplus, called savings, can be invested to increase productive capacity and allow the economy to expand. Production, therefore, has a "multiplier effect" whereby surplus goods, funds, or equipment tend to increase the surplus of the next production period. Conversely, any halt of production by natural calamity, strike, or disease interrupts the expansion of the economy; thus in addition to the immediate loss due to interruption of output, there is the long-term and greater loss due to failure of the economy to expand at the usual rate.

For example, in the U.S. in recent years there has been an annual rise in the gross national product of about 3%. This has been accompanied by a rise in the cost of living, which, while of course inflationary, clearly reflects a higher living standard for most of the citizens and a greater prosperity for the nation. At times during which the economic boom was progressing there was substantial (4–5%) unemployment. It was pointed out by labor and government economists that failure to provide full employment reduced the rate at which expansion could occur, and that because of a lag in employment, the unprecedented prosperity was not as great as full employment could make it. Missing from the economy were the billions of dollars that 7 million unemployed could have been earning, spending, and investing. Graphically (see Fig. 2–5), the difference between a rising gross national product with unemployment and the same item with full employment could be seen as a widening gap between what the nation's economy could perform and what it was in fact performing.

This kind of gap or loss, so familiar to critics and defenders of the midcentury government fiscal policy in the U.S., has not been mentioned by malariologists. Their estimates of the actual annual loss of about 2 billion dollars are probably conservative, considering that such loss is multiplied by the consequent failure to expand the economies of the countries where the immediate loss is felt. The cost

---

* "Malaria: The World's Most Expensive Disease," *Med. J. Australia*, May 10, 1958, p. 640.

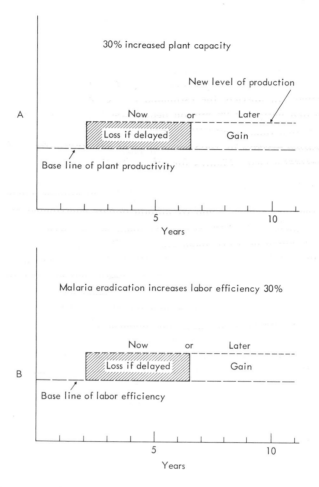

FIG. 2–5. Comparing the "multiplier effect" in economics with the effect of malaria eradication in general. (A) Increased plant capacity produces increased output over the entire time thereafter, and failure to increase plant capacity delays such output. (B) Early eradication permits early gain in labor efficiency with increased prosperity. Failure to eradicate sacrifices years of such prosperity.

of malaria is a multiplying cost, a loss of potentialities, like the cost of childhood diseases, malnutrition, and lack of education.

But if a balance sheet must be drawn, perhaps a minimum figure—$1,700,000,000 a year—can be used in the cost column, leaving out potentialities, even the monetary, involved in such an annual sum, and neglecting entirely at this point the intangibles, the "human cost." Against the annual loss of a little less than 2 billion dollars, the cost of eradicating malaria may be realistically projected.

COST OF ERADICATION

As we have seen, eradication involves research, training, investigation, planning, attack by teams of experts and laborers, surveillance after attack, and public health maintenance. By 1958, the cost of some eradication programs could be accurately stated. In Mexico, the estimated cost of a five-year plan to eradicate malaria from that nation of 26 million people was $21,000,000. Malaria losses in Mexico were at that time about $18,000,000 a year. For India, the estimated total cost of eradicating malaria by 1965 was $114,000,000. Indian losses due to malaria were $400,000,000 a year. Figures for other situations are available from United Nations reports and other sources. Actual expenditures in dollars for the malaria eradication program, worldwide for 1958, included 39 million dollars contributed by world or regional agencies, and 61 million dollars contributed by governments of nations in which eradication was being planned or carried out. The United States had contributed by 1957 about $30,000,000 to world antimalaria programs, or about 0.18% of U.S. foreign aid. In 1958, the U.S. contributed more than $25,000,000. Based on the increased cost of imports and lost sale of exports, the annual cost of malaria to the U.S. is estimated at $300,000,000 (Russell). It can be seen from a glance at these figures that the annual cost of malaria is perhaps twenty times the budgeted amounts for malaria eradication. Considering the exposed population to be 1,400,000,000 persons, it can be estimated, however roughly, that over a six-year period the probable cost of eradicating malaria from the world is about half a billion dollars. Whether the latter estimate is in error, even by a factor of 2 to 4, does not change the obvious conclusion: eradication of malaria is an investment that would be paid back in less than a year and would thereafter earn from 100% to 500% per annum.

THE STRUGGLE SO FAR

The investment in malaria—basic research, hospitalization, treatment, control, and eradication—has been substantial already. The modern history of malaria is a story of courage, imagination, and persistence, exhibited by workers at many times and places. It is impossible to evaluate monetarily the investment that these men have made, but nowhere in the history of medical science can be found an effort more worthy of recognition and recall than that of the malariologists.

Laveran, a French army surgeon, discovered the malaria parasite in 1880. He saw exflagellation, the movement of microgametes separating from the micro-gametocyte, in a wet blood film. In 1875, Kelsch, a German, had seen pigmented bodies in the blood of malaria patients. Earlier, in 1847, Meckel, also a German, recognized that the pigmentation of the liver and spleen was due to malaria, and in 1753, Torti, an Italian, had distinguished malaria from other fevers by its susceptibility to cure by quinine, a drug introduced still earlier by the wife of the Spanish viceroy in Peru. Her physician had learned of it from the Indians, who used it in the crude form of bark of the Cinchona tree (named for the lady, the

Countess of Chinchon). Boyd (1930) has stated that the introduction of quinine into Europe was probably the most important event in malariology for two thousand years.

The last decades of the 19th century saw many discoveries in malariology. The Italian workers were active. Golgi, in 1886, described the blood forms of *malariae* and probably *vivax* malaria; Marchiafava and Celli, in 1885, named *Plasmodium malariae* (hence the taxonomic credit "Marchiafava and Celli" after the genus *Plasmodium* as well as after the above species). In 1890, Grassi and Felletti named *P. vivax*. In 1898, Grassi and others demonstrated the transmission of human malaria by anopheline mosquitoes.

Scientists of other nations during the same period also made essential contributions. In 1891, Romanowsky, a Russian, began using the buffered methylene-blue eosin stain that gave rise to the family of blood stains (Wright, Giemsa, etc.) so useful in differentiating the parasites. Ross, an Englishman working in India under peculiarly adverse conditions, demonstrated the transmission of bird malaria by culicine mosquitoes; his works were published in 1898, the year of Grassi's similar discovery regarding human malaria. Manson, an Englishman, and Kilbourne and Smith, Americans, in the 1890's had pointed unmistakably to the possibility of arthropod transmission. Manson had worked on the helminthic disease, filariasis; Kilbourne and Smith had studied the transmission of the protozoan disease, piroplasmosis, by ticks. In addition to the aforementioned French, Spanish, Peruvian, German, Italian, English, Russian, and American contributions, there have been valuable studies by the scientists and medical men of other nations. Malaria, a disease afflicting mankind, has been attacked since the early days of medical science on an international front.

Recent attacks, in the 20th century, have been international not only in personnel but also in support and structure, culminating in the present cooperative effort to eradicate malaria from the world.

It is a matter of some interest that the first successful eradication program was worked out in the United States (a nation whose political and economic structure bore some resemblance to the structure which the united nations of the world may one day adopt, as communication lines shorten and the need for international cooperation becomes ever more apparent). It was recognized that the tremendous problem of malaria in the United States required more than the individual efforts of state public health departments. The mosquito knew no state boundaries, and travel by the people in the U.S.A. was not restricted in any way by law. If any state chose to control malaria, its efforts would be nullified by the uncontrolled malaria of a neighboring state. Therefore regional organizations for study and control had to be set up.

Under the direction of the U.S. Public Health Service, aided by the advice of such organizations as the Rockefeller Institution and the National Malaria Society, and utilizing the suprastate operations of the Tennessee Valley Authority, malariologists attacked malaria in the United States on a broad front.

Until 1930, the attack was sporadic and exploratory. Military experiments in Cuba and the Panama Canal Zone had demonstrated that malaria could be effectively controlled by existing methods. Around 1915 the U.S. Public Health Service made demonstrations in selected communities in North Carolina and Mississippi, again showing the effectiveness of drainage, larvicides, and screening. During World War I, large numbers of troops were concentrated in camps in malarious parts of the U.S. The Public Health Service protected these troops and improved the health in areas surrounding the camps by control measures. Unfortunately, the immediate effect of such demonstrations, as Boyd (1941) has remarked, was to impress the public with mosquito control concepts, rather than with malaria control (which involves the ecology of certain species of *Anopheles* only). Consequently, many communities in the 1920's attempted to reduce the mosquito population indiscriminately, with little effect save an impression on the public that such projects were costly. New techniques, however, came into use, along with ever-improving knowledge of the important anophelines, so that during the 1930's malaria in the U.S. declined.

The new techniques were spraying with the larvicide "Paris Green" (a copper and arsenic compound), regulation of the water level of man-made reservoirs to prevent increase in anophelines, and, finally, the wide-spread application of DDT. These larvicidal factors, plus important improvements in the diagnosis and treatment of malaria patients, were extremely effective. The steady rise in standards of living with increased industrialization and improved educational facilities contributed substantially to the decline of malaria. Malaria disappeared from the United States during the 1940's.

Today malaria is disappearing from the world. The credit for the approaching victory in this war against a great affliction belongs to many; and no one can sift the biographies, the scientific reports, the cold data of surveys and exploratory studies, ranging in time from the mid-nineteenth century to the present—over a hundred years of varied and dedicated efforts—with much hope of being able to say "these were the leaders, these the heroes, these the casualties" of this war. The present dramatic trend toward complete eradication of malaria is the fruit of the past. It is also a hopeful sign for the future, for now when international tensions are high, with the possibility of cataclysmic war still present, international efforts to control malaria are showing unmistakable signs of success. The antimalaria program serves as an example of world cooperation, of how much the medical and scientific resources of the world can accomplish when they are mobilized to seek a common goal.

The most recent spectacular advances against malaria have been led by a new agency, the World Health Organization of the United Nations. In 1949, one year after its founding, WHO chose as its major objectives the control of three diseases: tuberculosis, venereal disease, and malaria. Among antimalaria organizations already in existence was the Rockefeller Foundation, which since 1920 had been supporting cooperative programs throughout the world, especially programs emphasizing epidemiological study, control demonstrations, training, and the establishment of official bureaus and institutes. Other organizations among nations have made effective contributions. The Pan-American Scientific Congress (1916), the Malaria Commission of the Health Organization of the League of Nations (1923 to 1939), the United Nations Relief and Rehabilitation Administration (1943 to 1946, effective especially in Greece and Sardinia), and the Institute of Inter-American Affairs (1942), all aided in various important ways. But the continuing stage of the battle is dominated by United Nations agencies WHO, UNICEF, and PASO (the latter an autonomous organization serving as regional offices for WHO in the Americas). These agencies are cooperating with each other and with the various national governments to eradicate malaria.

Money for the vast program comes from three primary sources: the United Nations, the privately endowed and regional organizations, and the participating governments. The latter have provided about 60% of the total. Russell (1958) estimated some of the amounts spent by the international and regional agencies as follows: UNICEF had spent about 34 million dollars, for insecticides, equipment, and transportation. UNICEF's 1958 malaria budget was 8 million. The WHO special account (Malaria Eradication Special Account—MESA), which accepts gifts from national governments, received from the U.S., in 1951, 5 million dollars. The United States, through its International Cooperation Administration (ICA) had spent about 9 million dollars for malaria in 30 countries by 1958. ICA budgeted 23 million for malaria eradication in 1958. The various phases of eradication mentioned earlier are financed by the above contributions.

In 1962, Alvarado and Bruce-Chwatt drew a balance sheet of success measured against need in the malaria eradication program. About one and one-half billion people live in areas where exposure to malaria does or can occur. (These exposed populations are in areas where climatic conditions and biological factors, specifically tropical to temperate climate and the presence of anophelines, make malaria a threat. The actual occurrence of malaria is not necessary to the concept of "exposed population.") In other words, about 45% of the people on the earth were the potential victims of malaria in the 1940's when the major program started. Of these, about 300 million lived in malaria-free regions. By 1961, eradication programs were operating or being completed in countries having exposed populations of about 710 million. Planning, in that year, had been officially begun in other countries with exposed populations totalling nearly 200 million. About 223 million had no protection, nor was any planned at that time. Russell (1958) listed the following countries as having nearly or completely achieved malaria eradication: Argentina, British Guiana, Chile, France (Corsica), Cyprus, French Guiana, the Gaza Strip, Italy (Sardinia, Sicily included), Martinique, Mauritius, the Netherlands, Puerto Rico, Reunion, Rumania, United States, and Venezuela. While this book is being written other countries should doubtless be added to that list. It can be stated that by 1961 malaria incidence had dropped from the 300 millions of the 1940's to about 100 millions. There is every reason to believe that the decrease is continuing.

The outlook for the future is indeed bright. It has been pointed out that when victory is in sight and only maintenance of earlier positions is needed, the chief obstacle to success in eradicating malaria is the temptation to relax vigilance, to tighten purse-strings, and to allow the enemy to creep back into its former strongholds. The enlightenment of governments and the education of the people to this continuing need for sound public health practices are necessary for the complete eradication of malaria.

The implications of such a victory are obvious. First, a "condition of man," such as the disease called malaria, can be corrected by combined and coordinated effort. Second, international cooperation among both scientists and technicians, and among governments as well, is necessary for such an effort to succeed. Third, there are many other "conditions of man," such as overpopulation, poverty, hunger, ignorance, pestilence, even war itself, that may be subject to the same kind of effective attack used against malaria. *Study* by men of skill and good will in many lands—*planning* by cooperating governments, with the advice of an international agency, supported by the necessary funds—*attack*, using the knowledge and the plans to change or abolish the "condition"—*surveillance*, by trained and dedicated specialists, to prevent recurrence of the danger—these are the steps by which mankind may advance over obstacles that today seem insurmountable, as did malaria only yesterday.

SUGGESTED READINGS

ALVARADO, C. A., and L. J. BRUCE-CHWATT, "Malaria," *Sci. Amer.*, May 1962, pp. 86–98.

BOYD, M. F., *Malariology*. Saunders, Philadelphia, 1949.

CHANDLER, ASA C., *Introduction to Parasitology*, 9th ed. Wiley, New York, 1955.

COVELL, G., G. R. COATNEY, J. S. FIELD, and J. SINGH, "Chemotherapy of Malaria," WHO Monograph Series No. 27, 1955.

EXPERT COMMITTEE ON MALARIA, 6th, 7th, and 8th reports, WHO Technical Report Series, Nos. 123, 162, and 205. 1957, 1959, 1961.

FAUST, E. C. and P. F. Russell, *Craig and Faust's Clinical Parasitology*, 7th ed. Lea and Febiger, Philadelphia, 1964, pp. 232–282.

MACDONALD, G., *The Epidemiology and Control of Malaria*. Oxford University Press, London, 1957.

"Malaria: The World's Most Expensive Disease," *Med. J. Australia*, May 10, 1958, p. 640.

RUSSELL, P. F., L. S. WEST, R. D. MANWELL, and G. MACDONALD, *Practical Malariology*, 2nd ed. Oxford University Press, London, 1963.

RUSSELL, P. F., "Man against Malaria—Progress and Problems," Rice Institute Pamphlet 45(1) : 9–22, 1958.

SHEPARD, M., "The Battle for Health—and Dollars," *St. Louis Globe Democrat*, Feb. 12–20, 1956.

CHAPTER 3

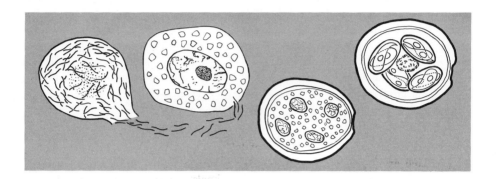

OTHER SPOROZOA

The sporozoan genus *Plasmodium,* while by far the most important member of the class, is not the only sporozoan seriously affecting human welfare. Two other groups of Sporozoa, the order Coccidia and the family Babesiidae of the order Haemosporidia, cause widespread disease among domestic animals and poultry. Before describing these parasites and the diseases they cause, a discussion of the significance of veterinary parasitology would be helpful.

Domestic animals, which constitute a major portion of the world's wealth, suffer from various diseases, all of which diminish to some extent the prosperity of mankind.

In 1957–58, the Food and Agriculture Organization of the United Nations estimated the annual world production of animal products used as food to be about 300 million tons (of milk, meat, eggs, and animal fats). Other valuable products, of course, include hides and wool. Animals perform useful work, the value of which is large yet difficult to estimate. Perhaps the worth of agricultural and domestic animals can be best examined in relation to certain regions and areas.

In the United States, where machinery has almost entirely replaced animals as a source of labor and where a disproportionate share of the world's protein is produced and consumed, economic animals fall into two major categories, products and pets. The meat and dairy products industry, for example, yields annually about 80,000,000 tons of milk, meat, and animal fats. American chickens laid nearly 62 billion eggs in 1961. U.S. farmers had 59 million hogs. Intermediate between the "product" and the "pet" concept are such useful animals as horses and certain kinds of dogs, the breeding, care, and training of which employ a substantial number of persons and require many thousands of acres of fine land. House pets are the stimulus for a fantastically large industry in the U.S. In 1961, for instance, dog food cost consumers $360,000,000, which is more than the amount spent for canned baby food or several other standard commodities.

In contrast to the United States are some of the underdeveloped nations, where agricultural production results chiefly in plant carbohydrate and levels of diet are perilously near deficiency. In nations such as Bolivia, Pakistan, and Mexico, domestic animals are used as a source of labor, and those animal products which are marketed are renewable resources, like wool or dairy products. Because it is inevitably wasteful to use vegetable food for animals which are then to be eaten by man, nations of low productivity or development cannot afford the production of meat in their agriculture. It goes without saying that in these countries pets are not a major economic factor. But the share which animals play in the underdeveloped regions is nevertheless important. The 45 million water-buffalo in India, the 6 million mules and asses in Mexico, the nearly one million camels in Saudi Arabia and northern Africa, and the 6 million llamas and vicunas in Bolivia and Peru, for instance, indicate the substantial role animals play in the agricultural resources of those lands. Moreover, wealth is not measured in the same way by the prosperous as by the poor, and the value of the thoroughbred horse in Kentucky may be in some manner comparable to the value of the burro in Durango, Mexico.

Animal disease is a major loss affecting man's agricultural wealth.

The 1961 FAO Animal Health Yearbook lists approximately 150 diseases of veterinary animals, with data on incidence and distribution of each within six major geographical divisions of the world. Among these diseases are well-known conditions such as anthrax, hoof and mouth disease, bacterial and virus diseases of poultry, and various diseases caused by animal parasites. The latter cause about 20% of the diseases affecting animals; if diseases transmitted by arthropod parasites are added, the parasite-caused diseases of stock amount to almost one-third of the total.

A recent attempt (U.S.D.A., Agriculture Handbook No. 291, 1965) to evaluate losses to livestock in the United States due to internal parasites arrives at a grand total of $340,206,000 annual loss. Annual losses due to insect pests and parasites affecting livestock and poultry come to $877,850,000. Projecting these figures onto a worldwide instead of national scale would multiply these losses three or four times. Thus in both numbers of diseases and extent of losses the parasite caused diseases of farm animals are extremely important. It is difficult to say whether these diseases are comparable to other diseases in both number and importance, for while a few are spectacular (fowl coccidiosis, trypanosomiasis, the cattle fevers), others are insidious. The latter diseases (including many helminthiases) cause "unthriftiness," a condition in livestock that resembles rather closely the debilitated state of victims of the chronic parasitisms of man, such as malaria, hookworm disease, and schistosomiasis. Unthriftiness is probably a major aspect of the diseases of animals vital to an economy, but of all aspects it is the most difficult to evaluate. Like labor loss as an increment to the cost of manufactured goods, unthriftiness in animals is an addition to the cost of husbandry and reduces the margin between subsistence and surplus.

In addition to the economic, man is affected by animal disease in other ways. Domestic animals serve as reservoirs for certain human infections, making the diseases difficult to control or eradicate. Animals serve also as intermediate hosts for some parasites. Moreover, on occasion certain dangerous pathogens are accidentally acquired from animals; zoonosis (the name given to man's diseases which are acquired from animals) is a particularly difficult problem, and involves several important parasites.

Protozoa cause various animal diseases. Although these will be discussed in the chapters on human diseases having similar causes, it may be useful to list them here in a chapter entirely devoted to some veterinary protozoa.

In the class Mastigophora, the family Trypanosomidae contains trypanosomes which cause severe disease in horses, camels, cattle, sheep, goats, dogs, and swine. The genus *Leishmania* in this family affects dogs. Other flagellates are the trichomonads, which cause one form of contagious abortion in cattle, and a curious organism, *Histomonas meleagridis*, which lethally affects turkeys.

The class Sarcodina is unimportant in veterinary medicine, although the amoebas living in man can be reared in some laboratory animals for study.

The Ciliata contain *Balantidium coli*, a relatively harmless parasite of swine but one which affects man rather severely if he acquires it. Thus balantidiasis is a zoonosis.

The Sporozoa, of course, include the agents of bird malaria.

Other sporozoans cause two expensive diseases of domestic animals. Coccidiosis, while significant in the health of various mammals, is chiefly a disease of poultry. It can destroy whole flocks, and, more often, can mean the difference between successful and unsuccessful production of meat and eggs. Piroplasmosis, the group of cattle fevers, has been a factor of historical importance, because it has interfered with the raising and shipping of cattle in such widespread regions as South Africa and Texas; although control of arthropod vectors has greatly improved the situation, the cattle fevers are still a constant threat to cattle production in many places.

Toxoplasma gondii is a parasite of doubtful affinities; recent studies seem to support its inclusion in the Class Sporozoa, although its position within this very large and heterogeneous group has not been established. Judging from serological surveys in several parts of the world, *Toxoplasma* is a common parasite of man. It lives within cells of the reticuloendothelial system, multiplies by binary fission, and spreads in clinically severe cases to various other tissues and organs where it causes lesions. Toxoplasmosis of the newborn is a severe and often fatal disease apparently acquired *in utero* from the infected but symptomless mother. Other forms of transmission have been demonstrated in the laboratory, but the epidemiology of toxoplasmosis in nature is obscure. Any warmblooded animal may harbor *Toxoplasma* and may serve as a reservoir for human infection.

COCCIDIA

Members of the subclass Telosporidia (order Coccidia) cause severe diseases of poultry, livestock, and furbearing animals. All the important Coccidia belong to the suborder Eimeriida, characterized by a life cycle in which mature gamete-forming cells do not associate in pairs (syzygy of the protozoologists) as they do in certain other orders.

Typically the life cycle has three distinct phases: a dispersion phase, a growth phase, and a sexual phase.

The dispersion phase involves an oocyst. This is a spheroidal mass, consisting of a highly resistant outer shell or membrane and one or more sporocysts. The latter contain one or more sporozoites. In some species the sporocyst is absent, and an oocyst contains the sporozoites. The oocyst is formed as a result of sexual fusion of gametes in the host body; development of the sporocysts and sporozoites usually occurs after the oocyst has been discharged from the host. The oocyst is capable of withstanding extremes of heat and cold, and is resistant to attack by most sterilizing agents. Oocysts can remain infective in the soil for more than 15 months. Their size, from slightly over 10 to about 30 microns, makes them easily transportable as dust or in water. They are ingested by accident with the food of their host animals. Since coccidial infections are chronic in at least some carriers, a steady source of oocysts maintains the environmental supply at a level sufficient for transmission.

The growth phase occurs in the tissues of a host and accounts for various pathological effects. When an oocyst is ingested it passes down the digestive tract and ruptures from chemical or mechanical causes. The sporocysts containing the sporozoites are liberated, and the sporozoites emerge into the intestinal lumen. The parasites invade the intestinal epithelium, and at the expense of host cells they grow either within the epithelium itself or in subepithelial layers of the intestine. The growing parasites split into many small forms (by schizogony, into merozoites); as a result the host cell ruptures, and the merozoites are free to enter new cells. After several repetitions of this growth and splitting cycle, the parasites become transformed into the sexual stages. The asexual or growth stages are the most damaging, for they often cause the loss of much intestinal lining from hemorrhage and secondary infection.

The sexual stages begin with the development of recognizably different cells, the gametocytes. Male cells or microgametocytes are rounded cells which become multinucleate, with hundreds of small, distinctly stained nuclei migrating toward the cell membrane. These nuclei elongate to give the gametocyte a fibrous or

FIG. 3–1. Life cycle of *Eimeria tenella*. Oocysts (A) are voided with host feces in an unripe state. Outside the host, oocysts form within them four sporocysts, each with two sporozoites. After hatching of the oocyst in a new host's intestine the sporozoites (B) enter epithelial cells of the cecum. The infected cells migrate to the submucosa where they hypertrophy (C). Within the cells hundreds of merozoites (D, E) form through schizogony. These enter new cells, and the schizogonic cycle is repeated. After several such cycles, merozoites enter epithelial cells and develop into microgametocytes (F) and macrogametocytes (G). Microgametes are liberated to fertilize macrogametes, and the latter become zygotes which develop into oocysts and enter the fecal mass to be discharged.

striped appearance. Eventually the gametocyte and host cell rupture, and microgametes (tiny, biflagellate motile forms) emerge. Meanwhile the female cell or macrogametocyte has developed, with a characteristically distinct, central nucleus and massive accumulations of granules next to the cell membrane. Fertilization of the macrogamete by a microgamete is followed by formation of the tough, resistant oocyst wall. It is believed that only the zygote, or fertilized macrogamete, is a diploid organism, that is, it possesses a double set of chromosomes. Certain observations on related protozoa suggest that this is true. Further development probably involves, therefore, reduction in the number of chromosomes as the sporocysts and sporozoites are formed by repeated divisions of the zygote.

Figure 3–1 illustrates the coccidian life cycle of *Eimeria tenella* as outlined above.

Before proceeding to particular coccidioses, let us examine certain relationships between the life cycle of the Coccidia and the diseases which they cause. First, it is extremely difficult to prevent coccidiosis in flocks of poultry, herds of livestock, or pens of furbearers; the oocysts are durable, very small, and easily ingested. Second, because these organisms multiply rapidly during their growth phase, they often cause severe and even fatal disease; however, nearly all hosts of Coccidia have the ability to develop eventual resistance, and the survival of both host and parasite depends on this resistance. Low-grade, continuing production of oocysts by symptomless carriers of Coccidia probably ensures, in nature, the infection of new hosts by very small numbers of oocysts and gives these hosts a chance to develop resistance without being overwhelmed. Crowded and unsanitary conditions, however, produce an opposite effect, with epidemics of acute disease producing vast numbers of oocysts, which in turn threaten new generations of hosts.

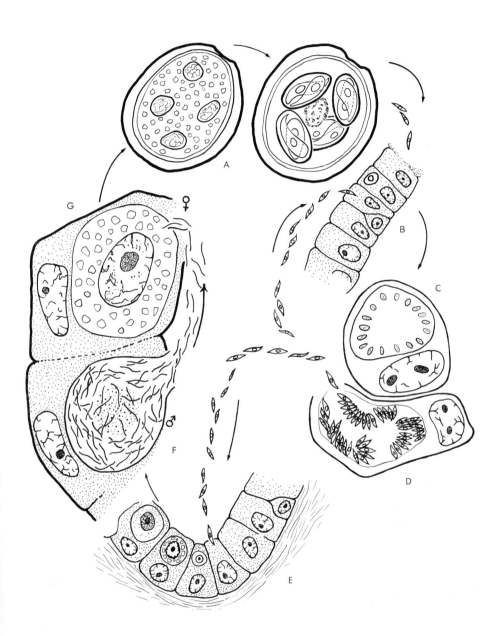

Fowl coccidiosis is a worldwide disease of poultry, including chickens, turkeys, pheasants and quail, pigeons, ducks, and geese. The pathogens are generally specific for each species of host, and different species of Coccidia cause distinctly different symptoms. Over twenty species of Coccidia are known to infect fowl. The three best known fowl coccidioses are those produced in chickens by the respective species *Eimeria tenella, E. necatrix,* and *E. acervulina.*

Eimeria tenella causes cecal coccidiosis. Its sporozoites invade the ceca (intestinal pouches) of young chickens and rapidly multiply. It has been estimated that each sporozoite can produce 900 first-generation merozoites, and that each of the latter can produce 350 second-generation merozoites. Thus over two million second-generation merozoites can result from the ingestion of one oocyst (with its 8 sporozoites). This vast proliferation causes a regularly progressing disease, as increased areas of cecal lining become affected. Five days after infection the droppings of the bird become hemorrhagic. Marked bleeding occurs by the fifth day, and oocysts are found in the feces by the seventh day. Meanwhile the bird loses appetite and energy. About 90% of the mortality occurs during the first seven days, and birds that survive this period generally recover. But recovery is usually not truly complete. Cecal coccidiosis causes permanent impairment of the function of the cecum, and at least 10 weeks are required before infected birds become comparable to uninfected birds in weight; infected birds never become as efficient at egg-laying as noninfected birds.

Eimeria necatrix causes chronic intestinal coccidiosis of chickens. This parasite invades the lining of the small intestine (rather than the ceca), but its cycle and development are otherwise very similar to those of *E. tenella.* Several generations of merozoites occur characteristically in the submucosal layers, where infected epithelial cells have migrated. One striking difference in development between *E. necatrix* and *E. tenella* is the irregularity of the former; generation time is not comparable for the progeny of different oocysts within the same host, and study of infected tissue may show gametocytes present at the same time with first generation merozoites. After several generations of merozoites, however, sexual stages invade the cecal cells, where gametocytes develop. The course of disease is similar to that of *E. tenella* infection, but is somewhat milder, and may in some chickens go unrecognized. Its economic effect, even with its rather low mortality, is quite severe, since unthriftiness in surviving birds may be more costly than the brief expense of rearing young birds to the time of death. (A somewhat similarly grim economic factor applies to debilitating disease in man and this factor has been publicly recognized in discussions of biological warfare, where the disabling of an enemy population is judged more disruptive of an enemy economy than outright killing). *E. necatrix* presents the same problems of control as *E. tenella;* indeed, the saturation of the chicken's environment by oocysts of the former may be relatively complete and continuous, because of the somewhat longer duration of active infection by *E. necatrix.*

Other fowl coccidioses are characterized by the features of the two diseases discussed above. The parasites invade and destroy the epithelium of a particular part of the intestine (Fig. 3–2). They are specific for particular hosts; that is, the coccidians of chickens do not infect turkeys, pigeons, or ducks. The symptoms of the disease are both local (loose or bloody stool) and general (droopiness, listlessness, often a fatal debilitation), and are due, of course, to the pathological lesions and to the effects of these on the whole body's well-being.

Immunity develops slowly, but a bird that survives coccidiosis is totally resistant to reinfection with the same species of parasite. This fact may be advantageous in preventing severe coccidiosis; low-level infections, as shown experimentally, produce mild disease which, upon recovery, confers immunity. Although severe coccidioses leave lasting damage (intestinal scarring or adhesions; impaired intestinal function), mild coccidioses may provide benefit by giving protection against severe disease. Coccidiacidal drugs may be used in very low concentrations as part of

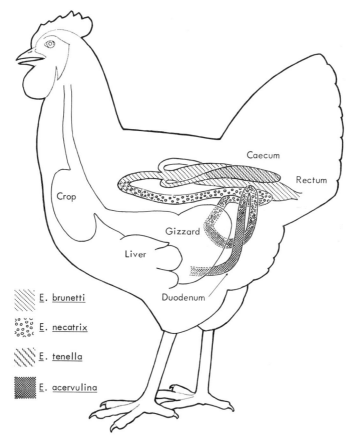

FIG. 3–2. Location of damage by several species of *Eimeria* in the fowl. Each of the four species occupies a different habitat in the fowl's intestinal tract, although some overlap the habitat of others, as *E. brunetti* with *E. tenella* and *E. necatrix* with *E. acervulina*.

the feed regularly given; apparently such drugs reduce the number of viable sporozoites reaching the tissues, and thus reduce the severity of infections while permitting immunity to develop. Combined with frequent cleaning of the pens or houses in which birds are kept, chemical coccidiostasis is effective in controlling disease and minimizing loss. Treatment of fowl coccidiosis is essentially useless once symptoms have developed. It has been suggested that the flock or population should be considered as a unit. Once symptoms of coccidiosis have appeared among its members, treatment of the flock with drugs and sanitary measures should "cure" the flock, although sick individuals would not be saved. But the best treatment is preventive: a sanitary environment, good diet, and judicious use of stasis-inducing drugs.

Coccidiosis in mammals, while not as serious or costly a condition as it is in fowl, is widespread. Furbearers, cats and dogs, rabbits, swine, sheep and goats, cattle, and horses, as well as other animals of economic importance, can harbor coccidial infections.

Although at least four species of *Isospora* are known from foxes, and two more, plus two members of *Eimeria*, from mink, these parasites are probably completely unimportant in the wild and have been of little significance in fox or mink farms, where the raising of animals on wire mesh prevents dangerous fecal contamination of the environment.

The coccidians of cats and dogs cause a typical coccidiosis, sometimes severe. Like fowl coccidiosis, the disease may be self-limiting, and confers immunity upon the host. Canine and feline coccidiosis occurs in from 2 to 20% of animals examined in various surveys in the United States.

Domestic rabbits suffer from disease caused by six species of coccidians. Hepatic coccidiosis of rabbits is a chronic disease caused by *Eimeria stiedae*, which invades the bile ducts and liver, causing lesions and hypertrophy. Intestinal coccidiosis, due to various other species of *Eimeria*, causes loss of appetite, diarrhea, and roughness of coat. The attack ranges in severity from almost symptomless to fatal. Coccidiosis is a major cause of death in domestic rabbits.

Coccidiosis of swine is due to at least five species of *Eimeria* and one of *Isospora*. This disease has been shown experimentally to cause diarrhea, loss of appetite, and failure to gain weight. Since the symptoms of even acute attacks do not include bloody stools, the disease may often be overlooked, especially in its milder forms. The importance of porcine coccidiosis is difficult to estimate.

More than eight species of *Eimeria* are found in sheep and goats. The importance of coccidiosis in these animals depends on the conditions under which the animals are kept. When lambs are held on feed-lots, outbreaks of coccidiosis occur; the result is scouring (diarrhea), loss of weight, and sometimes death. In certain regions (parts of the USSR, for instance) winter quarters for livestock tend to concentrate coccidial cysts from carrier hosts and effect transmission easily.

Cattle suffer from coccidiosis due to ten species of *Eimeria*. The importance, pathology, and transmission are typical of the coccidioses of other frequently or temporarily crowded animals. Among calves severe outbreaks bring losses from death (from 2 to 40%) as well as from lowered growth rate. In cows, milk-production is affected by more or less chronic coccidioses.

Although horses are known to harbor coccidia, the disease is not of sufficient consequence to be recognized in leading works on veterinary parasitology.

From the above brief survey of economic coccidioses, it appears that these diseases are widespread, important, and difficult to treat or control. Coccidiosis affects the production of such valuable animals as poultry and other fowl, sheep and goats, and cattle. Estimates of the extent of damage to these products are not available except in isolated instances. Figures (see Ershov, 1956) have been given, however, for losses in poultry flocks (up to 100% mortality) and in wool or meat production in sheep (28% reduction in wool, or 15% reduction in weight of lambs). In 1965, the U.S. Department of Agriculture estimated the annual loss due to coccidiosis of poultry at about $45,000,000.

Certain biological peculiarities of the disease are also apparent. The organisms, *Eimeria*, *Isospora*, etc., have relatively simple life cycles, alternating very rapid growth by schizogony in a host's tissues with effective dispersion through the oocyst stage. The persistence of the schizogonic phase in a chronic or carrier condition ensures the continuous spread of oocysts, while the durable nature of the cysts themselves aids their survival as well as their scattering by biological and mechanical agents. The persistence of coccidiosis in symptomless flocks and herds threatens young animals of each new generation with severe epizootic disease. The prevalence of Coccidia among gregarious animals illustrates the principle (elaborated in Chapter 26) that simple means of transmission serve best where available hosts are numerous or crowded. Also, man's domestic animals, especially the herd animals, have not been domesticated long enough (10,000–20,000 years) for coccidioses, mild on the natural range, under the low population densities of these animals in nature, to have become adapted (less virulent in form) under the heavy infection rates made possible by controlled husbandry. If such evidence of genetic stability can be taken seriously (and the Coccidia, being haploid during nearly their entire cycle, would be expected to behave conservatively in evolutionary respects) then control of coccidiosis by stasis chemotherapy should not rapidly create drug-resistant strains. Available methods of control, based on stasis or low initial infection coupled with sanitary animal housing, should continue to be successful where seriously applied. Thus the problem may be unlike the malaria problem of the early 1950's, where rapid acquisition of insecticide-resistance by mosquitoes forced a change of plan from control to eradication; the coccidiosis problem may yield to control alone.

THE PIROPLASMS

A second group of sporozoan diseases is the cattle fevers (piroplasmoses). These cosmopolitan parasitemias are due to minute intracellular parasites resembling the malaria organisms to some degree, but unique in their biology, especially transmis-

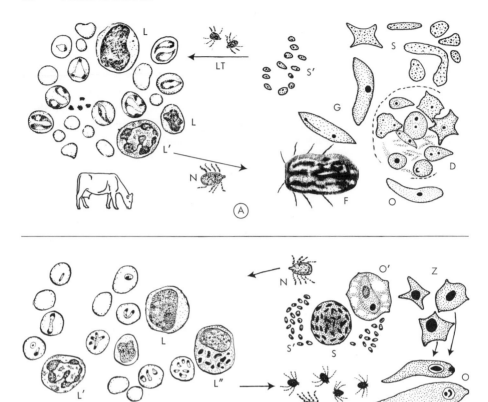

FIG. 3–3.   Life cycles of (A) *Babesia* and (B) *Theileria.* (A) At left are blood cells from a heavily infected animal. Typically paired forms, as well as other stages, are seen presumably dividing. Cells labeled (L) and (L′) are lymphocytes and heterophile leucocyte, respectively. A nymph or adult of the tick, *Boophilus,* ingests such blood. At right, stages in the tick are seen as follows: (G) blood forms, greatly enlarged, after escape from red cells into the tick's gut; (O) motile zygote, "ookinete," from the ovum of a mature female tick (F); (D) development of sporoblasts in tissues of a larval tick (LT); (S) "sporokinetes," motile forms from the larval tick's salivary glands; (S′) sporozoites from saliva of a larval tick (LT). (B) At left are blood cells. One of these (L″) is a lymphocyte filled with proliferating *Theileria* cells. The red blood cells show stages suggesting division, although division occurs primarily in lymphocytes. A nymph, larva, or adult tick may ingest blood, and development to infectivity occurs in that tick without the necessity (as in *Babesia*) for egg transmission. At right are (Z) zygotes from the tick gut; (O) motile zygotes or "ookinetes" from the tick body cavity; (O′) ookinete in a salivary gland cell; (S) multinucleate sporoblast from tick salivary gland; (S′) sporozoites from saliva of tick larva, nymph, or adult. [After Hall, 1953.]

sion. Members of the genus *Babesia* become congenital parasites of their vectors (various ticks), for within the female tick they multiply, invade the ovary and eggs, and thus are carried by the next generation. This delayed or protracted stay in the intermediate host once made very difficult the elucidation of the cycle, and still presents special problems of control. Other features of the cattle fever organisms are their variable effects on particular cattle or groups of cattle, protective premunition in chronically exposed animals, and the striking differences in virulence among geographic or even local strains of the parasite.

Babesia and *Theileria* cause cattle fever. The former has a somewhat more complicated life cycle than the latter (see Fig. 3–3). Essentially, *Babesia* multiplies asexually in the erythrocytes of the bovine host, undergoes sexual fusion in the gut of a tick, and develops from the motile zygote into numerous schizogonic stages in the tissues of the engorged tick; some of these stages then invade the tick's ovaries, where they survive within the eggs, and eventually produce fragments (sporozoites) which invade the salivary glands of the young ticks of the next generation. *Theileria*, on the other hand, multiplies asexually in the lymphocytes and endothelial cells of the lymph nodes of its bovine host, producing minute invasive forms which enter the erythrocytes but do not reproduce. These forms, taken into a tick's gut, are believed to undergo sexual fusion. At any rate, fission products (sporoblasts and sporozoites) are found in the tick's saliva, with which they are injected into the blood of a new host. Apparently *Theileria* cannot pass directly from one generation of the vector to another.

The habits of the vector are important. Adult cattle ticks of the genus *Boophilus* do not leave the bovine host except for oviposition and subsequent death. The larval ticks provide the only means of transmission, and therefore in the *Babesia* life cycle in America, where *Boophilus* is the vector, there is an obligate delay in transmission between the ingestion of piroplasms by a female tick and the injection of sporozoites by one of her offspring. It was noted earlier that pastures or feed lots, rather than infected cattle themselves, were the instruments of infection. Thus the cycle of *Babesia* requires two generations of ticks for transmission, and the parasites may be latent in a pasture for an entire season. The *Theileria* cycle, on the other hand, utilizes the habit, common in many kinds of larval ticks, of dropping off a host in order to moult. The next larval stage finds a new host, and thereby transmits the disease. Theileriasis, therefore, is characterized by a contagionlike transmission, with fairly rapid spread and with only short obligate lapses (the moulting period) between tick engorgement from an infected host and subsequent infection of a clean host.

The symptoms of babesiasis include elevated body temperature, depression, reduced appetite, jaundice, and hemoglobinuria. Death, the usual outcome in freshly exposed herds, occurs about a week after symptoms first appear. Theileriasis has a slightly different course because of the primary involvement of lymphatic cells and tissues. Swollen lymph glands, edema of various organs, and pronounced leukopenia (or lymphocytopenia) are observed. Liver damage occurs in both forms of piroplasmosis. Pulmonary edema is an important cause of death in theleriasis, whereas kidney damage and anemia are usually associated with babesiasis. Both diseases, however, are general in the sense that damaged erythrocytes and other cells upset the balance which normally exists among organs and systems. Diagnosis depends on finding characteristic parasites in the blood. Mixed infections probably occur frequently, making diagnosis an inexact procedure.

No treatment is entirely effective, although it is practical at times to treat an exposed herd by injecting all animals with a specific therapeutic. Treatment of individual sick animals should be supportive, chiefly, while allowing the drug to impede the parasite's progress. Good food, shelter, and rest are indicated.

Control of cattle fevers means tick control. As explained in Chapter 22, cattle ticks are obligate ectoparasites which must engorge between moults and before laying eggs. Pastures or feed-lots infested with infective ticks, either second-generation larvae harboring *Babesia* or larval ticks recently infected with *Theileria,* are dangerous to cattle or other susceptible stock, and should not be used. *Theileria* infection cannot last beyond the life span of the infected ticks, which are in a moulting phase, and must either feed within a definite period or die. *Babesia* infection may be latent in pastures for a longer time, since the entire period of egg-laying and larval emergence as "seed-ticks" must precede infection of a new host. The eggs and young ticks may persist up to 18 months, under favorable climatic conditions.

On the basis of the above information, two control measures are recommended: (1) rotating pasture use to permit infected ticks or their descendants to die out, and (2) removing ticks from cattle to prevent spread of infection. Various systems of dipping the animals in tick-destroying solutions will reduce substantially the number of ticks and prevent the spread of disease. In the United States, where once the entire southeastern half of the country was an endemic focus of babesiasis, the disease has disappeared, since the vector, *Boophilus annulatus,* has been effectively eradicated. Quarantine regulations between the United States and Mexico prevent reentry of cattle ticks and fever.

Before leaving the piroplasmoses, it should be pointed out that the eventual control of these diseases must come from understanding factors of immunity, host-parasite interaction, and the biology of the vectors.

Ideally, it should be possible to produce a vaccine which when injected into susceptible animals would stimulate the production of specific antibodies similar to the ones which undoubtedly exist in carriers. The difficulty with this prospect is that piroplasms exist in numerous strains, and a successful vaccine would have to contain antigen from all common strains, and would have to anticipate antigenic substances of new or mutant strains. Apparently, as the host recovers, it produces in each infection the antibody necessary to protect against only the currently infecting pathogen. Exposure to "foreign" strains of piroplasm results in a disease of which the severity is unrelated to prior infection with another strain.

The host-parasite interaction establishes in a herd relative safety from the piroplasms which occupy the herd's normal range. Probably such safety results from a historical or evolutionary process in each instance. Initial contact between the parasite and a susceptible herd kills off many animals, but it establishes some as low-level carriers, immune to damaging attacks thereafter so long as some level of infection persists. It is probable that variation among both parasites and hosts is leveled off or reduced by these processes of death and immunity, the more virulent strains of piroplasm dying out with the more susceptible hosts. Thus in time a herd of cattle becomes selectively resistant to a piroplasm, while the piroplasm becomes selectively mild to that particular herd. The parallel with human malaria is clear.

CONCLUSION

Coccidiosis, a group of diseases of poultry and other animals, is due to the growth and multiplication of intracellular parasites resembling the malaria organism in the details of their reproduction, but requiring no intermediate host for transmission. Durable and ubiquitous oocysts ensure that species of *Eimeria* survive in barnyards and chicken pens, and require that the poultryman seek to protect his flock by limiting the number of oocysts which can effectively infect susceptible birds, since it is usually impossible to eliminate these parasites completely. The concept of stasis leading to immunity has been applied successfully, and this method of controlling coccidiosis, while usually not completely successful, is practical and economical. Perhaps it should be said that under ordinary poultry-raising conditions a few coccidia are better than none.

The piroplasmoses of cattle have both historical and present interest. Historically they provided the first evidence of transmission of protozoa by an arthropod, a discovery which led to the relatively prompt control and eventual eradication of American babesiasis (Texas cattle fever) through attack upon the single-host tick which transmitted this disease. Present interest centers around two phenomena. The first of these is the difficulty of controlling many of the vectors of certain *Babesia* and *Theileria* species; such a challenge to entomologists requires that further research on tick life cycles, reservoir hosts, etc., be carried on. The second phenomenon is the plasticity of the host-parasite relationship in piroplasmosis; mild to virulent strains of parasites are of common occurrence, and immunity, perhaps an incipient racial or herd immunity, seems to protect the cattle of each region from disastrous epidemics of their local form of disease. No general principle of piroplasm control can be formulated until more is known about the many variables that characterize these diseases.

SUGGESTED READINGS

Ershov, V. S. (ed.), *Parasitology and Parasitic Diseases of Livestock.* State Publishing House for Agricultural Literature, Moscow, 1956, pp. 461–478. (English translation, Israel Program for Scientific Translations.)

LeClerg, E. L. (ed.), "Losses in Agriculture," Agriculture Handbook No. 291, U.S. Govt. Printing Office, Washington, 1965.

Levine, N. D., *Protozoan Parasites of Domestic Animals and Man.* Burgess, Minneapolis, 1961.

Morgan, B. B., and P. A. Hawkins, *Veterinary Protozoology.* Burgess, Minneapolis, 1949.

Neitz, W. O., "Classification, Transmission and Biology of Piroplasms of Domestic Animals," *Ann. N.Y. Acad. Sci.,* 64:56–111, 1956.

CHAPTER 4

BLOOD AND TISSUE
FLAGELLATES

INTRODUCTION

Chagas' disease, oriental sore, forest yaws, kala azar, African sleeping sickness, nagana, mal de caderas—these are some of the diseases caused by the blood and tissue flagellates of man and his domestic animals. The symptoms and pathologies of these diseases are as varied as their names, but the organisms which cause them are surprisingly similar to each other. They are members of a single family, the Trypanosomidae, and belong to only three genera: *Schizotrypanum, Leishmania,* and *Trypanosoma.* Almost without exception the species of these genera are transmitted by insects; all are parasites of the tissues of vertebrates, and all undergo a transformational life cycle in which some stages common to all the species can be recognized.

This group of parasites illustrates remarkably well such principles as (1) the concept of zoonosis (human disease derived from other animals), (2) the evolution of parasitism (the derivation of parasites from free-living but preadapted

49

species), and (3) the idea of physiological variation with a minimum of morphological change (the existence of races and strains within a species, or a species concept based on host-relationships). The last principle has also interesting taxonomic implications, for it helps to explain how these parasites are so widespread in the world, and how they have such a variety of hosts and modes of transmission yet remain, in appearance, so nearly identical to each other as to make the recognition of species by ordinary means almost impossible. But before we examine the above principles (which are helpful in understanding parasitism), we shall learn the basic facts about the parasites and the diseases they cause, for the latter are both interesting and important.

CHAGAS' DISEASE

Chagas' disease is a common, serious, incurable (as of this writing) condition of people of the poorer classes in South America, Central America, and Mexico. (See Fig. 4–1.) Perhaps 35,000,000 people live in areas where the disease occurs. It exists in animals in the southwestern United States, but has been reported only once from a person in that area. It is a classic example of a zoonosis. Chagas discovered in Brazil, around 1910, that bugs of the family Reduviidae (see Chapter 23) frequently harbored trypanosomes in their intestines. Trypanosomes, slender flagellates with a distinctive undulating membrane ending in a free flagellum (see Fig. 4–4), were known at that time as agents of disease in Africa, and Chagas reasoned that the blood-sucking triatomids (kissing bugs) might be the transmitting agents of an unrecognized disease in South America. He searched for the parasites in vertebrate hosts, and eventually found them in armadillos, as well as in human autopsy tissues. It became evident that the parasite multiplies in both kissing bug and mammal. During defecation the bug deposits the parasite at the place where biting occurs (usually the face, hence the name kissing bug or barbeiro), and subsequently the victim rubs the parasite into the puncture wound. The parasites are carried by the blood, but come to rest, possibly by phagocytosis, in various organs.

The blood, in fact, seldom has many of these parasites in it. Before the development of serological diagnostic tests (precipitin reactions and complement fixation, see Chapters 6 and 27), Chagas' disease was commonly diagnosed by xenodiagnosis (literally, strange diagnosis), in which an uninfected kissing bug was allowed to engorge on the patient's blood, and then, after several weeks' time, was examined for the many trypanosomes which would be found in its intestine if the bug had become infected. Xenodiagnosis is still widely used. The life cycle of *Schizotrypanum cruzi* is outlined in Fig. 4–2.

A frequently damaged tissue is the muscle of the heart. South American parasitologists have demonstrated in certain regions a strong correlation between abnormally frequent heart disease and high frequencies of infected kissing bugs. The parasites, particularly harmful to children, cause severe heart damage, fever, and edema (swelling of subcutaneous tissues, an allergic or allergylike reaction).

FIG. 4–1. Distribution of Chagas' disease. Dots indicate areas of human infection. Infected armadillos, etc., have been found over a wider range. [After Faust and Russell, 1964.]

FIG. 4–2.   Life cycle and pathogenesis of *Schizotrypanum cruzi*. The organs (A) frequently affected by Chagas' disease are shown stippled. A heart muscle cell (B) is filled with leishmania forms of the parasite. A thatched hut (C) suitable for breeding and shelter of triatomid bugs (F) is typical of Chagas' disease regions. Armadillos and other animals (D) serve as reservoirs for the parasite (E) shown in five stages as follows: (1) leishmania form, dividing into (2) leptomonads, which give rise to (3) crithidias, and (4) metacyclic trypanosomes in the body of the insect vector, with (5) the definitive form found in the blood of a vertebrate host. [(E) after Chandler and Read, 1961.]

About 10% of the infected children die from acute Chagas' disease. Adults suffer from a chronic infection which interferes with normal activity and shortens life. Up to five million people in Brazil are probably afflicted with Chagas' disease. South into Argentina and Chile and north into Mexico, millions more are infected. This disease is important not only because of the large number of victims but also because of the generally "poor prognosis." This medical term means "small likelihood of recovery or cure." In this case it applies to the public health problem of Chagas' disease as well as to individual cases, for the disease flourishes where people are poor and ill-housed. As a result of the failure of population control, there is unfortunately no bright prospect of removing poverty or improving housing in the vast "tropical slum" that is spreading over much of Latin America. Of course, the situation is not hopeless, for means of treatment, perhaps mass treatment, may be discovered at any time; in fact there are some partial solutions to the public health problem of Chagas' disease.* Reduction of contact between triatomids and human hosts is achieved in Brazil by plastering the cracked walls of mud huts with cow dung and sand, a cement used for many centuries by South American oven-birds. This practice seals up the hiding places of kissing bugs and limits their association with humans. It is a truism that infection rates depend on the frequency with which contact occurs, and that antitriatomid measures will save many lives and prevent other lives from being short and miserable.

TRYPANOSOMIASIS

African sleeping sickness is the human form of trypanosomiasis, a group of important diseases of domestic animals in all tropical regions of the world. These diseases almost certainly originated in Africa, for there the large grazing animals have the

---

* It was reliably reported, when this book was in production, that a toxin had been isolated from *Schizotrypanum cruzi* which, when inoculated into susceptible laboratory animals, caused the latter to develop a protective immunity. Perhaps this discovery has already relieved the "grave" public health "prognosis."

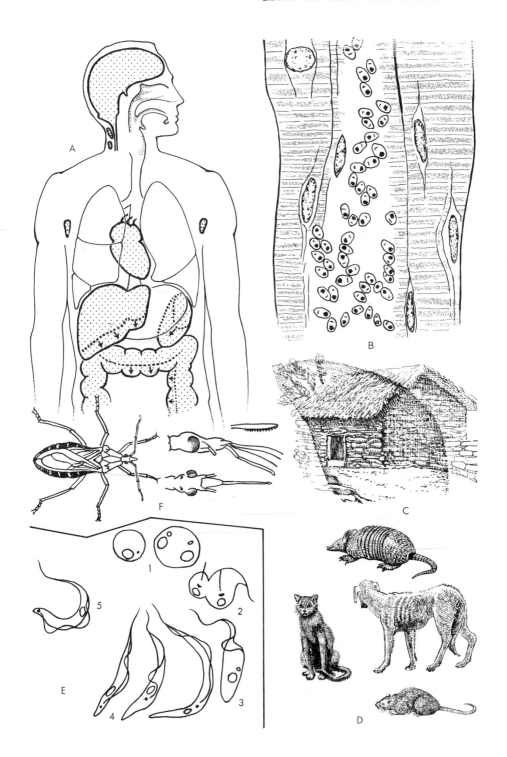

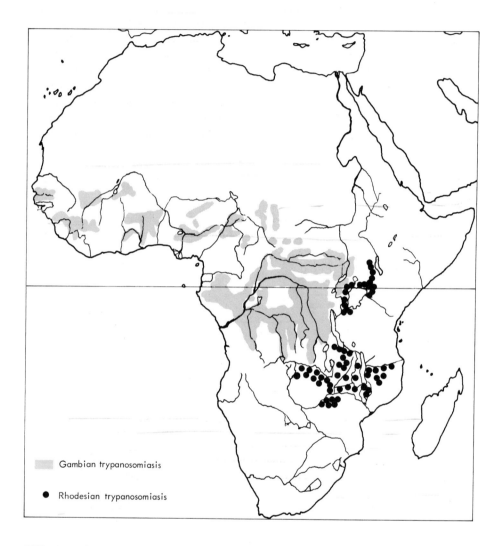

FIG. 4–3.   Distribution of African trypanosomiasis. Dots indicate Rhodesian sleeping sickness; shaded areas indicate Gambian. Trypanosomiasis in domestic animals exceeds the range indicated above. [After Faust and Russell, 1964.]

parasites in their blood, but are not harmed by them. Man, as well as the pigs, goats, and cattle which accompanied him in his prehistoric invasions of Africa, became a new host and found himself not at all well adapted to survive the new disease. In recent history large areas of Africa have been depopulated by trypanosomiasis from time to time. The opening up of trade to Africa, and especially the development of trade routes across land infested with the tsetse fly transmitter of trypanosomiasis, stimulated the spread of the disease and resulted in terrible epidemics in the 19th century. Moreover, even where people can safely exist because of conditions that interfere with frequent transmission of trypanosomes to humans by tsetse flies, nevertheless cattle and other domestic animals are exposed to trypanosomiasis and frequently become infected. Hence it has been said that one-fourth of the total area of Africa is barred to agricultural development by the presence of trypanosomiasis. Figure 4–3 shows the extent of the disease in Africa.

African sleeping sickness exists in two forms, both of which are severe and usually fatal. One, Gambian or western sleeping sickness, requires up to seven years for the terminal symptoms to develop, while the other, Rhodesian or eastern sleeping sickness, is usually fatal within 3 or 4 months. The life cycle of *Trypanosoma gambiense* is shown in Fig. 4–4.

The first symptom of infection is an itching, irritated lesion, the bite of the tsetse fly. The infected bite differs from the ordinary bite by becoming a rather large, dark-red, circular spot. Fever and headache follow in a few days. Recurrence of these symptoms at periods separated from each other by weeks to months brings on increasing weakness accompanied by a characteristic enlargement of lymph glands. Trypanosomes can be found in the blood during this period but may be quite scarce. At a relatively late stage in the disease (a few months in the Rhodesian type or several years in the Gambian) the parasites invade the central nervous system. Pathological changes in the brain, specifically an accumulation of cells ("round cells") against the walls of the blood vessels and an increase of cells in the cerebrospinal fluid, characterize this late stage of trypanosomiasis. A symptom at this time is the classical sleepiness. Increasingly frequent episodes of drowsiness merge into a terminal coma, from which the victim cannot be wakened. During the neurological phase of sleeping sickness various symptoms of brain and nerve damage become apparent. Insanity, convulsions, and tremors frequently occur. Extreme emaciation, probably due to indifference to food, is also characteristic. While some victims may recover from the early, febrile stage of the disease, there is no direct evidence that this ever happens; most authorities believe that untreated African trypanosomiasis is invariably fatal. Diagnosis of this disease results from discovery of the trypanosomes in fluid from the host. During the early stages the blood contains the parasites; they are usually found in such small numbers, however, that special diagnostic methods are required to demonstrate them. Concentration of parasites by sedimentation with the centrifuge may be used. Another kind of detection is animal inoculation, since the trypanosomes of sleeping sickness multiply rapidly in laboratory animals. In late stages of the disease, cell counts from the spinal fluid are suggestive (but they may suggest

FIG. 4–4. Life cycle of *Trypanosoma gambiense*. (A) Blood forms infective to tsetse fly: (1) undulating membrane; (2) nucleus; (3) kinetoplast (blepharoplast and parabasal body). (B) Dorsal and lateral view of resting tsetse fly, *Glossina* sp. (C) Diagram of tsetse with development of trypanosomes: (1) ingested forms found in foregut and (2) in midgut; (3) attached crithidias in lumen of salivary gland; (4) infective metacyclic trypanosomes in salivary gland and proboscis. (D) The actual host and various potential or reservoir hosts are shown. *T. gambiense* is thought to be a man-adapted strain of *T. brucei*, common in wild and domestic animals in Africa.

other cerebrospinal diseases as well), and trypanosomes may be found in the spinal fluid. Fortunately, treatment is effective. Intramuscular injections of Antrypol (Bayer 205) or Lomidine (or other pentamidines), or a combination of these two kinds of drugs which interact beneficially, can cure nonneurological cases rapidly and permanently. They have the great advantage also of prophylaxis, that is, protecting for several months an uninfected person from acquiring the infection. Unfortunately, treatment of the neurological stage of the disease is quite difficult. Arsenical drugs must be used. These are dangerously toxic and frequently ineffective as arsenic-resistant strains of trypanosomes occur. Control and prevention, as well as eradication, of human trypanosomiasis in Africa can be attempted with some hope of success, provided socioeconomic conditions make such attempts possible. Local eradication has been accomplished by mass treatment of the people with curative and prophylactic drugs, accompanied by tsetse control by trapping of the flies, clearing of fly-breeding areas, etc. (see Chapter 25). In the Gambian disease, such programs can be effective, for *Trypanosoma gambiense* is essentially a man-adapted strain or species and is probably not common in animal reservoirs. *T. rhodesiense*, on the other hand, is probably a geographical variant of *T. brucei*, which occurs in the large game animals of West Africa. Such a zoonosis is impossible to eradicate, and the human population must be protected by chemotherapeutic prophylaxis and fly control, a continuous program that cannot be relaxed.

Unable as yet to control the human trypanosomiases, man faces the same difficulties with these diseases in his livestock. In fact, certain areas of Africa are withheld from human use by the presence of pathogens of livestock just as surely as by the presence of human sleeping sickness. These areas are the tsetse lands, where wild game serves as a reservoir for the trypanosomes of man and his domestic animals, and thus where neither man nor his domestic animals can survive. It is not necessary to elaborate here the relationships between disease, disease of livestock, full use of land, stable prosperity, a healthy political and sociological climate, and indeed the welfare of nations remote from the African tsetse country. These relationships, while characteristic of most problems of health

and economics, are obvious and demonstrable in the case of the blood flagellates of domesticated animals. The following discussion will deal first with the tsetse-transmitted trypanosomiases, then with those transmitted by other means.

In central Africa livestock is often considered to be afflicted with "fly disease." Regions infested with the tsetse fly (see Chapter 24 for details of this fly, its breeding and biting behavior, and the ways of controlling it) are generally the homes of several species and many strains of pathogenic hemoflagellates, some or all of which may be present in the same herd of cattle, or even in the same host. Under such conditions, symptoms, like the pathogens, may be mixed, and a broad term like "fly-disease" may be more accurate and descriptive than the names of specific diseases.

Trypanosoma vivax is widely distributed in Africa. (It should be noted, however, that *T. vivax* is also found in South America, where the tsetse is absent.) Like most hemoflagellates, this parasite varies, presumably in genotype, from area to area; it is severely pathogenic to cattle in West Africa, mildly pathogenic in South Africa, and pathogenic chiefly to sheep in parts of South America. The symptoms are emaciation and weakness without loss of appetite. Perhaps because of the wide distribution of *T. vivax*, its transmission by different kinds of flies, and the variation in its effects on the final host, this parasite has been considered by some to be a group of species. Several of its strains bear specific names, for which some justification probably exists.

Trypanosoma congolense, together with the related parasites which comprise its group, causes severe disease in cattle, sheep, goats, and pigs. It lives as a harmless parasite in the blood of wild animals in tsetse-infested areas. *T. ruandae, T. somaliense*, and several other forms described in the literature may be strains or local varieties of *T. congolense*. Since such descriptions are based on the appearance, shape, and form of blood parasites and since these characteristics may vary widely within the same host individual, it is not easy either to recognize established species or to define new ones. *T. simiae*, which resembles *T. congolense* in general but on the average is longer than the latter and somewhat more variable within a single host, causes an extremely acute disease in pigs.

Trypanosoma brucei, named for the famous investigator of trypanosomiasis and brucellosis, is pathogenic chiefly to dogs, horses, asses, and camels, rather than to ruminants. The disease is similar in symptoms to the condition called "surra" in India, "mal de caderas" and "derrengadera" in South America, and "murrina" in South Africa. It differs from these diseases chiefly in mode of transmission, since the latter are spread by various bloodsucking insects other than tsetse flies.

The non-tsetse diseases are caused by *Trypanosoma evansi* or a number of related species. Among these are *T. berberum* of Algeria, *T. marocanum* of Morocco, *T. annamense* in Annam, *T. soudanense* in Egypt and Algeria, *T. hippicum* in Panama, *T. venezuelense* in Venezuela, *T. equinum* in Argentina, and *T. ninae kohl-jakimovi* of central Asia. The fact that the *evansi*-caused diseases closely resemble those caused by *T. brucei* in the tsetse areas suggests to some parasitologists that members of the *evansi* group represent variant strains derived from the *brucei* group, which, by becoming adapted to mechanical transmission by various insects in which no cyclical stages are found, have successfully invaded areas beyond the tsetse country in Africa. Transmission is practically limited to warm humid periods, suitable for survival of the parasites in or on the proboscis of the mechanically transmitting arthropod. Reservoirs of infection are cattle and wild game, which serve as symptomless carriers.

Trypanosoma equiperdum of horses requires no intermediate host whatsoever. It is transmitted during copulation, spreading from the genital mucous membrane of one host to another. It may be transmitted occasionally by biting flies. The parasite resembles *T. evansi* very closely in size and shape. The disease, called "dourine," runs a characteristic course somewhat different from the course of other trypanosomal diseases of stock. First there is edema of the genitals, with some discharge from the vagina of affected females. Some areas or spots of the mucous membranes of vulva and penis lose pigmentation. After about a month urticarial plaques (a rash characterized by raised, clearly demarcated, circular flat areas under the skin) appear, disappear, and reappear on the sides of the body. Eventually neuromuscular symptoms supervene, and in certain groups of muscles paralysis develops and eventually becomes general. Mortality varies among breeds and among areas. In some regions dourine is very mild, while in others up to 70% fatalities occur. Apparently the disease is milder in the cooler latitudes. There are no reservoir hosts; the chronic or carrier infections of normal hosts serve the function of maintaining the disease.

Treatment of the various trypanosomiases of livestock depends on several factors, including the value of the particular animal affected, the risk of spreading infection, and the effectiveness of available therapeutic measures. Injection with various drugs is effective in treatment of valuable livestock, such as camels or horses. Prophylactic doses of such drugs may protect animals or herds temporarily exposed to infection, and may prevent considerable loss. In regions bordering on areas where *evansi* infection or dourine is common, it may be wise to destroy infected animals to prevent further spread into uninfected flocks. One serious risk of drug therapy is the creation, by inadequate dosage, of drug resistant strains of trypanosomes. The drug dimidium bromide used in combination with Antrycide shows promise of being prophylactic and curative of animal trypanosomiasis. Control, however, aimed at reduction in tsetse flies or other transmitters and at recognition and elimination of reservoirs and carriers, seems to be a better long-range plan than continued attempts at treatment.

Certain miscellaneous hemoflagellates should now be mentioned. *T. lewisi*, of rats, is of interest because as a laboratory animal it has yielded valuable information on immunity. The discovery of ablastin by Taliaferro in the twenties led to many experiments, and possibly to an understanding of the nature of relapse in several diseases (see Chapter 27). *Schizotrypanum cruzi* occurs frequently in dogs, where it causes a condition similar to visceral leishmaniasis (see below). *T. melophagium* of sheep is very common and is probably worldwide, but the degree of parasitemia is so slight that blood culture is necessary for diagnosis, and the organism is probably not pathogenic. It is transmitted by the sheep ked *Melophagus ovinus*. *T. theileri* of cattle is another widespread but nonpathogenic trypanosome.

Veterinary trypanosomiasis in general resembles the human diseases caused by the hemoflagellates; however, in the animal forms the variety of causative organisms is much greater, and the distribution is more nearly cosmopolitan. Whereas only two trypanosomes in the tsetse country of Africa and one more in triatomid-infected areas of South America cause human disease, the animal diseases are caused by (1) the *vivax-brucei-congolense-simiae* complex of tsetse-borne organisms in a restricted African zone, (2) the *evansi* complex of mechanically transmitted trypanosomes more widely distributed as surra, mal de caderas, etc., and (3) the contagious species *equiperdum* capable of occurring wherever horses are bred or transported. The diseases characteristic of these parasitisms are in general like those of man, that is, usually gradual in onset, sometimes fatal in outcome, and often debilitating, running a chronic course. Also, the epidemiological problems are basically similar, involving arthropod transmission and the existence of numerous reservoirs as well as symptomless carriers. Treatment is still not completely satisfactory; in fact it is probably futile in Chagas' disease of man or dogs and is of doubtful value in trypanosomiasis of cattle. Apparently control aimed at eradication of trypanosomiases of livestock is a reasonable approach just as it is for the disease in man.

Perhaps it is practical to view trypanosomiasis as a condition—not alone of man or his stock, but of man's life and livelihood. Trypanosomes in nature represent a great potential for harm. They are like such elemental forces as windstorms, vulcanism, or floods. From time to time these forces become organized into great catastrophes. It was once thought that nothing could be done but to yield to such forces and attempt to survive. Gradually man has learned to predict hurricanes and typhoons, to chart the seismic faults which produce earthquakes and volcanic eruptions, and to build dams and reservoirs to prevent floods. As more becomes known about trypanosomiasis in all its forms, epizootics can be predicted and prevented, vectors can be destroyed or avoided, and treatment of carriers can remove the threat from the immediate herd to the background of game animals or the untreated herds of distant neighbors. Thus the condition described above will yield to scientific discovery and to the common sense and common effort of stock breeders and public health officials all over the world.

LEISHMANIASIS

Leishmaniasis is a group of diseases caused by organisms closely related to *Schizotrypanum* and the trypanosomes. The classification and evolution of these parasites will be discussed later in this chapter following a description of the diseases.

Cutaneous and mucocutaneous leishmaniasis are two forms of a rather similar condition caused by the presence and multiplication of the nonflagellated stage of the parasites in cells near the surface of the body.

The classical lesion of cutaneous leishmaniasis is the oriental sore. The relatively dry sore, about 5 to 15 cm in diameter, requires several months from inoculation to development; it persists for several months, heals spontaneously, and results in lasting immunity against reinfection. Classical sores are found from the Mediterranean region to India and southwestern Asia. Parts of Africa are also involved. Another form of oriental sore is relatively wet, requires a shorter incubation time than the classical sore, and usually involves lymph nodes as well as the initial site where the transmitting sand fly has injected the parasites. These two sores are epidemiologically, as well as pathologically, distinct; the former is probably a truly human condition (although dogs are susceptible), in which the parasites are well adapted to human skin and the human host is reasonably tolerant and ultimately resistant to the parasite. The wet sore is probably a zoonosis, since it occurs in rural areas where ground squirrels and gerbils maintain it in their territories and where it is transmitted by a *Phlebotomus*, which lives with these burrowing rodents.

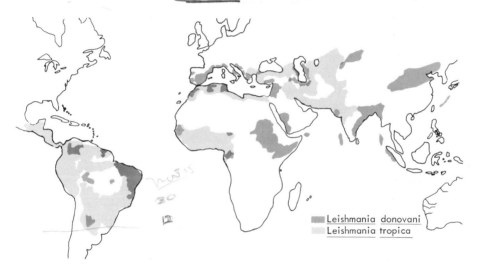

Leishmania donovani
Leishmania tropica

FIG. 4–5. Distribution of leishmaniasis. South and Central America contain foci of both visceral leishmaniasis and classical oriental sore, although the common leishmanial infections in the New World are probably strains of *L. tropica*. [After Faust and Russell, 1964.]

Mucocutaneous leishmaniasis is characteristically a neotropical zoonosis. The various forms it takes are evidence that the parasites exist as local or geographical strains. In the forest of the Amazon Basin a particularly harmful strain causes destruction of the membranes and underlying hard tissues of the nose and mouth. This condition, called "espundia" and "forest yaws," is usually acquired far from human habitation by workers or explorers in the deep forest. Like other leishmaniases, it is probably transmitted by sand flies, but these insects, in the case of

Fig. 4–6. Strains of *Leishmania tropica*. Vectors, known reservoir hosts, and human victims of cutaneous leishmaniasis appear in their geographical settings: in South America, mucocutaneous infection; in Mexico, chiclero ulcer; in Egypt, multiple facial lesions; and in India, the classical "dry" sore.

"espundia," must always acquire their parasites from some wild animal of the jungle, as yet unidentified. Less severe New World cutaneous leishmaniases occur in the Andean highlands (Peruvian "uta"), Central America ("buba" or "pian bois"), and Mexico ("chiclero ulcer"). Probably the Old World oriental sore has also been introduced into the American tropics, and lesions resembling this classic disease, that is, not involving mucous membranes and not characteristically attacking the ears (as in Mexican chiclero ulcer), can probably be attributed to oriental or Mediterranean strains of the parasite. The world distribution of leishmaniasis is shown in Fig. 4–5.

Leishmania tropica (Fig. 4–6), the various strains of which can be recognized only by the reactions of their hosts, yields readily to chemotherapy. Antimony tartrate, applied locally after removal of scabs and crusts, is effective in healing ordinary leishmania ulcers. In mucocutaneous disease and when lymph nodes are involved, injection of antimony compounds is indicated. Because secondary infection by bacteria, fungi, or treponema may be present, the use of antibiotics and antiseptic precautions should be considered. It is interesting to note that preventive inoculation was first employed against oriental sore. Mothers in ancient times deliberately infected their children with material from active sores in order to prevent the later acquisition of a disfiguring facial ulcer. The relationship between this age-old practice and the modern use of vaccination is well known to medical historians.

Visceral leishmaniasis, while caused by an organism quite similar to the above mentioned *Leishmania tropica* and transmitted similarly by sand flies, is a more serious disease than even the painful and sometimes destructive cutaneous leishmaniases. The parasites *Leishmania donovani* (see Fig. 4–7) are injected by the bite of sand flies. Once in the skin, these parasites are engulfed by phagocytes of the connective tissue. The phagocyte usually involved is the macrophage, a large, lymphoid wandering cell. Instead of being destroyed by the phagocytes (which usually function by neutralizing or dissolving foreign organisms), the leishmania bodies reproduce to pack the infective cell, which bursts. In this way the parasites

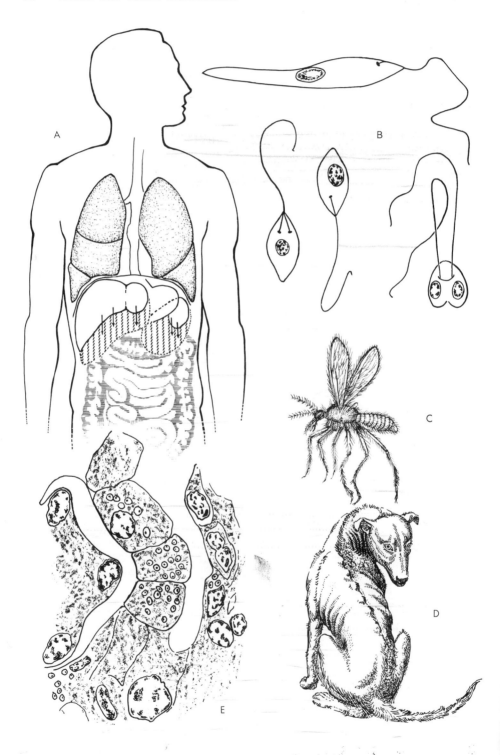

FIG. 4–7. Life cycle of *Leishmania donovani*. Enlargement of liver and spleen is indicated by arrows in (A). Leptomonad forms of the parasite in culture (B) or in the sand-fly vector (C) are transmitted by bites to man or (sometimes) a reservoir such as the dog (D). Multiplication of leishmania forms in the macrophages of liver (E) or spleen cause the disease known as kala-azar.

spread from phagocyte to phagocyte. During a period ranging from a few weeks to a year or more this slow spread among the phagocytic cells continues. Some monocytes get into the blood stream via the lymphatics, and are carried to many organs. Clinical or symptomatic kala-azar develops when the phagocytes of liver, spleen, kidneys, and bone marrow become parasitized. Irregular fever, weakness, and headaches occur; the spleen swells greatly, liver function is disrupted, and, worst of all, the normal pattern of the blood cell population changes. In response to the destruction of a great number of monocytes, many monocytes are produced by the blood-forming red bone marrow apparently at the expense of other defensive cells, such as neutrophiles (polymorphonuclear leucocytes). This reduction in the number of all the white blood cells except monocytes, a condition called "monocytosis with leucopenia," seriously affects the body's defense against all sorts of invading organisms, and the victim of kala-azar usually dies from some infection which he would normally be able to resist. One of the unsolved mysteries about visceral leishmaniasis, however, is the apparent variability in response to this disease even in people living in the same area. Most of those who are actually ill die if they are untreated; but others undoubtedly acquire the disease yet suffer very little from it and become permanently immune upon recovery. That this is true is demonstrated by the severity of the occasional epidemics and the fact that the morbidity rates are higher among children than among adults. The only explanation can be found in the theory that each epidemic leaves many survivors who have become immune. Treatment of kala-azar, as in cutaneous leishmaniasis, is quite effective. Antimony compounds are injected intravenously or intramuscularly; pentavalent compounds such as Neostibosan and Solustibosan are most effective and best tolerated. Supportive or symptomatic treatment by good nursing care, enriched diet, and even blood transfusion is important in severe cases. After recovery, a skin condition called dermal leishmanoid sometimes results. White patches (depigmented areas) appear in the skin about one year after recovery. The appearance of these patches is followed by dermal nodules in which leishmania organisms occur within the monocytes. Since the organisms can be found in recovered cases in internal organs (by spleen or bone-marrow biopsy, for instance), it is believed that "post kala-azar dermal leishmanoid" demonstrates effective, but not complete, immunity to the parasite. Just as does *Leishmania tropica*, so *L. donovani* exists in several geographic strains, each somewhat different from the others in its host relationships. Thus in China, kala-azar has an important reservoir in dogs, which suffer from extensive cutaneous lesions. Sand flies easily obtain the parasites from surfaces, and thus transmit the disease to humans. In India, on the other hand, kala-azar has no zoonotic features, for it is transmitted primarily from person to person by means of sand flies living in moist, vegetation-shaded areas near houses. Epidemics of visceral leishmaniasis have occasionally occurred, with infection rates seemingly far higher than man-to-man sand fly transmission or acquisition from animal reservoirs would explain. It has been shown that ingestion of the organisms can cause kala azar, and it may be assumed that a considerable number of human cases may result from contact of contaminated food and water rather than from insect vectors. In the Mediterranean region, dogs are important reservoirs; children in this area are more frequently infected than adults. In the Sudan, domestic vectors are important, and the disease is both extremely virulent and difficult to cure. In South America, the disease shows similarity to several of the above varieties; it is clinically similar to the Indian form, but is found in dogs and other reservoirs as it is in China. It seems reasonable to believe that South American kala-azar, like some forms of South American dermal leishmaniasis, may be an imported disease —possibly a mixture of several exotic strains. The significance of geographical variation will be discussed later, when the classification of trypanosomes and leishmanias will be considered. The importance of visceral leishmaniasis, a very severe disease capable of epidemic outbreaks and quite persistent in the many regions where it affects millions annually need not be emphasized.

SOME PRINCIPLES OF PARASITISM

The dangerous parasites (the trypanosomes and leishmanias) discussed above show remarkable abilities of an evolutionary nature. This aspect of parasitism, discussed in general terms in Chapters 26 and 29 and implied, of course, in the host relationships of *Plasmodium* (above), explains rather well many puzzling facts. Some puzzles about the hemoflagellates are their curious metamorphic cycles, their unusual modes of transmission, their host relationships among vertebrates, and their variation in virulence and pathological effect.

The hemoflagellates go through two or more of four distinct stages in their life cycles. One of these stages is a simple flagellate form called the leptomonas stage. It has a characteristic elongate, tapered body, central nucleus, subterminal parabasal body or kinetoplast (granule or anchor for the flagellum), and a terminal flagellum. Another stage, called leishmania, is nonflagellate. The body is more or less rounded and has a nucleus and kinetoplast, with a short fibrous structure attached to the latter. The two other stages are flagellate and in addition are equipped with an undulating membrane which connects part of the flagellum to the side of the body. The so-called crithidia stage has a flagellum and membrane extending only halfway along the body; the nucleus and kinetoplast are displaced somewhat toward the rear of the animal, and the undulating membrane and flagellum lie in the anterior region. The **trypanosoma** stage has a complete undulating membrane, with the kinetoplast posterior to the nucleus. These four forms are named for certain genera of hemoflagellates in which they occur. Because it is somewhat confusing to use the same word for a form and a genus, we should remember that the name of a genus (unlike ordinary words, such as crithidia or leptomonas, which refer to structural entities or forms) is always capitalized and must be either italicized or underscored. The genera of the family Trypanosomidae are defined not only in terms of the forms found in their life cycles but also in terms of their host relationships. Thus the familiar genera discussed above are defined as follows:

Trypanosoma, with the trypanosoma form in the blood or other tissues of a vertebrate, and the leptomonas, crithidia, and leishmania forms either absent or developing into trypanosoma forms in an intermediate host;

Schizotrypanum, with both trypanosoma and leishmania forms in the tissues of vertebrates, and the other stages, including trypanosoma, in the intestine of a reduviid bug;

Leishmania, with the leishmania form in the tissues of vertebrates, and only the leptomonas and leishmania forms in the intestine of insects.

(A few other genera are defined in Chapter 29 as parasites of invertebrates or of plants exclusively. See especially Fig. 29–1.) As is pointed out in Chapter 29, this group of parasites shows by its present-day range of hosts (plants, insects, vertebrates) and by the life cycles of its more complex genera (*Trypanosoma, Schizotrypanum, Herpetomonas*) that these parasites of vertebrates are probably descended from parasites of invertebrates. The similarities of the genera *Phytomonas* and *Leptomonas* (of plants and insects, respectively) suggest an ultimate origin of this group of protozoa from free-living or saprophytic species.

Moreover, in the genus *Trypanosoma*, the present-day list of means of transmission reveals one or more evolutionary relationships. *Trypanosoma brucei*, as well as the very similar *T. gambiense* and *T. rhodesiense*, requires a period of time between one vertebrate host and the next during which the parasites move into the tsetse's stomach, multiply there, and then invade the salivary glands, where metacyclic (terminal stage) trypanosomes develop. These metacyclic forms are injected into the next host. *Trypanosoma vivax*, an important parasite of all large domestic animals and nonpathogenic only in dogs, undergoes multiplication and cyclic transformation in tsetse flies; however, this change occurs only in the proboscis, for it is unable to develop in the tsetse stomach or intestine. *T. vivax* can also be transmitted mechanically, and it exists outside tsetse areas. In much of Africa and South America this trypanosome is transmitted by tabanids and by stable flies of the genus *Stomoxys* (see Chapter 24). *Trypanosoma evansi*, on the other hand, does not develop cyclically in any insect vector but is mechanically transmitted by the contaminated proboscis of various biting insects, including the tsetse fly. Tabanids (see Chapter 24) are important vectors of this trypanosome of camels. *Trypanosoma equinum*, parasite of horses in Africa and South America, is transmitted mechanically also. (It is noteworthy to mention that each of the above two parasites can be transmitted by vampire bats, in the blood of which the trypanosomes live and multiply.) Another parasite of horses, *T. equiperdum*, has no intermediate host, for it is transmitted during copulation. These examples from the genus *Trypanosoma* illustrate a possible evolutionary series. Although it cannot be demonstrated whether loss of a vector or intermediate host is primitive or modern, reason clearly supports the view that trypanosomes, at least those tsetse-derived forms discussed above, show progressive independence of the insect host with complete freedom, in the case of *T. equiperdum*, of any intermediate host at all. The importance of these variations in vector relationships is not so much that they show speculative trends or directions, but in the fact of parasitic ingenuity (in the evolutionary sense) which they so clearly demonstrate.

It is not surprising, therefore, that within each species of *Trypanosoma* there are genetic strains or varieties. There is good reason to consider *T. gambiense* to be a man-adapted strain of *T. brucei*, for only minute distinctions such as those based on quantitatively estimated morphological measurements, the requirements and tolerances of the two forms in laboratory culture media, and their respective virulence for laboratory animals (far from obvious differences) can be made between these parasites of domestic animals and man. The susceptibility of certain wild reservoirs of *T. brucei* to *T. gambiense* also suggests, of course, a common origin for these species. *T. rhodesiense* is believed by many authorities to be a recent man-adapted form of *T. brucei*; its extraordinary virulence and sporadic occurrence in thinly populated areas are suggestive of a zoonosis—an invasion of man by a parasite normally adapted to some other host. Even within well-defined species, such as *T. congolense*, strains or varieties have been recognized. Particularly important are the strains that occur in drug-treated herds, strains which survive such drug treatment and threaten to replace the drug-susceptible parasites. Drug-resistance is another example of rapid and continuing evolution among these protozoan parasites.

Variation in virulence and variation in tissue and organ specificity are observed in all hemoflagellates. The difference between western and eastern African sleeping sickness, between wet and dry forms of oriental sores, between the kala azars in India and in China, and among the many varieties of New World leishmaniasis, to say nothing of the different kinds of animal trypanosomiases noted above, all point unmistakably to continuing evolution. Therefore, what value can be derived in naming these parasites, that is, assigning them to species? Two reasons can be given. First, it is obviously valuable to have names for definable entities, and it may be argued that a scientific name which can be standardized internationally is more useful than a local, geographical, or clinical name. Thus it might be helpful to accept *Leishmania brasiliensis* for *L. tropica* in South America, or even to divide *L. tropica* still further. Second, physiological differences are just as real as, and probably more important than, the morphological differences upon which species are usually erected. Indeed, it is possible to define the hemoflagellates most accurately in terms of such differences. As we have seen, the *brucei* complex of trypanosomes is divisible by host relationships (*brucei* vs. *gambiense*, *rhodesiense*) and by virulence (*gambiense* vs. *rhodesiense*) as well as by geography (eastern-vs.-western sleeping sickness). Thus in spite of the undoubtedly close relationship within these three groups of organisms, as is evident in the morphological continuities among them, there are sound biological bases for separating them into taxonomic categories. Isolation and behavioral differences here form the bases of species recognition. The changes which all species in these groups of parasites are undergoing may be more sudden and perhaps less predictable than evolutionary change elsewhere, but species are no longer believed by any biologist to be eternal; species are the present approximate definitions of real groups of animals or plants, named for convenience and for communication. Since it is useful to have names for various hemoflagellates, no further justification in naming them is needed.

The preceding pages of this chapter dealt with the hemoflagellates as agents of disease and as interesting examples of evolutionarily fluid categories, the present-day species. The remaining portion will consider these parasites as a series of problems or work for the future.

PROBLEMS AND PROSPECTS

A primary problem in trypanosomiasis and leishmaniasis alike is the way in which these parasites harm their hosts. This problem is not unique to these pathogens; microbiologists and physiologists have sought with little success for an understanding of the general mechanisms of pathogenesis. Bacterial toxins, of course, are known to injure hosts. Host tissues may be lysed (dissolved) by histolytic enzymes produced by a variety of agents, including the pathogenic amoeba. Host response, the immune reaction itself (see Chapter 27), may be harmful. But there are many diseases—those caused by the neurotropic viruses, by the man-adapted bacteria such as leprosy and tuberculosis bacilli, by treponemes of syphilis and yaws, and by many other pathogens—where the way in which the pathogen injures the host is not known. Lack of this knowledge hinders treatment, especially that which is supportive or symptomatic.

The fact that individuals vary in their response to trypanosomiasis and leishmaniasis suggests that the parasites are on the evolutionary verge of being tolerated. Perhaps knowledge of factors involved in the natural tolerance possessed by some individuals among a susceptible population will suggest ways to increase tolerance, i.e., to render victims of sleeping sickness, kala-azar, etc., symptomless.

Although the possession of toxins by hemoflagellates has not yet been demonstrated (see footnote on page 52), mechanical injury may be attributed to both *Schizotrypanum cruzi* and the *Leishmania* species, for these are intracellular parasites which severely damage host cells as well as tissues. Thus the heart muscle, in Chagas' disease, and the reticuloendothelium (system of monocytes, macrophages and lymphoid cells) in visceral leishmaniasis, both of which are invaded by the parasites, fail in function; chronic heart disease in the one case and greatly lowered resistance to miscellaneous infection in the other result directly from the parasites' action. Obviously, any cure for these diseases must include destruction or inhibition of the parasites.

Laboratory studies now being carried out on metabolism and specific blood and tissue alterations in infected hosts should prove fruitful. Although results so far have been rather inconclusive, studies on glucose metabolism of infected hosts have disclosed profound alteration in blood sugar levels, in the last stages of sleeping sickness, for example. More refined comparisons between the biochemistry of normal and diseased hosts can be expected to result in better understanding of metabolic factors in pathogenesis.

Therapy itself is a promising field of study. The discovery that two trypanosomicidal drugs mentioned earlier (Antrypol and the pentamidines) have a mutually reinforcing effect is a finding of great practical value. It would be desirable, of course, to know the exact mode of action of drugs in current use, for from such knowledge might come new treatments which would take full advantage of as yet unsuspected specific effects upon the parasites. Of course, this deficiency in knowledge of how therapy works is rather widespread in all fields of medicine; the subtle distinctions that make it possible to destroy a tissue parasite without killing the neighboring host cells have their counterpart in the problem of cancer therapy, which probably can be solved only when more is known about growth, development, and differentiation—one of the truly basic problems in the study of life itself.

Immunity to hemoflagellates has been studied perhaps longer than any other aspect of these diseases; the observations, as we have seen, go back beyond the early days of modern medicine (Jenner's time, circa 1800) all the way into prehistory and the shrouded origin of protective inoculation with oriental sore. Observations on Gambian sleeping sickness suggest "self-cure," or spontaneous recovery, in some cases. Also, the extremes of variation in host response to trypanosomiasis in general suggest strongly that not only the parasites but also the hosts are varied in constitution; at least some of the hosts are capable of resistance which is probably immune in type. Specifically, an immune reaction is known in *Trypanosoma lewisi* of rats, and in *T. duttoni* of mice (Fig. 27–3). After an initial rise in the number of trypanosomes in infected animals, the population becomes stable and dividing forms are absent. Morphological variety, characteristic of growing populations of trypanosomes, is replaced by morphological uniformity. Then the population slowly declines as phagocytes, aided by opsonins, destroy the trypanosomes. *T. lewisi* or *T. duttoni* infections in guinea pigs, however, run a different course. The parasites never lose their variability in form, while their numbers fluctuate sharply as the host produces antibodies against each new peak in parasite growth. Although the guinea pig quells several such population surges, eventually the host weakens and dies. It is now known that guinea pigs lack an antigrowth substance (named "ablastin") which is part of the rat-mouse antibody against *T. lewisi* and *T. duttoni*, respectively, and we can reasonably conclude that this substance, by preventing further reproduction in the parasites, gives time for the attack by phagocytosis to be effective, and, of course, prevents the emergence of phagocyte-resistant strains.

On the basis of what is now known about immunity to hemoflagellates, it is probably too early to hope for practical immunization tomorrow or next year. The goal is worth working for, especially since there are so many weaknesses in other approaches to the prevention, control, and eradication of the trypanosomiases and leishmaniases.

Control of hemoflagellates presents some peculiarly difficult problems. Three factors in kala-azar—transmission, treatment, and reservoir hosts—must be considered. In India, where reservoir hosts are not important, treatment of human cases and destruction of sand flies alone can greatly reduce the amount of visceral leishmaniasis. In the countries around the Mediterranean, however, where dogs are frequently carriers of *L. donovani*, nothing short of eradication of this disease from man and dog would probably work. Human trypanosomiasis in Africa presents two contrasting pictures of control. Gambian sleeping sickness is thoroughly man-adapted; prophylaxis by drugs (pentamidines), if applied to all persons living in tsetse areas, would greatly reduce the numbers of available trypanosomes and might completely stop transmission. Tsetse control would have a very helpful effect, too. Rhodesian sleeping sickness, on the other hand, is a true zoonosis; control of that disease would mean extermination of the reservoir hosts (the great game animals) or the tsetse fly. The wearing of suitable clothing to prevent bites from the tsetse and prophylactic therapy to kill the invading parasites themselves seem to be rather effective measures if properly applied. But control of the size of the *T. rhodesiense* populations in the wild is impossible. Control of oriental sore is perhaps not so pressing a problem as that of some other diseases because of the rather effective treatment for this unpleasant but not necessarily disabling condition. Infection may be prevented by avoiding sand flies, but of course that cannot be done by the very poor in a tropical slum, or by chicle hunters in a Central American rain forest. Because reservoirs exist for all kinds of *Leishmania tropica*, it is unlikely that control of these parasites will be effective. In Chagas' disease, the three factors—treatment, transmission, and reservoirs—must be dealt with. Unfortunately, treatment is unknown. Transmission by triatomid bugs can be stopped by a simple process, i.e., ridding a house of cracks in which the bugs can hide and breed. Reservoirs, however, are so varied and so common that there is no hope of eradicating Chagas' disease. Perhaps if treatment can be developed and housing conditions improved in tropical America, *Schizotrypanum cruzi* can be driven back into the animal community of opossums and armadillos where it existed before man arrived in the Americas.

The problems sketched above show the difficulties which confront victory over the hemoflagellates of man and beast. The future of these problems cannot be foreseen. A look at another problem may help to show the hemoflagellates in fair perspective and may relieve somewhat the pessimism of the preceding pages.

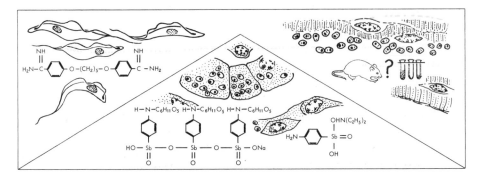

FIG. 4–8. Therapy in trypanosomiasis and leishmaniasis. (Left) Typical trypano-somes with the formula of a common remedy, pentamidine. (Center) Leishmanial infection with two drugs, stibamide glucoside (Neostam) and diethyl amino-p-aminophenyl stibinate (Neostibosan, Bayer 693). (Right) Rat and test tubes symbolizing continuing research in therapy of Chagas' disease.

Malaria (see Chapter 2) is viewed by many as a problem being solved. Malaria eradication has been successful in many regions where this disease was once overwhelmingly the major health problem. Stable malaria of the jungles and rain forests, where extremely primitive people still live in a stone age that is barely more accessible to modern man than the stone ages of the past, seems an unlikely prospect for eradication. Perhaps stable malaria presents the same epidemiological difficulties as a zoonosis. But the conquest of malaria in the populated lands is certain, because we have the scientific and technical tools to accomplish it and because the leaders of mankind wish it.

We may now consider the hemoflagellates as a problem similar to malaria. What are the prospects for solution? Without reiterating what has already been stated in this chapter, we can remind ourselves that there are several kinds of trypanosomiasis and leishmaniasis and that each has its own peculiar difficulties. These difficulties include treatment (of Chagas' disease), vector control (of all, but particularly of the African trypanosomiases), and the existence of animal reservoirs (of most forms of leishmaniases, Rhodesian and veterinary trypanosomiasis, and Chagas' disease). Research in all these areas is active (see Fig. 4–8). Improvement in chemotherapy is unpredictable, and Chagas' disease may be conquered by drugs, by immunization, or by strict vector control. Tsetse trapping has been used to reduce transmission of Gambian sleeping sickness; studies of insect behavior may soon reveal tropistic factors in tsetse flies that will doom these vectors. Systemic insecticides, which kill arthropod parasites and predators upon cattle, may eventually be made suitable for controlling the transmission of animal trypanosomes. Research should be encouraged in these and many other directions; there is plenty of money in this world's affluent society, and the explosion of knowledge is occurring in parasitology as well as in space science. We should be confidently impatient to remove the hemoflagellates from tomorrow's brave new world.

SUGGESTED READINGS

Expert Committee on Trypanosomiasis, 1st Report, WHO Technical Report Series No. 247, Geneva, 1962.

Faust, E. C., and P. F. Russell, *Craig and Faust's Clinical Parasitology*, 7th ed. Lea and Febiger, Philadelphia, 1964, pp. 104–166.

Hoare, C. A., "The Epidemiological Role of Animal Reservoirs in Human Leishmaniasis and Trypanosomiasis, *Veterin. Rev. and Annot.*, 1:62–68, 1955.

Hoare, C. A., "The Relationship of the Hemoflagellates," Proc. 4th Int. Cong. Trop. Med. and Malaria, U.S. State Dept., Washington, 2:1110–1116, 1948.

Hutchinson, M. P., "The Epidemiology of Human Trypanosomiasis in British West Africa." *Ann. Trop. Med. Parasit.*, 47:156–182, 1953, 48:75–94, 1954.

Levine, N. D., *Protozoan Parasites of Domestic Animals and Man.* Burgess, Minneapolis, 1961.

Morris, K. R. S., "New Frontiers to Health in Africa," *Science* 132:652–658, 1960.

Noble, E. R., "The Morphology and Life Cycles of Trypanosomes," *Q. Rev. Biol.*, 30:1–28, 1955.

Ormerod, W. E., "The Epidemic Spread of Rhodesian Sleeping Sickness, 1908–1960," *Trans. Roy. Soc. Trop. Med. Hyg.*, 55:525–538, 1961.

Williamson, J., "Chemotherapy and Chemoprophylaxis in African Trypanosomiasis," *Exptl. Parasit.*, 12:274–322, 1962.

CHAPTER 5

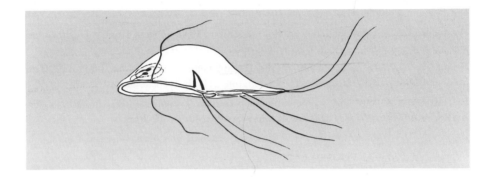

OTHER FLAGELLATES

While the blood and tissue flagellates just discussed are by far the most important members of the class Mastigophora in terms of human and animal welfare, there are other flagellates of considerable significance to parasitology.

LUMEN FLAGELLATES OF MAN

Only two of the species of flagellate protozoa inhabiting the alimentary, urinary, or genital ducts of man cause disease. Of these, *Trichomonas vaginalis* is found in the vagina of women and the urethra and prostate gland of men, while *Giardia lamblia* (or *G. intestinalis*), occurs in the duodenal region of the intestine. The harmless flagellates all occur in the lower intestine except *Trichomonas tenax*, which lives in the mouth. Flagellates related to the human forms are found in many other vertebrates, including domestic animals and fowl. Many of these parasites can be reared successfully in foreign hosts; all except *Giardia* can be rather easily maintained in pure, axenic (bacteria-free) culture. These facts indicate considerable adaptability of the lumen-dwelling flagellates. It is noteworthy to mention that such tolerance to environmental variation does not permit variation in habitat within the host, for the three species of *Trichomonas* in man cannot be made to live in organs other than the mouth, vagina, or colon, respectively, although each species has been transferred successfully to a corresponding site in a nonhuman host.

The significance of the nonpathogenic flagellates is merely diagnostic and epidemiological. It is possible to confuse the cysts or trophozoites of several of these with pathogenic forms, and the laboratory technician must be aware of this danger. Also, the incidence of any of the intestinal flagellates is an index of the sanitation level in the community. Along with data on the frequency of pathogenic parasites, these data help to support recommendations by public health officials for improved sanitary practices. Before discussing the pathogens, therefore, we shall look briefly at five common commensal flagellates of man.

NONPATHOGENIC FLAGELLATES

Retortamonas intestinalis belongs to a family of flagellates, the Bodonidae, which include free-living, parasitic, and coprophagous (feeding on fecal matter) species. Among the latter are members of the genus *Bodo*, which are occasionally reported from stool specimens but are known to be contaminants rather than derived from the intestine. *R. intestinalis* is quite small, only 4–9 microns long (see Fig. 5–1), with one anterior flagellum and one trailing flagellum. The cysts are even smaller, about 4–7 by 3–4 microns. *Retortamonas intestinalis* is never common anywhere but has been found all over the world. A species, *R. sinensis*, possibly identical with *R. intestinalis*, has been reported from China. The rareness of these parasites in man is thought by some authors to be evidence that *Retortamonas* infections are accidentally acquired from some nonhuman reservoir.

Enteromonas hominis, often called *Tricercomonas intestinalis* (Fig. 5–1), is small (4–10 by 3–6 microns) and pear-shaped. Like other members of its family (Tetramitidae), it possesses four flagella—one trailing and the others projecting anteriorly. The cysts (6–8 by 4–6 microns), when mature, contain four nuclei. *E. hominis* is rather common in the tropics; it is found in nearly 70% of the population in certain parts of Egypt but has been considered rare elsewhere. Probably this parasite is often overlooked in routine fecal examination.

Chilomastix mesnili (family: Chilomastigidae) is, like other members of the family, superficially similar to the tricercomonads, in having four flagella—three projecting and one trailing—arranged in the tricercomonad way. *Chilomastix* and its related genera, however, have a prominent cytostome (Fig. 5–1) within which the trailing flagellum lies. They thus appear to have only three flagella. Trophozoites of *C. mesnili* are quite variable in size; they range from 6 to 20 microns in length, depending on their state of growth and activity. Cysts are 7–10 by 4–6 microns. They have a thick wall, and usually only a single nucleus is present. *C. mesnili* occupies the cecal region of the large intestine in as much as 10% of the population.

FIG. 5–1. Intestinal and other lumen-inhabiting flagellates of man. (A) *Retortamonas intestinalis*. (B) Cyst of same. (C) *Chilomastix mesnili*. (D) Cyst of same. (E) *Trichomonas hominis*. (F,G) *Tricercomonas intestinalis* and cyst. (H,I,J) Cyst, dorsal view, and lateral view, respectively, of *Giardia lamblia*. (K) *Trichomonas vaginalis*. (L) *Trichomonas tenax*. All are drawn at the same magnification, about 2300×.

The three species of intestinal flagellates just discussed are always more common in diarrheic stools than in fresh ones. This fact has led some workers to attribute dysenteric symptoms to some pathogenic effects of the flagellates. Other workers believe that the diarrheic conditions, of whatever cause, probably stimulate growth of the harmless commensals, and their large number in stools is a result, not a cause, of the diarrhea.

Transmission of the above species is by cyst. Obviously, general sanitation reduces the incidence of cyst-transmitted parasites. There is no reason to be concerned about these commensals, however, except for the fact that their presence indicates fecal contamination of food or water—a serious enough risk of infection with harmful parasites.

Trichomonas hominis is the intestinal member of the three species of *Trichomonas* found in man. Members of the family Trichomonadidae have 3 to 5 flagella and an attached flagellum with an undulating membrane. They have a conspicuous axostyle, a tapered rod extending posteriorly through the body from the region of the basal granules. A cytostome has been described (Fig. 5–1) for members of the genus *Trichomonas*; phagocytosis and perhaps pinocytosis (ingestion of liquid droplets) may occur at the body surface. *T. hominis*, with usually 4, sometimes 3 or 5, anterior flagella, is 5–14 by 7–10 microns in size. It lives in the upper (cecal) region of the human colon, where it moves about very actively; this is based on observations of living specimens from diarrheic stools. In formed stools, *T. hominis* rounds up and cannot easily be recognized. Unlike the other intestinal flagellates, it does not form cysts. Transmission, however, seems to occur frequently, since the parasite is relatively common; it is found in about 1% of the inhabitants of the United States, but is far more common in tropical lands, where incidence is nearly 12%. Children under 10 are more often infected than adolescents or adults. The rounded fecal forms can survive for a day or more in water or food, or presumably in the stomachs of flies. These forms are also capable of passing unharmed through a human stomach and small intestine if ingested with semiliquid food or water. There is no evidence of pathogenicity; *T. hominis*, occasionally present in rather large numbers in loose stools, probably responds to unusual intestinal conditions but does not contribute to such pathologies.

Trichomonas tenax lives in the mouths of many human beings (4 to 53%). This organism is of medium size (5–12 microns in length), has 4 free flagella, and has

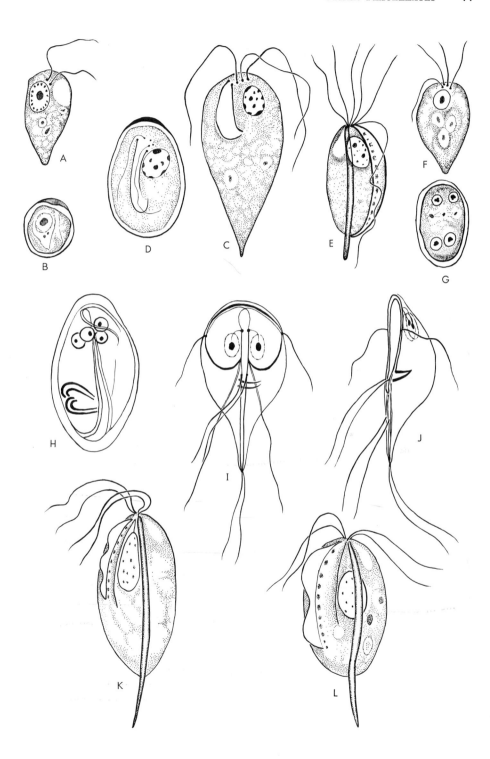

a relatively short undulating membrane, which does not reach the posterior end of the body. It feeds on bacteria, organic debris, and cell exudates in various crypts and cavities of teeth and gums. While diseased mouths often harbor large colonies of *T. tenax*, there is no reason to believe that this parasite contributes to the bad conditions. Occasional involvement of the respiratory tract has been noticed. Transmission of this cystless organism is either direct (by kissing or coughing) or through contaminated utensils. This parasite, like the intestinal trichomonad, survives outside the body for some time.

TRICHOMONAS VAGINALIS

A third trichomonad of man is *T. vaginalis*. Because this is an annoying pathogen affecting up to 25% of all women and perhaps a similar number of men (although evidence for the latter figure is scanty), it will be given considerable attention in these pages.

Trichomonas vaginalis is larger than *T. tenax* or *T. hominis* (7–23 by 5–12 microns). The undulating membrane barely reaches half the length of the animal. The cytoplasm is richly granular. The cytostome is inconspicuous. No rounded up forms have been observed. The parasite characteristically inhabits the lining of the vagina and sometimes invades the urethra and associated glands. In men, it lives in the urethra and prostate gland. Unlike *T. tenax* and *T. hominis*, *T. vaginalis* is quite delicate; it dies within 40 minutes in water and is easily killed by desiccation or by environmental temperature no higher than 40°C. Most investigators believe that the primary mode of transmission is by sexual intercourse. In the crowded post-war conditions of Europe, however, transmission of *T. vaginalis* occurred through contaminated wash cloths, bathing water, etc. Hopefully, since the life of refugees and concentration-camp internees is not normal to the human species, this woman-to-woman transmission was unusual. Ordinarily incidence of trichomonad vaginitis is directly related to the degree of promiscuity in sexual behavior. Although it does not carry the stigma of shame or embarrassment conferred by syphilis and gonorrhea, this is a venereal disease.

The effects upon women are relatively severe. The parasites cause irritation and sloughing of the vaginal epithelium, infiltration of the area with leucocytes, increased secretion of fluid, and discharge of purulent material which may greatly irritate the vulva and the urethral orifice. The condition of the vagina is altered as the surface becomes swollen and inflamed, and the normal acidity of the vagina is changed toward neutrality. As the disease progresses, symptoms become milder. The purulent discharge changes to a frothy white material consisting largely of leucocytes and trichomonads. In some persons each menstruation brings on a recurrence of acute symptoms, which subside during the intermenses. In males, there are usually no symptoms. However, trichomoniasis has been found to be associated with about one third of nonspecific cases of urethritis in men, which suggests more than a chance relationship.

The true frequency of *T. vaginalis* in men is unknown. Estimates usually indicate 4–15%, which may be lower than the actual figure because asymptomatic diseases are hardly ever diagnosed or treated. There is no *a priori* reason to suspect that women acquire trichomoniasis more frequently from men than the latter acquire it from women.

Treatment of vaginal trichomoniasis is not very effective. It can be cured, however, with the combination of (1) local application of drugs (by suppositories or douches containing such drugs as iodine or arsenic derivatives, plus bactericidal accessories such as sulfonamides), (2) induced changes in the vaginal pH by introduction of acid-forming bacteria (Döderlein bacilli) usually present in the healthy vagina, and (3) orally administered systemic drugs, usually antibiotics. Treatment of male patients is by urethral irrigation and oral medication.

Male patients, although usually without symptoms, should be treated if there is any hope of controlling this disease. Otherwise married women under treatment must inevitably become reinfected by their husbands. Like the other venereal diseases, trichomoniasis flourishes because of the secrecy that prevails among many so-called civilized peoples concerning sexual matters. Whatever may be the supposed value of sexual suppression in public or private morality, or in religion, this cultural practice is entirely harmful from a public health point of view. It encourages the spread of venereal disease, obstructs the efforts of epidemiologists seeking to find and treat existing cases, and prevents many individuals from seeking help for painful and often dangerous infections.

GIARDIA LAMBLIA

Giardia lamblia (or *Giardia intestinalis*) is a parasite of the human small intestine, especially the duodenum. (See Fig. 5–1.) Members of the family Hexamitidae, to which *Giardia* belongs, are characterized by having two nuclei and three or four pairs of flagella, arranged bilaterally. The trophozoites of *Giardia* are memorable objects which appear like the caricature of a sparsely bearded face. From the host's point of view, the most significant feature of this organism is a shallow sucking disc on the anterior ventral surface, by which the animal attaches itself firmly to cells of the host's intestinal epithelium. *Giardia* may literally cover the inner surface of the upper small intestine and cause strange pathological effects (see below). *Giardia* forms cysts, the stage usually seen in fecal examinations. These are ellipsoidal, have a well-separated cyst membrane, contain four nuclei when mature, and show several bundles of curved axonemes (roots of the flagellar system). The cysts form as the liquid content of the bowel thickens and becomes dehydrated. Cysts are usually not present in unformed or diarrheic stools, and trophozoites are absent from normal feces. *Giardia* is difficult to culture, but recent attempts using a yeast-enriched medium are believed to have been successful. As in amoebiasis or enterobiasis, transmission occurs by various cyst-contaminated objects, food, and water.

The effect of *Giardia* upon the host may be very slight, but severe illness may occur. The latter is due to the blanketing effect of vast numbers of trophozoites upon the wall of the upper small intestine. There an important function of digestion —fat absorption—is directly interfered with, so that a fatty or oily diarrhea (steatorrhea) ensues. Since important vitamins (fat-soluble carotins, for instance) are prevented from reaching the host, secondary effects are usually present. Duodenal irritation, dehydration from the steatorrhea, loss of weight, and poor appetite are some of the symptoms of a *Giardia* infection. The bile duct and gall bladder are sometimes affected also. Nevertheless, in most individuals, giardiasis does not produce any symptoms; probably the many parasites are somehow prevented from reaching vast proportions, and no significant blockage of fat absorption occurs.

Incidence of *Giardia* in man is clearly related to age. Young children, as they grow older, show a rise in frequency of infection, a frequency which peaks at puberty. Then there is a sharp decline, until about one-fourth the maximum incidence is reached. Adults in the temperate zone have about 2% incidence. Adolescents or preadolescents in the tropics show 25% giardiasis. The figures for children indicate that *Giardia* infection is contagious and that increased opportunity with age results in a rise of incidence. The loss of *Giardia* infection after puberty suggests some late-developing immune response. Another explanation is change in physiology of the host, or in personal habit or diet, which is a purely speculative idea but is perhaps related to the data on pinworm incidence (see Chapter 18) and other such conditions. A local epidemic of giardiasis affecting all age groups was observed in 1954–55; this incident raises interesting questions, including the possibility that *Giardia* exists in a number of geographically isolated strains.

Treatment of symptomatic giardiasis is usually successful with Atabrine (an antimalarial drug, one of the acridine dyes). Treatment in the absence of symptoms may have some public health value but has not been proposed, since *Giardia*, like temperate-zone *Entamoeba histolytica* (see Chapter 6), is usually a harmless commensal. Moreover, even in the most sanitary surroundings existing today, much can still be done to improve sanitation and prevent transmission of infectious pollutants of water, food, and various objects. Thus giardiasis, like enterobiasis, mild amoebiasis, and many other conditions not strictly labeled "diseases," will probably fade away with progress in sanitation.

FLAGELLATES OF VETERINARY IMPORTANCE

Two flagellates of veterinary importance remain to be mentioned, *Histomonas meleagridis* and *Tritrichomonas foetus*.

Histomonas meleagridis causes the disease of turkeys and other fowl called "blackhead," which refers to the dark discoloration of the head seen in some, but by no means all, cases. Such discoloration is not peculiar to this disease alone. The parasite is a flagellate which throughout most of its life cycle resembles an amoeba; it possesses as many as four flagella and exhibits flagellate movement only in the lumen of its host's cecum or intestine, a place from which the parasite is absent most of the time. In the liver tissue or wall of the cecum, where the organism proliferates and causes extensive pathological lesions, it appears entirely amoeboid; the only evidence of its flagellate nature is a small extranuclear body, probably the blepharoplast. The organism forms no cysts or other resistant stage capable of survival outside the host; its transmission depends upon either the immediate ingestion of droppings (feces) of infected birds by new hosts or the inclusion of the parasite in the durable eggs of the cecal nematode, *Heterakis gallinae*. The latter means of transmission is quite unusual, if not unique, for a protozoan parasite. Only *Dientamoeba fragilis*, an amoeba of man showing some affinities for flagellates, may have a similar transmission in the eggs of the pinworm (*Enterobius vermicularis*), although this is not by any means an established fact.

The disease, blackhead or histomoniasis, varies in its severity. Young turkeys are apparently least resistant and almost always suffer fatal sickness. Older turkeys suffer less, and chickens, very little. The latter are important reservoirs or carriers, and may be the source of epizootics in turkey flocks. Symptoms of acute disease appear one to two weeks after ingestion of the parasites. The birds seem droopy, and sulfur-colored droppings may be noted. Death may occur within a few days, and mortality may reach 100%. In flocks of young turkeys 50% mortality is usual. Less than 25% of the older birds die.

The organs affected are the liver and one or both ceca. As the parasites invade the tissue, they pass between the cells, at first ingesting small particles, presumably fragments of host cells. Later, as they multiply and as the infected tissues become distorted and damaged, the parasites become clear or hyaline and are without food vacuoles. Presumably their nutrition is now osmotic. In the liver, many necrotic spots are seen scattered throughout the organ. The cecum becomes ulcerated and may rupture. Usually a hard, tough plug deposited within the cecal lumen is laid down concentrically by material exuded from the wall. Extensive scarring results in permanent damage to the cecum even if recovery occurs. It is interesting to note that in chickens various immune reactions, including infiltration of affected areas with phagocytic cells, result in nearly complete restoration of damaged cecal tissue. Treatment is not effective.

Control and prevention of blackhead disease depend on recognition of two facts mentioned earlier—the spread of the parasite in fresh droppings of infected birds, and the persistence of the parasite in *Heterakis* eggs. If young turkeys (the stage in which the disease is most severe) are kept off the ground where older turkeys have run and if they are separated from chickens (which are very commonly carriers), they have a good chance of escaping the disease. It is also important to keep the ground clean in order to minimize the chance of rapid spread, via droppings, once an outbreak occurs. Since blackhead seldom appears except as a result of the presence of infective feces or nematode eggs, it should be possible to raise *Histomonas*-free flocks. Good management requires several pens or pastures, which must be kept clean and which can be rotated alternately. Various mechanized systems of rearing turkeys have been highly successful. Of course, cost of special brooders, wire-floored pens, etc., must be weighed against profit, and simpler methods in many cases may turn out to be profitable in spite of higher incidence of disease.

Tritrichomonas foetus is another flagellate pathogen of agricultural importance. It causes one form of infectious abortion in cattle. (Another is the bacterial "Bang's

Disease" or "contagious abortion," which is also very serious.) The parasite is one of a large group of lumen-dwelling flagellates which bear three or more flagella and have the body stiffened with a rodlike axostyle. In *T. foetus*, there are three free flagella, and a fourth is attached to the body by a prominent undulating membrane. It is similar, of course, to other members of the genus and is best recognized by being recovered from the characteristic lesions or tissues of its host. This flagellate is found in the preputial cavity of the infected bull and in the uterus and vagina of infected cows. In severe trichomoniases, the epididymis, seminal vesicles, and testes of the bull may harbor parasites, and the cow's uterine tissues may be invaded along with the tissues of an embryo or fetus which may be present. Trichomoniasis is usually mild and chronic in bulls and severe but sometimes self-curing in cows, although it usually results in failure of conception or the death and abortion of a fetus. Cows may retain an infection in the form of vaginitis or they may recover to become immune to further infection. Bulls probably never lose the infection, and it is the best practice to destroy such animals. This disease is spread by copulation, and thus, like dourine of horses, is a venereal disease. Treatment is ineffective. Apparently, as in human trichomoniasis of the vagina, the secretions of the mucous membranes are not susceptible to saturation with systemic medicines or drugs, and direct applications to the affected tissues fail to reach all the parasites. Fortunately, prevention of venereal disease of cattle is easy. Since the only source of infection of cows is an infected bull, examination and elimination of such sources is a practical way to protect other cattle.

CONCLUSION

The first part of this chapter called attention to the flagellate protozoa of the human intestinal and genital ducts. Most of these protozoa are harmless commensals. Some, like the small *Retortamonas, Enteromonas, and Chilomastix*, are relatively uncommon, may be easily overlooked, and are of interest chiefly as sources of confusion in the diagnosis of giardiasis. *Trichomonas hominis*, a harmless inhabitant of the bowel, has two interesting features: (1) its ability to survive for many hours outside the body without the protection of a cyst membrane, and (2) its relationship with two other trichomonads, inhabitants of mouth and genital tract, respectively. The evolutionary relationships of these three quite similar parasites might be speculated upon. *Trichomonas vaginalis*, whatever its origin, is now frankly a venereal parasite. Yet its occasional transmission by contaminated objects such as wash cloths shows its epidemiological flexibility, and indicates its relationship to the more easily transmitted mouth and bowel trichomonads. The disease caused by *T. vaginalis* is quite serious, yet control depends on changes in culture (indeed, changes in a moral postulate of Western culture) which are unlikely to be brought about. *Giardia lamblia*, usually a harmless feeder upon mucous exudates of the small intestine, occasionally becomes so numerous that severe nutritional disorders arise. Treatment of giardiasis is effective, but control of this cyst-producing, highly contagious parasite probably depends on further sanitary improvement, even in the more civilized parts of the temperate zone. It thus appears thât the intestinal flagellates and their relatives are well adapted to inhabit man and to move successfully from habitat to habitat. The fact that they occasionally cause disease is not important enough to bring about their eradication so long as easier and more serious problems remain unsolved.

The second part of this chapter has dealt with two flagellates of domestic animals, *Histomonas mealeagridis* and *Tritrichomonas foetus*. The former, chronic or mild in chickens, causes devastating epizootics in turkey flocks. The curious transmission of *Histomonas* through the eggs of *Heterakis*, a nematode, ensures continuity and dispersion of this protozoan; infection by ingestion of feces, however, is the cause of fatal spread throughout flocks of young birds. In contrast, *Tritrichomonas* is transmitted among cattle by copulation. Although dangerous, tritrichomoniasis can be prevented by destruction of infected bulls. Each of the flagellates of domestic animals presents a problem of some seriousness and difficulty.

SUGGESTED READINGS

FEO, L. G., "The Incidence and Significance of *Trichomonas vaginalis* Infestation in the Male," *Am. J. Trop. Med.*, 24:195–198, 1944.

LEVINE, N. D., *Protozoan Parasites of Domestic Animals and Man.* Burgess, Minneapolis, 1961.

MOORE, S. F., and J. W. SIMPSON, "Trichomonas Vaginitis. An Emotionally Conditioned Syndrome," *Southern Med. J.*, 49:1495–1501, 1956.

O'DONOVAN, D. K., J. McGRAITH, and S. J. BOLAND, "Giardial Infestation with Steatorrhea," *Lancet*, ii, 4–6, 1942.

PETERSON, J. M., "Intestinal Changes in *Giardia lamblia* Infestation," *Am. J. Roentgenol.*, 77:670–677, 1957.

TRUSSELL, R. E., *Trichomonas vaginalis and Trichomoniasis.* Thomas, Springfield, Ill., 1947.

CHAPTER 6

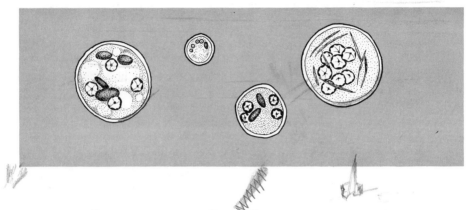

AMOEBIASIS

INTRODUCTION: PARASITIC AMOEBAS

The amoebas living in man and other hosts are less obviously parasitic than the other important parasites. They resemble free-living amoebas in appearance and behavior. Anyone who has observed a free-living amoeba, surely individually a most adaptable animal, can imagine how such an animal could thrive, given a few protective changes, within the digestive tract of man. The bacterial flora of mouth and colon provide ample food, and the homeostatic processes of the host maintain this environment relatively constant in temperature, pH, pressure, and viscosity. Only minor adaptations seem necessary for such parasitism and commensalism, and these adaptations probably are represented in two differences between some parasitic and most free-living amoebas. First, most parasitic amoebas form cysts with tough coverings within which the animal lies dormant and is prepared for transfer to new hosts. (It is noteworthy to mention that some free-living amoebas also form cysts, which permit their transmission from one wet season to another.) Second, some parasitic amoebas can actively invade the tissues of their host and multiply there at the host's expense. This seems to be a true adaptation to parasitism, for similar invasiveness is not found in free-living amoebas.

Other adaptations to parasitism should be considered. The temperature of a mammalian host requires, of course, considerable tolerance of temperature change on the part of a hypothetical invader from the free community of pond or stream; yet a parasite of reptiles, *Entamoeba invadens*, is quite similar to the pathogenic *E. histolytica* of man, suggesting the possibility of transition from cold environment to poikilothermic ("cold-blooded") host as a first step in progress toward a warm-blooded host. There is also the possibility of a very long association of amoebas with vertebrates, an association that began in the Mesozoic before homoiothermy ("warm-bloodedness") had ever arisen. Another obstacle to successful parasitism is the chemical hostility of the digestive tract to naked protoplasm. Such enzymes as are found in the stomach and small intestine do not occur in the environment of a free-living amoeba. But parasitic amoebas actually are not found in the regions of active digestion; they grow only in the mouth or colon and the rectum, or sometimes in the tissues. Those which must reach their habitat by passing through the stomach and small intestine are always protected by a cyst (except the mysterious *Dientamoeba fragilis*, which may perhaps be transmitted through the eggs of *Enterobius*, a nematode). The rich fermenting medium in which amoebas of the colon live is perhaps not more toxic than various stagnant and polluted pools in which many kinds of free-living protozoa abound. The amoeba of the mouth is transmitted by kissing; it forms no cysts, nor does it need them. The universal risk in parasitism is the problem of transmission from one rich environment to another and seems to have been easily solved by those enteric amoebas whose life cycles are known; the parasites even increase in number while encysted, and thus they get a reproductive boost during the critical period of transmission.

The amoebas living in man (see Fig. 6–1) are nearly always harmless. *Entamoeba gingivalis* inhabits the crevices between the teeth at the borders of the gums, where it ingests bacteria, particles of food, and sloughed epithelial cells. Because its population is greater in mouths with tooth decay than in healthy mouths, some workers conclude that this amoeba increases decay or injures mouth tissues, but this view is probably not supportable on present evidence; *E. gingivalis* thrives in practically all persons who acquire it. It is successfully and frequently transmitted by oral contacts, at least in those cultures (the West) where such contact is permitted. *Entamoeba coli* is the largest intestinal amoeba of man. It lives in the colon, where it ingests bacteria and forms cysts with eight nuclei. Of course, young cysts of *E. coli* must be distinguished in diagnosis from the important 4-nucleate *E. histolytica* cysts, but their identification can be based upon the structure of the nucleus (eccentric endosome in *E. coli*, Fig. 6–1 C2) and on the usual presence of splinterlike (contrasted with rounded) "chromatoid" bodies in the *E. coli* cyst. Also, of course, transitional stages like 4-nucleate *coli* cysts are relatively rare among the numerous completed cysts. Other harmless amoebas in the human intestine are the rather small forms, *Iodamoeba butschlii* and *Endolimax nana*. An amoeba of doubtful pathogenicity and perhaps doubtful relationships is *Dientamoeba fragilis*. Unlike the other enteric amoebas, this is normally binucleate.

Also its nuclei are quite different from those of the other amoebas. The nuclei contain a central group of granules—probably condensed chromosomes—and there may be a fibril joining the two nuclei; such a structure is similar to the intracellular fibrils seen in some flagellates but is never observed in the typical amoebas. *Dientamoeba fragilis* is sometimes very numerous and has been implicated in diarrhea or mild dysentery. Its pathogenicity is not well established, but should be investigated further. The life cycle of this organism is unknown. Because in structure it resembles in some ways the flagellated amoeba, *Histomonas meleagridis* of fowl (see Chapter 5), and because the latter parasite is transmitted within the eggs of a nematode, it has been suggested and possibly demonstrated that *Dientamoeba fragilis* may enter the ovary of the common pinworm, *Enterobius* (see Chapter 18), and become incorporated in the eggs of this ubiquitous parasite. The last amoeba to be mentioned in this list is *Entamoeba histolytica*, which causes the dangerous disease or syndrome known as amoebiasis. *E. histolytica* will be discussed throughout the rest of this chapter.

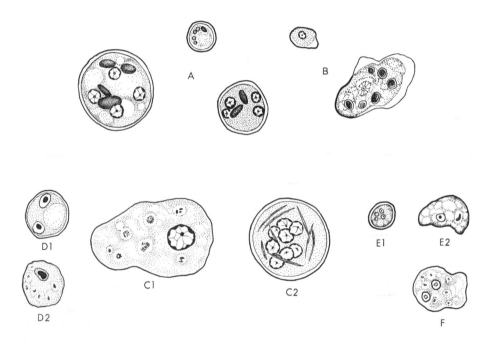

FIG. 6–1.   Amoebas living in the human intestine. (A) Cysts of *Entamoeba histolytica* of different sizes. (B) Trophozoites of *E. histolytica* from small and large races. (C1) Trophozoite and (C2) cyst of *Entamoeba coli*. (D1) Cyst and (D2) trophozoite of *Iodamoeba butschlii*. (E1) Cyst and (E2) trophozoite of *Endolimax nana*. (F) Trophozoite (cysts unknown) of *Dientamoeba fragilis*. All are drawn to the same scale, about 1070×. [After Hall, 1953.]

ENTAMOEBA HISTOLYTICA AND AMOEBIASIS

Entamoeba histolytica is found throughout the world. Its cysts can be recovered from the feces of 10–25% of populations in temperate zones, and up to 100% of certain populations of the tropics. However, climate is not the only important factor, for there is close correlation between poor sanitation and high amoebic incidence. Yet symptomatic disease does seem to be related to climate. While some hosts of *E. histolytica* in temperate zones are known to suffer chronic symptoms, practically all the really serious amoebiases either occur in the tropics or are acquired there. For this reason certain workers believe that there are two races of *E. histolytica:* (1) a pathogenic variety endemic to the tropical regions, and (2) a harmless variety native to the temperate zones. The harmless race, called *Entamoeba hartmanni,* has been given specific rank by some. Workers also believe that the pathogenic race has two phases or forms: (1) a so-called "minuta" or small-cyst form which is a lumen-dwelling commensal feeding upon the colonic bacteria, and (2) a "magna," or large-cyst form which is an active invader of tissue and the cause of the many kinds of clinical amoebiasis. The above separation of *E. histolytica* into two species, one of which has two races, may seem rather complicated; actually there is reason to think that *E. histolytica* is composed of many races or varieties, just like every other organism which has been sufficiently studied. A further complication in the picture of *E. histolytica* as a pathogen or a group of pathogens and commensals is the great variation that occurs in host responses to amoebic infection. The spectrum of symptoms described by various medical writers is evidence of a spectrum of host reactions, which range from possible refractoriness (to infection) through tolerance of the parasite as a commensal, to susceptibility to various degrees of invasion from minute submucosal lesions to massive liver or lung abscesses. Symptoms may be nil, may range from vague to severe discomforts, may be rapidly fatal dysentery, or may reflect the peculiar location of abscesses far from the intestine. In laboratory animals a similar range of difference has been found; kittens are susceptible and are severely affected, while rats and monkeys respond in a manner more like that of people, with considerable resistance and varying degrees of tolerance.

Naturally, therapeutic measures must be prescribed to fit particular cases, for the intestinal parasites are not susceptible to the same drugs that destroy the tissue invaders, and vice versa.

Control measures also are made difficult by the above features of amoebiasis, because the question of pathogenicity, dependent on the identification and definition of truly pathogenic strains, should be settled before heavy investment is made in public health surveys, mass treatment, or the like.

In short, *Entamoeba histolytica* presents very difficult problems of recognition (taxonomy), therapy, and control, because of its variability, the complex and rich environments which nourish it, and the differences among its hosts and victims.

FIG. 6–2. Pathology of amoebiasis. Ulcerated colon [gross features (A) and histological section (B)] permits amoebas to enter venous circulation (C) and be carried to various organs, such as liver, lung, and brain, where erosive abscesses [a liver abscess (D)] form.

PATHOGENESIS

Much is known about the way in which *E. histolytica* damages its host. Presumably the harmless forms—*E. hartmanni* to those who consider these a distinct species— can live indefinitely in the large intestine, feeding on bacteria, multiplying, producing cysts, and causing no recognizable damage either to the host in general or to his tissues in particular. While capable of living with very little damage to the host, however, the pathogenic forms usually pursue a well-defined course. Moving freely over the surface of the mucosa, the amoebas secrete proteolytic enzymes, some of which have been isolated and tested. These enzymes, while probably used chiefly within the amoeba during digestion, are capable of dissolving the surfaces of intestinal cells and thus providing pathways for invasion.

(The actual methods of invasion are not yet known. Evidence from experiments with bacteria-free hosts strongly suggests that in the absence of a normal intestinal flora the amoeba alone, at least in some strains, is unable to penetrate the intact mucosa. Thus the role of the above-mentioned histolytic enzymes in invasion is still undefined.)

Once amoebas are within the tissues of the submucosa, they become larger, engulf bits of cellular debris as well as whole erythrocytes, and divide rapidly. Their enzymes break down neighboring host tissue, and thus a hollow necrotic area develops in the form of a small, closed ulcer or local abscess. The intestine may have many such lesions, usually very small and often discoverable only during autopsy. In moderate to severe cases, however, the amoebic ulcers spread, some- times joining each other, and rather large pieces of epithelium break away to leave open pits (see Fig. 6–2). These damaged areas are subject to invasion by other pathogens, such as bacteria. Repair of ulcerated bowel lining eventually occurs, but the flexible, absorptive mucosa is usually replaced with fibrous scar tissue. Sometimes this tissue partially constricts the intestine, blocking peristalsis and interfering with normal function. The above tissue damage and repair are charac- teristic of primary amoebiasis. Less characteristic is the condition of "amoebic dysentery." In such cases the process of erosion is so rapid that much of the intestinal lining is lost, and hence there is discharge of blood, mucus, and amoebas into the lumen. Such dysentery resembles bacillary dysentery in severity, in the wasting, emaciating effect it has on its victims, and in the pain and rectal straining which are present; but amoebic dysentery seldom is accompanied by the fever

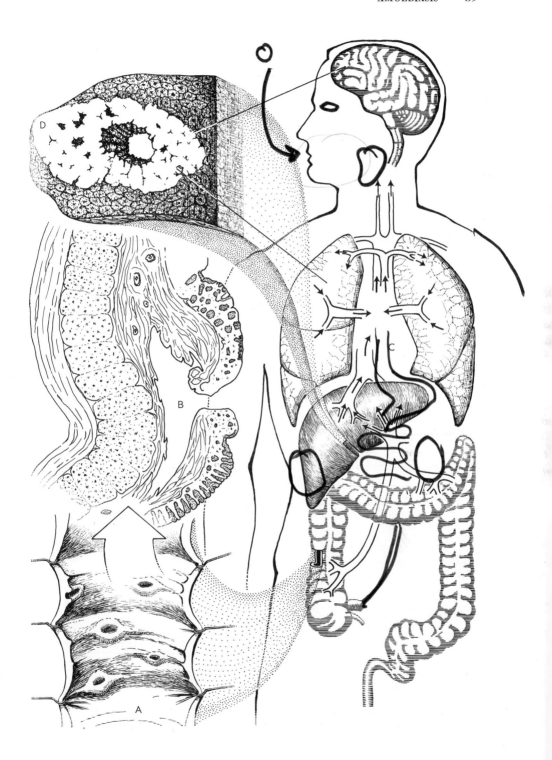

which is always part of bacillary dysentery. Recovery from amoebic dysentery is of uncertain duration, with recurrence of severe symptoms years after the earlier illness. Secondary amoebiasis is due to transportation of amoebas via circulation from primary abscesses in the intestine to other tissues. The liver, lungs, and brain develop amoebic abscesses in the given order of frequency. A liver abscess (see Fig. 6–2 D) consists of a hollow eroded region, containing a viscous fluid mass of dead amoebas, blood, and tissue detritus. Around the necrotic center of the abscess the liver tissue is full of amoebas which actively invade healthy tissues as the amoebas multiply. No fibrous envelope forms around such an abscess, and it spreads steadily with age. Amoebic abscesses are usually bacteria-free. Lung abscesses may develop directly from liver abscesses through the spread of the latter across the diaphragm. Brain abscesses result from amoebas which have become lodged in the brain and have multiplied. Brain abscesses are less common than lung or liver. Other sites of amoebic infection have been reported. Recovery from surgery is sometimes hampered by amoebic involvement of incisions and sutures.

Apparently the only defense the body has against invasive amoebiasis is phago-cytosis. Abscesses contain a large number of leucocytes which have engulfed amoebas, and systemic (or secondary) amoebiasis usually produces a raised leucocyte count. In some individuals and with certain races of amoeba, this de-fense is so weak that abscesses form and grow in spite of leucocyte activity. The variations in pathogenicity have been attributed to various causes—differences in race or strain of pathogen, differences in inherent susceptibility of host, and dif-ferences in internal environment. The latter seem to be particularly important. Attacks of bacillary dysentery in a number of cases were followed by amoebic invasion, which suggests very strongly that pathogenic changes in the intestinal lining make this tissue susceptible to entrance of amoebas. Chemical and physical irritants may also be effective. Certain bacteria seem to be necessary for amoebic virulence, even if the bacteria themselves are harmless. Thus synergistic (mutually energizing) relationships between amoebas and other inhabitants of the intestine are an important part of pathogenicity. Diet, in both experimental animals and humans, has a definite effect on tissue invasion. As mentioned earlier, specific differences in host response may have their counterparts in individual differences among members of the human species. In short, the many and diverse pathological conditions caused by *Entamoeba histolytica* are due to many diverse causes, a few of which are to some extent understood.

DIAGNOSIS AND TREATMENT

Diagnosis and treatment of amoebiasis are unusual in two respects: (1) the diffi-culty and costliness of diagnosis, and (2) the variety of drugs used in treatment. Primary amoebiasis cannot be recognized by symptoms, but must be diagnosed by recovery of cysts or trophozoites of *Entamoeba histolytica* from fecal specimens. Even with modern, rapid methods of making slides, the preparations must be made by a trained technician with a good laboratory, and the slides must be examined by a parasitologist. Cysts of *E. histolytica* are not easy to distinguish from other amoebic cysts, and a novice may make rather gross errors of identification. The other amoebas in man (as discussed earlier) are important chiefly because of the risk of confusing them with *E. histolytica* and prescribing unnecessary, expensive, and perhaps dangerous treatment. If characteristic cysts or amoebas (see Fig. 6–1) are found, then diagnosis is positive and treatment can be started. But if no amoebas or cysts are seen in one examination, there is still a very good chance that the patient may have amoebiasis; it is estimated that for a reasonably sound negative diagnosis up to six examinations, spaced over several days, may be necessary. Other methods of diagnosis, such as serological tests and inoculation of culture media with fecal samples to permit multiplication of scarce amoebas are of doubtful utility and are at least as difficult and expensive as microscopic examination of stained smears. Upon positive diagnosis treatment can be started. Certain drugs are effective against intestinal amoebas but not against systemic forms, and the reverse is also true. Thus a combination of drugs should be used. Among drugs effective against intestinal forms are various antibiotics, as well as certain arsenicals and iodine compounds. The antibiotics are particularly useful for treating acute dysenteric phases of amoebiasis. Their effectiveness against chronic infections seems not to be very high, however, and the arsenicals and iodine formulations are used for complete cure of intestinal amoebiasis. Systemic amoebiasis does not respond to either antibiotics or arsenicals and iodines. However chloroquine, a well-known antimalarial drug, is very effective against liver and other abscesses. A serious obstacle to successful treatment is the uncertainty of post-treatment diagnostic check-up. Thus the difficulty of diagnosis is a serious factor both at the beginning and the end of a clinical course of amoebiasis; perhaps this fact is reflected in the large number of drugs which have been tried, since no drug is susceptible of absolute proof of efficacy.

It should be clear to the reader of this chapter that amoebiasis is still an unsolved problem. In addition to the difficulties in diagnosis and treatment just mentioned, there are also the other difficulties discussed earlier: (1) questions of the extent of morbidity among the millions harboring *Entamoeba histolytica*; (2) problems of taxonomy, especially the matter of pathogenic races of varied virulence, and the probability that *Entamoeba hartmanni* is actually a distinct species, nonpathogenic, and very common; (3) differences in susceptibility of human racial groups as well as individuals. In order to suggest certain approaches for solving some problems, let us consider other aspects of amoebiasis—resistance and immunity. At the end, the importance of amoebiasis will be estimated. The last point, of course, is crucial to the decisions which must be made on future expenditures of effort and money to control amoebiasis.

RESISTANCE AND IMMUNITY

It has been pointed out that resistance to infections with *E. histolytica* varies among individuals and perhaps among genetically different groups. Data supporting this statement come from epidemiological surveys over many years. An interesting study made by Meleney in the 1930's suggested that age is a factor in resistance.

While age is a well-known factor in resistance to many diseases, Meleney's data are rather peculiar, resembling the incidence curve of schistosomiasis (Chapter 9). Both curves show a rise during childhood, a leveling off at maturity, and a decline in old age. The schistosomiasis curve has been interpreted very reasonably as evidence of two factors at work. The childhood rise in incidence is due to repetition of exposure, since each immersion of feet or body in cercarious water increases the chance of cercarial penetration, that is, of infection. The incidence curve, if dependent on exposure alone, should rise steadily, because exposure is repeated throughout life. Since the curve actually flattens out at maturity and declines in later years, it is reasonable to conclude that a second factor—resistance, immunity, "self-cure"—is involved. Experimental work on schistosomiasis in animals does indeed point strongly to an immunity factor. A third factor, the shortened life span of sufferers from clinically severe schistosomiasis, might also explain the lower incidence in old people, for they would be survivors of a population in which schistosomiasis is a contributing cause of death. Similar interpretation of the amoebiasis curve seems reasonable, at least for the data up to old age. The rise of amoebic infection during childhood is to be expected, since amoebiasis is acquired by ingesting cysts in water or food. Such cysts are sporadically available, and opportunities to acquire infection multiply with time. This explanation of the rise in incidence during youth is probably too simple, for the sanitation of the environment is an important and variable factor in amoebiasis. It might be argued, also, that young children are less sanitary than adults; but adults are probably the source of childhood amoebic infection, a fact which reduces somewhat the importance of hand-washing, indiscriminate tasting of objects, etc., by young children. A reduced rate of new infections might explain the leveling of the incidence curve at ages 30–50, if there were evidence of improved sanitation beginning during middle age. Other complicating factors are changes in mobility with age (an unknown for the populations studied, although Meleney dealt largely with rural people in the U.S.A.), and possible differences in diet (since diet is known to be a factor in resistance to amoebiasis). Recovery (loss of amoebas) is another possible explanation of the apparent shift in rate of acquired infection. Obviously, a simple incidence curve may be the result of a variety of interacting forces. The similarity of the age-incidence relationship in amoebiasis to that observed in schistosomiasis, and attributable in the latter to immunity, is worth noting, for there is in these data at least a suggestion that human immunity to amoebiasis may play a role in protecting some members of exposed populations.

Immune responses do occur. One of the methods sometimes used in diagnosing amoebiasis is the complement fixation test, which may be explained briefly as follows. Many substances, when introduced into the body, cause specific antibodies to be formed. These antibodies are capable of combining chemically with the foreign substances (antigen), in a way that makes the foreign substance susceptible to various defensive mechanisms of the body. A common and probably universal reaction following antibody-antigen combination is the union of the latter with "complement," a substance found in the blood. Complement, plus antigen-antibody, seems to be a second step necessary for the various protective reactions of the body. The latter reactions may be phagocytosis (ingestion of foreign material by certain cells of the blood and connective tissues), lysis (destruction of the foreign substance), precipitation (taking the foreign substance out of solution), agglutination (causing the foreign substance to form large aggregates, which can then be phagocytosed), or perhaps all of these reactions. The only property of antibody that we are concerned with here, however, is its power to combine with, i.e. "fix," complement. Since it is possible to use test-tube (serological) techniques to determine whether, in fact, a given serum contains a particular antibody-antigen combination by testing the ability of that serum to "fix" complement, such a test can be used to discover immune responses. It should be mentioned here (as will be discussed more fully in Chapter 27) that the presence of antibody as revealed by complement fixation is not evidence of the effectiveness of the immune response. In fact, complement fixation has long been used for the diagnosis of syphilis, a disease against which immunity has little or no effect. In amoebiasis, the complement fixation test reveals that antibodies are formed in response to the presence of *Entamoeba histolytica*. As would be expected, the level of antibody is relatively high in systemic amoebiasis (where the amoebas are actually within the tissues of liver, lung, etc.) and practically absent in intestinal amoebiasis. In the latter case, the fact that immune reactions occur at all is rather good evidence that some lesions—"pinpoint ulcers"—may be present in supposedly intestinal amoebiasis. It is generally believed that antigen must enter the blood or other tissues in order to stimulate antibody formation.

One further fact about amoebiasis suggests the probability that the immune response may be effective to some degree. It has been observed that members of populations in the tropics often show more tolerance for amoebiasis than do visitors or immigrants from temperate zones. For instance, the incidence of clinical amoebiasis in English soldiers serving alongside Indian soldiers in the 1930's was higher than in their Indian comrades. The fact that the Indian soldiers had spent all their lives in regions where clinical amoebiasis (hence the virulent strain) was common, while the English had not, suggests at least two explanations for the differences in their response to infection. First, it is possible that an evolutionary mechanism has been working for centuries to eliminate highly susceptible members of the Indian population, and thus a relative resistance to amoebiasis may have become part of the Indian racial heritage. It is also possible that the Indian soldiers had nearly all suffered many subclinical infections, and had been repeatedly reinfected; therefore they maintained a degree of protective immunity of the premunition kind (see Chapters 2 and 27). Of course, both of the above explanations may be true. Leading workers on amoebiasis are not willing to deny the possibility that protective immunity may exist, but the weight of authoritative opinion seems to favor the various factors which have been discussed—strain differences among amoebas, individual differences both genetic and environmental among hosts—as being more important than immunity in the amoebiasis host-parasite relationship.

IMPORTANCE OF AMOEBIASIS

Up to now, questions which have been raised can be answered only by continued research into clinical and biological amoebiasis. It should not be assumed that because such questions have been asked they will be answered. Even in today's affluent world, there are conflicting demands for the use of money for various worthwhile enterprises. And the world's affluence has never extended far enough in terms of well-educated minds. Thus it is necessary to consider proposed programs in terms of their potential value to individuals, to nations, and to the world.

What is the importance of amoebiasis? A partial answer is supplied by the number of individuals who harbor *Entamoeba histolytica*. One authority estimates 400,000,000, but he properly adds that 80 per cent of these are without symptoms. As we have seen, the symptoms of chronic amoebiasis are so vague that the term "without symptoms" is not very meaningful. Yet one can assume a figure close to 100,000,000 for the actual sufferers from amoebiasis, a group which includes persons dying of acute dysentery or invasive abscess as well as those having vague abdominal pains, headache, chronic fatigue, or other nonspecific symptoms of ill health.

One basis for estimating the importance of a disease, once the number of victims is known, would be the cost of diagnosis and treatment. We have seen that diagnosis requires more than one fecal examination. A common method of examination is to take a minute sample of feces, spread it thinly on a slide, fix the material with a coagulant (fixing-solution such as mercuric chloride, alcohol, and acetic acid), stain it, and examine it for amoebic cysts or trophozoites. The time required to make such a preparation and examine it is measured in hours. The technician able to make such a diagnosis (or the technician-parasitologist team) is a highly trained specialist. The cost of one such examination in the United States is about ten dollars. An average of four to eight such examinations per patient would be usual, for after-treatment checkup is important, and at least two examinations before treatment, on the average, would be required. Other expensive diagnostic methods, especially concentration techniques, are often necessary. The cost of amoebicidal drugs and doctor's fees, as well as hospitalization for treatment in some cases, should also be considered. None of the figures is a firmly established amount, but by multiplying the number of potential patients (100,000,000 mentioned above) by, say, a modest twenty dollars, one reaches the rather large figure of two billion dollars as the medical cost of amoebiasis. This rather naive method of estimation is not intended very seriously, for it is unrealistic as to the majority of patients, who cannot afford such treatment, and it is visionary as to the treatment, which is not available on such a scale. However, there is probably some connection between medical costs of a disease and the economic and other kinds of damage due to it. The problem of amoebiasis is important.

However, in comparison with some other parasite-caused diseases—malaria, schistosomiasis, trypanosomiasis, for example—amoebiasis is perhaps not critically important. We have seen how malaria, while on the way to eradication or at least containment within tropical rain-forest areas, still disables over one hundred million and kills perhaps one million annually. In Chapter 9, we shall see how schistosomiasis can hardly be contained and how it affects vast multitudes. We recognize easily the importance of the hemoflagellates to man and beast in Africa and to millions of people in the slums of South America. Amoebiasis is not, perhaps, in a class with these destroyers and disablers. If the importance of a particular disease were the sole argument for spending money for research on that disease, however, certain great advances would not have been made. One of the best reasons for working on a problem is its difficulty. Hookworm in the United States during the early twentieth century was perhaps not a major health problem; indeed, the public refused to recognize it as a problem (Chapter 15). Hookworm was (as it still is in much of the world) a very difficult problem, because its victims were poor, ignorant, and unsanitary, and also because the political subdivisions of the United States made certain general welfare programs difficult to apply in spite of the fact that the bad health of part of the nation affected the well being of all the citizens.

But the solution of this problem, by farsighted planning, educational efforts, and wise cooperation between political units demonstrated a truth which has had almost unlimited power for good—the fact that teaching by example, with modest financial aid, can multiply in effectiveness. The methods of the Rockefeller Foundation in hookworm control have been adopted in many international situations, with great success at present and promise for the future. The amoebiasis problem is very difficult. Its solution, which is possible if energetic research is continued, should point the way to solving other difficult problems. Improved sanitation, necessary for control of amoebiasis, would control many other parasitic, bacterial, and virus diseases. Education for amoebiasis control would add impetus to education in general and would promote the prosperity that only an educated people can achieve. Medical research in the immunology, diagnosis, and therapy of amoebiasis would undoubtedly lead to advances in related fields. And the study of the taxonomic problem of the races or species of amoebas involved in this disease should give information upon general matters of evolution, variation, and host-parasite relationships, all of which are subjects of fundamental significance in biological advance.

SUGGESTED READINGS

BURROWS, R. B., and M. A. SWERDLOW, *"Enterobius vermicularis* as a Probable Vector of *Dientamoeba fragilis," Am. J. Trop. Med. Hyg.,* 5:258–265, 1956.

FAUST, E. C., "The Multiple Facets of *Entamoeba histolytica* Infection," *International Rev. Trop. Med.,* 1:43–76, 1961.

HOARE, C. A., "The Enigma of Host-Parasite Relations in Amoebiasis," Rice Institute Pamphlet, 45(1):23–35, 1958.

McDearman, S. C., and W. B. Dunham, "Complement Fixation Tests as an Aid in the Differential Diagnosis of Extra-intestinal Amoebiasis," *Am. J. Trop. Med. Hyg.*, 1:182–188, 1952.

Meleney, H. E., "Community Surveys for *Endamoeba histolytica* and Other Intestinal Protozoa in Tennessee, First Report," *J. Parasit.*, 16:146–153, 1930.

Phillips, B. P., P. A. Wolfe, C. W. Rees, H. A. Gordon, W. H. Wright, and J. A. Reyniers, "Studies on the Amoeba-Bacteria Relationship of Amoebiasis. Comparative Results of the Intracecal Inoculation of Germfree, Monocontaminated and Conventional Guinea Pigs with *Endamoeba histolytica*," *Am. J. Trop. Med. Hyg.*, 4:675–692, 1955.

Rees, C. W., *Problems in Amoebiasis*. Thomas, Springfield, Ill., 1955.

Sawitz, W. G., and E. C. Faust, "The Probability of Detecting Intestinal Protozoa by Successive Stool Examinations," *Am. J. Trop. Med.*, 22:131–136, 1942.

Silva, R., R. Donckaster, and R. Valencia, "Estudio de algunos factores epidemiológicos que favorecerían las infección y reinfección por *Entamoeba histolytica*," *Bol. Chileno de Parasitologia*, 13:22–25, 1958.

Sodeman, W. A., and P. C. Beaver, "A Study of the Therapeutic Effects of Some Amoebacidal Drugs," *Am. J. Med.*, 12:440–446, 1952.

Van Thiel, P. H., "Réconciliation entre les conceptions concernant la biologie de l'*Entamoeba histolytica* en rapport avec la perturbation de l'équilibre entre l'homme et le parasite par le traitement médical," *Bull. Soc. Path. Exot.*, 51:824–829, 1961.

WHO, Epidemiological and Vital Statistics Dept., 8:127–131, 1955.

Wright, W. H., "The Public Health Status of Amoebiasis in the United States as Revealed by Available Statistics," *Am. J. Trop. Med.*, 30:123–133, 1950.

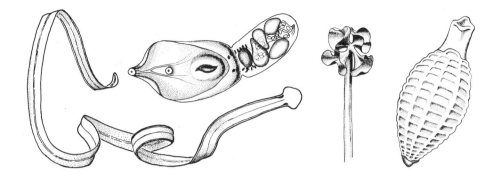

CHAPTER 7

INTRODUCTION TO
THE FLATWORMS

The phylum Platyhelminthes is important in several ways. While we are concerned necessarily with the parasites in this group and specifically with the parasites affecting man and his livestock, other biologists are interested in the flatworms as probable ancestors of the higher many-celled animals. Also recent psychological studies on the behavior of planarians (nonparasitic flatworms capable of regeneration) have revealed intriguing aspects of learning; an acquired ability, in these animals at least, seems to be transmissible by fragments cut from almost any part of a worm "trained" to run a maze.

Free-living flatworms belong to the class Turbellaria (see Fig. 7–1). These are ciliated, soft-bodied, hermaphroditic animals living in salt or fresh water (except for a few land-dwelling forms). They are carnivorous, being either predators on living prey or scavengers on the dead bodies of other animals. They possess three kinds of tissue, an outer layer of ciliated epidermal cells, a middle layer of muscular and excretory tissues, and an inner layer, either organized into an intestine or represented merely by specialized cells, of digestive tissues. A nervous system is

97

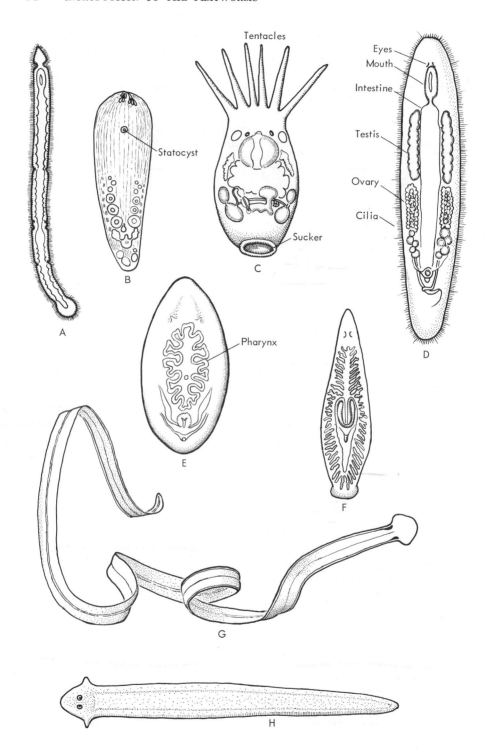

Tentacles

Statocyst

Sucker

Eyes
Mouth
Intestine
Testis
Ovary
Cilia

Pharynx

A

B

C

D

E

F

G

H

present with its chief ganglia concentrated as a brain in the anterior part of the body. Many Turbellaria have eyes, sensory hairs, olfactory or gustatory pits, and perhaps other sense organs. In some of the marine forms static organs (of equilibrium) are present. All Turbellaria reproduce sexually; the individuals exchange spermatozoa through a variety of copulatory organs. Few eggs are formed at a time, and the development from egg to adult is simple and direct. A number of Turbellaria reproduce asexually by a sort of segmentation and fission. The above-mentioned power of regeneration is the basis of such asexual reproduction as well as of the exceptional ability of these animals to withstand starvation by literally feeding on their own tissues. Starving planarians, for example, by means of phagocytic cells of their middle tissue layer, absorb and digest their own bodies and gradually become smaller until food becomes available. Some of these animals may become reduced to $\frac{1}{300}$ of their normal size. The regenerative and reorganizational capacities of turbellarians are probably significant in the evolution of the other classes of flatworms, the parasitic groups, for both trematodes and tapeworms are characterized by remarkable powers of asexual reproduction. In a sense, therefore, the Turbellaria may be said to be preadapted to a parasitic way of life in so far as reproductive potential is concerned.

The orders of the class Turbellaria illustrate the variety in this group and suggest a possible line of descent leading to higher forms.

The most primitive order is the Acoela. These marine Turbellaria are very small animals possessing a statocyst (balancing organ), sometimes possessing eyes, having a subepidermal nervous system (suggestive of coelenterate systems) or a more central system in the more complex forms, and lacking an intestine (hence the name Acoela). Food is captured by a mouth or simple muscular pharynx; it is received by a mass of digestive cells and, after fragmentation, digestion, and phagocytosis, is distributed to the cells and tissues by transportation and diffusion. There is no excretory system. Reproduction in the Acoela involves oocytes and ova, spermatocytes and spermatozoa; but there are no specialized female sex organs at all in most species, and in the males there is only a simple muscular penis by which sperm are injected hypodermically during copulation. Sperm wander through the middle tissues at random, fertilizing ova which leave the animal through the mouth or by rupture of the body wall. The primitive reproductive system just described is found in the simpler acoels; some have relatively complex systems foreshadowing the intricate organs of the higher Turbellaria.

FIG. 7–1. Representative members of the class Turbellaria. [After Hyman, 1951.] (A) Order Rhabdocoela (*Microstomum*). (B) Order Acoela. (C) Order Temnocephalida. (D) Order Rhabdocoela (*Macrostomum*). (E) Order Polycladida (*Hoploplana*). (F) Order Tricladida (*Bdelloura*). (G) Order Tricladida (*Bipalium*). (H) Order Tricladida (*Dugesia*).

The second order of Turbellaria (order Rhabdocoela) contains small animals with a simple, straight intestine (hence the name), an excretory system consisting of tubules ending in ciliated "flame cells" (see Fig. 8–1), and complex reproductive systems. Like all Turbellarians, the Rhabdocoela are monoecious (or hermaphroditic). While the order is large and complex, taxonomically, it will be sufficient here to call attention to two parasitologically significant facts. First, a number of rhabdocoels reproduce asexually most of the time (the common freshwater genus *Stenostomum*, for example), foreshadowing a major parasitic adaptation for fecundity. Second, many rhabdocoels are ecto- or endo-commensals (living upon or within other animals and sharing the food of the latter). Of particular interest in this respect is the curious group called Temnocephalida, which live attached to the surfaces of crayfishes of the southern hemisphere. These worms have anchorlike adhesive discs by which they cling to their hosts, and they are ciliated only when young. They will be mentioned again in connection with the evolution of parasitism.

There are three other orders of Turbellaria. The Alloeocoela are a varied group of small and mostly marine forms. The order Tricladida includes large turbellarians with three-branched diverticular intestine and prominent protrusible pharynx. The freshwater planarian, *Dugesia*, a familiar laboratory animal, is a member of this order, as are the very large and colorful land planarians sometimes seen in greenhouses and normally found in tropical rain forests. The order Polycladida consists of rather large marine forms ellipsoidal in outline, with a large, folded pharynx opening into a many-branched intestine. Some of them feed upon oysters, living within the shell in a curious relationship which is probably predacious rather than parasitic.

The above remarks about the class Turbellaria can serve only to call attention to a biologically fascinating group of primitive animals from which, possibly, the higher phyla have all evolved. The parasitological importance of the Turbellaria is also evolutionary, as will appear from the following discussion of the parasitic classes of flatworms, which should be viewed as derived from turbellarian ancestors of which they retain certain useful features, such as regenerative and reproductive capacity, and the ingenuity already exemplified by the diversity of the free-living flatworms.

FIG. 7–2. Representative Monogenea and Aspidogastrea. (A) *Udonella*, an ectoparasite upon parasitic copepods (Crustacea) attached to fish. [After Baer, 1951.] (B) *Aspidogaster conchicola*, subclass Aspidogastrea, parisitic in the heart chamber of freshwater mussels. [After Hyman, 1951.] (C) *Sphyranura*, from gills and surface of *Necturus*, an aquatic urodele amphibian. [After Hyman, 1951.] (D) Four holdfasts of Monogenea; (1) *Rajonchocotyle*, (2) *Sphyranura* [cf. (C) above] (3) *Gyrodactylus*, and (4) *Polystoma*. [After Baer, 1951.] (E) Two young *Aspidogaster*. [After Hyman, 1951.] (F) A young *Sphyranura*. [After Hyman, 1951.] (G) Larva of *Polystoma*, showing eyes, patches of cilia, and simple, undivided holdfast. [After Baer, 1951.]

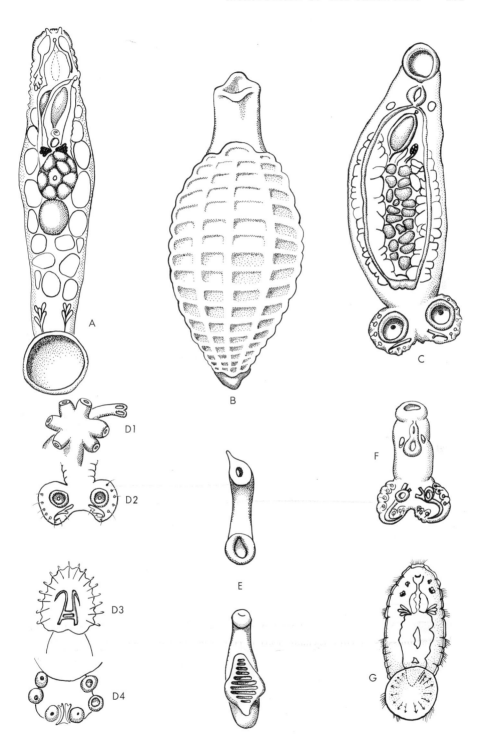

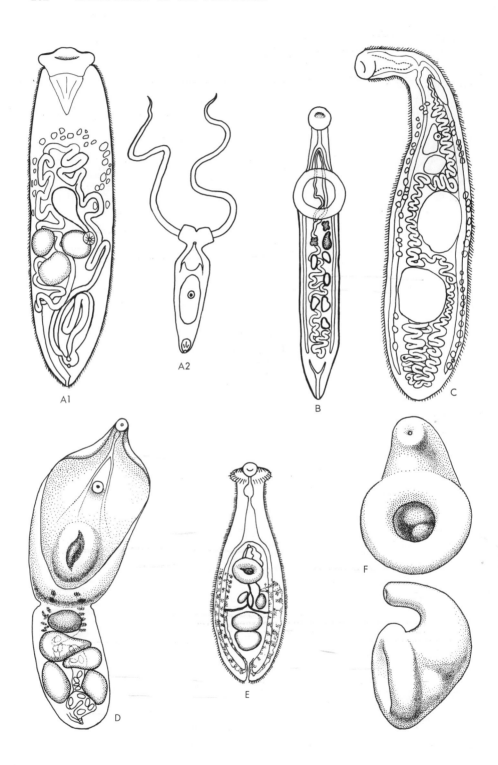

FIG. 7–3. Representative members of the subclass Digenea. (A1) Adult and (A2) cercaria from the family Bucephalidae of the "gasterostome" digenetic trematodes. [After Hyman, 1951.] (B) Adult, family Gorgoderidae, which include bladder flukes of frogs. [After Hyman, 1951.] (C) Adult, family Plagiorchidae, genus *Haemotoloechus*, a frog lung-fluke. [After Hyman, 1951.] (D) Family Strigeidae, genus *Neodiplostomum*. [After Noble and Noble, 1961.] (E) Family Echinostomatidae, genus *Echinostomum*. [After Chandler and Read, 1961.] (F) Family Paramphistomidae (*Paramphistomum*). [After Hyman, 1951.]

The class Trematoda more nearly resembles the Turbellaria than do any other flatworms. This class is divided into two or three major groups which probably represent quite distinct kinds of organisms (justifiably classes in their own right) —the subclasses Monogenea (ectoparasites with a simple life cycle), Aspidogastrea (endoparasitic with a simple life cycle) and Digenea (endoparasitic in vertebrates, with molluscs as obligate intermediate hosts, and polyembryony in the latter). The Monogenea (Fig. 7–2) are mainly parasites of fishes and amphibia; they have many of the characteristics of free-living animals (eyes, for instance) but possess well-developed hooks and suckers, and have cilia only in the young stages. They lay few eggs. The Aspidogastrea are not a well-understood group. They all possess a large, frequently subdivided, ventral sucker and have a simple saclike intestine. It is believed that they have a direct life cycle similar to that of members of the Monogenea and do not exhibit the polyembryony or asexual multiplication which occurs in the third subclass, Digenea. The best known aspidogastrid trematodes parasitize molluscs, but some species occur in reptiles, amphibia, or fishes. At least one life cycle involves a necessary intermediate host; members of the genus *Lophotaspis* develop as larvae in a marine snail but mature only in the intestine of turtles.

The subclass Digenea (Fig. 7–3) is by far the most important of the groups of trematodes. In fact, when most writers refer to trematodes, they mean "digenetic trematodes," indicating the thousands of species of flukes which infect man, his domestic animals, and innumerable kinds of wild vertebrates. The Digenea have a complex and remarkable life cycle. The eggs, passing out of the vertebrate host by any of a variety of routes, usually must reach water, wherein they hatch. The ciliated larva or miracidium which emerges (a recapitulation of turbellarian ancestry, no doubt) is able to swim, and may find and penetrate a mollusc, usually a gastropod. Usually only one species, or only a particular strain, of mollusc can serve as host. Once a suitable snail has been penetrated, the miracidium grows, losing its eyes and ciliated epidermis, until it becomes a sac (the sporocyst) filled either with many second-generation sporocysts or with larvae called rediae (somewhat wormlike, with a small vestigial gut, again reminiscent of Turbellaria). The

FIG. 7–4. Germ-cell cycle in digenetic trematodes. Each stage in the life cycle has body tissue which dies when that stage dies, as well as germinal tissue which gives rise to members of another generation. The solid objects represent germinal tissues, which are connected by arrows to their products. In the adult, germinal tissues of ovary and testis produce eggs and sperm, respectively, which unite to produce the zygote and miracidium of the next generation.

rediae escape into various tissues of the snail, where they grow, and become filled with tailed larvae, the cercariae. The cercariae are equipped with tails for swimming or a similar adaptation for "finding" a new host, with eyes and nervous system for directional response, and with the beginnings of sex organs for eventual reproduction. They emerge from redia or sporocyst and (usually) from the snail. Then in various ways they reach the final host. These ways include encystment on vegetation (the sheep liver fluke), penetration and encystment within a second intermediate host (*Clonorchis, Paragonimus*, and many other flukes of various animals), ingestion of the snail in species whose cercariae do not emerge, and, rarely, direct entry into the final host. In the final or definitive host, the cercariae grow into sexually mature, hermaphroditic (except in one family of blood flukes) individuals. Cross-fertilization is believed to be the usual method by which sperm and ovum unite. Then egg-laying commences, and the cycle is complete. (Details of several such cycles can be found in the chapters immediately following.) The several generations in the snail may be considered polyembryony, a successful device for rapidly increasing the number of cercariae (larvae infective to the vertebrate host). Also, the digenetic trematodes exemplify the germ line idea quite clearly, since each of the larval generations contains recognizable, distinct tissues which are somatic (mortal) and tissues which are germinal (immortal or destined to continue to reproduce). Figure 7–4 illustrates this fact.

From the above remarks about trematodes and turbellarians, it should be concluded that these animals are more diverse than the current scheme of classification suggests and that they are nonetheless rather closely related to each other by descent.

Members of the class Cestoidea (Fig. 7–5), while clearly flatworms, are quite distinct from the above four groups. They have no intestines, produce embryos with hooks, and are usually composed of chains of segments or proglottids. The class Cestoidea consists of two subclasses: the Cestodaria and the Cestoda. The first, the cestodarians, are a small group of rather curious forms which resemble the true cestodes chiefly because they have no intestine and develop from an embryo

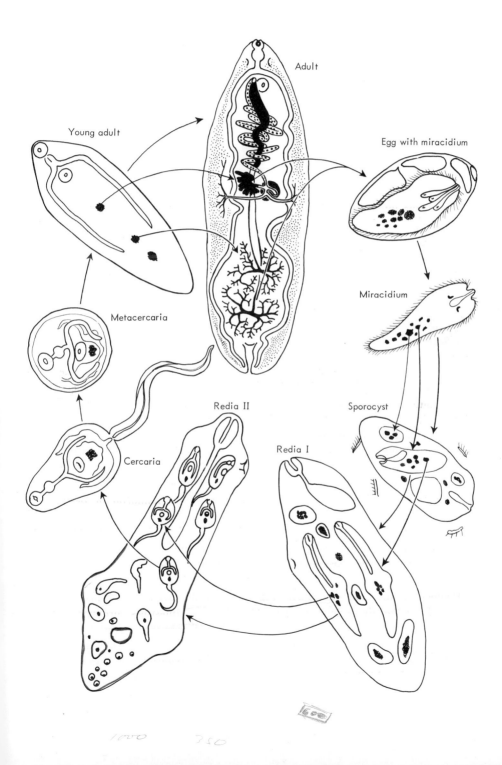

Adult

Young adult

Egg with miracidium

Miracidium

Metacercaria

Sporocyst

Redia II

Redia I

Cercaria

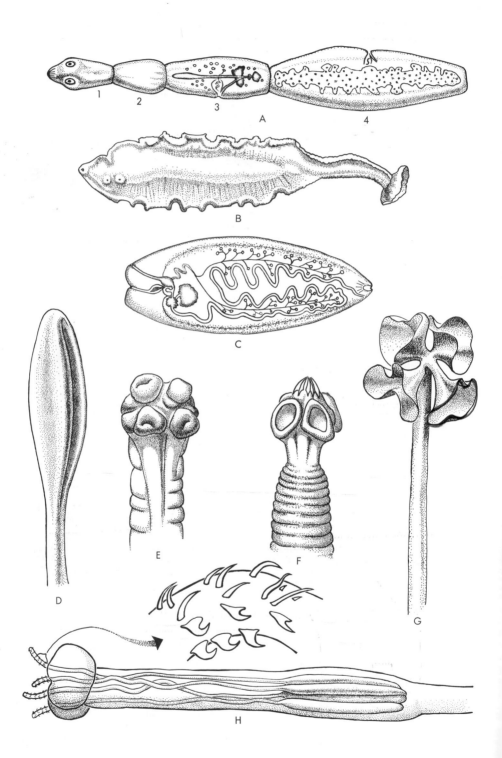

FIG. 7–5. Cestodes and Cestodarians. (A) Complete cestode, *Echinococcus granulosus*, from the intestine of a dog or wolf. (1) Scolex and neck, (2) immature proglottid, (3) sexually mature proglottid, and (4) gravid (egg-filled) proglottid. [After Chandler and Read, 1961.] (B) *Gyrocotyle*, subclass Cestodaria, from the body cavity of chimaeroid fishes. [After Wardle and McLeod, 1952.] (C) *Amphilina*, subclass Cestodaria, from the body cavity of the sturgeon. [After Wardle and McLeod, 1952.] (D,E,F,G) Scolices of *Dibothriocephalus*, *Ophiotaenia*, *Hymenolepis*, *Anthobothrium* of the orders Pseudophyllidea, Proteocephala, Cyclophyllidea, and Tetraphyllidea, respectively. [After Wardle & McLeod, 1952.] (H) Scolex and enlarged base of a proboscidial tentacle of *Grillotia*, a genus of the order Trypanorhyncha.

which bears hooks. These worms are unsegmented. They live in the body cavities of certain extremely ancient fishes. The rather robust worms of the genus *Gyrocotyle* occur in the chimaera, or ratfish, a primitive relative of the sharks and rays. The delicate worms which comprise the genus *Amphilina* live in sturgeons. Cestodarians live, not in the intestine, but in the peritoneal cavities of their hosts. Superficially, both *Gyrocotyle* and *Amphilina* resemble the plerocercoid larvae of certain true cestodes (subclass Cestoda). In fact these worms are considered by some authors to be neotenic (sexually precocious) larvae of extinct adults; this is a seemingly absurd idea until we consider that cestodes may well have inhabited certain large piscivorous reptiles of the Mesozoic seas and that such cestodes probably utilized large fishes as intermediate hosts. Since cestodarians are found today in the body cavities (not intestines) of two groups of very ancient fishes, the chimaeroids and the sturgeons, the suggestion that these strange worms are surviving larval stages of extinct forms seems plausible. This idea will be discussed further in the chapter on evolution.

The true cestodes are practically all segmented and have multiple reproductive units. These worms occur in five or more major groups which are roughly defined by corresponding groups of hosts. Thus the Cyclophyllidea are an order of tapeworms of birds and mammals (with a curious small family in amphibians); the Pseudophyllidea are found in fish and fish-eating hosts; the Proteocephala are worms of fishes, reptiles, and amphibia; and the Trypanorhyncha, as well as the Tetraphyllidea, are parasites of sharks and rays. The first two orders mentioned, Cyclophyllidea and Pseudophyllidea, are the only groups of tapeworms which concern man directly.

The cyclophyllidean cestodes are distinguished from others in several ways. The scolex or head possesses four cup-like suckers and often a terminal rostellum, a protrusible digging and anchoring organ armed with hooks. The strobila, or chain of segments, begins with a neck region just behind the scolex, where segmentation first appears as a series of transverse grooves at regularly increasing intervals. Posterior to this region are immature segments showing the anlagen (beginnings) of sex organs and gonads. Then come sexually mature segments. Usually the male system in these segments becomes functionally mature before the female. Characteristic of the Cyclophyllidea are relatively few testes, a compact yolk gland associated with the ovary, and a uterus which grows until it bursts since it is not equipped with a pore for egg-laying. Although each of these features may be found in other orders of cestodes, the above combination defines the Cyclophyllidea.

The Pseudophyllidea have grooves rather than suckers on the scolex, and there is no rostellum. Segmentation is much less regular than in the Cyclophyllidea; it sometimes occurs only after the strobila is quite large and often starts at more than one place, occasionally in the neck region or much farther down. The yolk glands of a mature segment are follicular and often widely scattered throughout the body. The uterus has a distinct pore through which eggs are laid.

While not very important in human or veterinary medicine, cestodes are interesting animals and will be the subject of a chapter in this book.

From the above description of the phylum Platyhelminthes one can derive a series of forms ranging from free-living to highly parasitic. The Turbellaria, well equipped for a predatory active life, have a few members which live as parasites in or on other animals. These parasitic forms lack cilia and may have holdfast organs; such features resemble the second class of flatworms, Trematoda. The latter, while all parasitic, range through several degrees of parasitic adaptations. Thus the ectoparasitic monogenetic forms resemble turbellarians in several ways. Members of the curious subclass Aspidogastrea, while living inside their hosts and having lost such evidences of freedom as eyes and locomotor apparatus, have a relatively simple and direct life cycle like that of Monogenea. Then there are the well-known flukes, members of the Digenea, in which the life cycle includes a highly fertile egg-laying adult generation and a polyembryonic, perhaps equally fecund, larval generation or series of generations; both are adaptations to a more precarious parasitism than is found in either of the first two subclasses.

In the next four chapters some important parasitic flatworms will be treated in detail. Liver flukes of man and other fish-eaters will introduce both the details of a trematode life cycle and the cultural factor in human disease. Blood flukes will illustrate a problem in disease control which so far has resisted all attempts at solution yet requires a solution as schistosomiasis spreads into new areas. Several other trematodes of man and his domestic animals will be mentioned. Then the cestodes will be examined in two ways: (1) as parasites of some medical and veterinary significance, and (2) as interesting animals showing particularly well some of the features of advanced parasitism.

SUGGESTED READINGS

HYMAN, L. H., *The Invertebrates, Vol. II: Platyhelminthes and Rhynchocoela. The Acoelomate Bilateria.* McGraw-Hill, New York, 1951, pp. 1–422.

CHAPTER 8

CLONORCHIASIS

INTRODUCTION

In the southeastern corner of Asia lies a long peninsula which extends southward from China and India. Formerly Indochina, the land now consists of the countries of Burma, Thailand, Laos, Cambodia, North and South Vietnam, and Malaysia. Since World War II, lands in Southeast Asia which once belonged to England and France have now become independent nations.

In the center of the peninsula lies Thailand (formerly Siam), a rich nation with a large alluvial plain in which some of the world's best rice is grown. The product is exported and returns wealth to the nation. Rice culture is irrigated farming. Across the flat country of Thailand are many canals. Since there are relatively few roads, people live on boats or in villages along the canals.

The educational level of the Thai people is relatively high. By 1954, 54% of the population over 10 years old were literate. (Compare Haiti, with something less than 10% literacy.) Buddhism is the chief religion; all young men of the Buddhist majority (90%) are expected to enter the priesthood for several months as part of their religious training. (Buddhism, of course, is a sophisticated and intellectual religion.)

109

About one-seventh of the land is devoted to agriculture, and about nine-tenths of that is used for planting rice. Other food products come from small grains, corn, and fisheries. Marine and freshwater fisheries produce about 250,000 tons (as compared with 6 million tons of rice). This implies a ratio (1/24) of protein to carbohydrate which is perhaps unrealistic; marine fisheries products are often exported, and seldom, at any rate, reach inland markets. Freshwater fish probably amount to not more than 70,000 tons. Since fish are the chief source of protein in the Thai diet, the protein-carbohydrate ratio may actually be close to 2%. (In the United States, the protein-carbohydrate ratio is about 40%.) Although this proportion of animal protein in human diet is no lower than that found in much of Asia, it represents a marginal amount and the bare necessity for normal growth, resistance to disease, and bodily repair.

The above seemingly unrelated facts—a naturally wealthy nation, literate people, far-from-modern transportation system, and protein-deficient diet—may be considered to be the actual and potential factors in the incidence and persistence of clonorchiasis, or human liver-fluke disease, which is widespread throughout Thailand. Let us see how custom and apparent necessity oppose the influences of education and modern science, and how the existence of a preventable disease depends on human culture.

Clonorchiasis affects approximately 19,000,000 people in the world. In Thailand, very heavy incidence has been reported in certain regions where surveys have found nearly 70% of the people infected; of the entire population of some 22 million, about one and one-half million people have clonorchiasis. Having a very peculiar distribution, the disease is limited to regions where three conditions occur simultaneously: (1) the pollution of fresh water with human excrement, (2) the presence of snails capable of harboring *Clonorchis* larvae, and (3) the custom of eating raw fish. People acquire clonorchiasis from larvae encysted in the flesh of fish. These larvae escape in the intestine, reach the liver via the bile duct, and feed upon liver tissue; pathological changes occur in proportion to their numbers and the duration of their residency. Over 5,000 living flukes have been removed from the liver of a single patient. This particular patient had died from the disease.

ANATOMY

The typical digenetic trematode has gonads (sex glands) of both sexes with the necessary accessories. It is therefore hermaphroditic. Adults of *Clonorchis sinensis* and *Opisthorchis viverrini*, the important liver flukes of man and other fish-eaters, have a female reproductive system that consists of a somewhat lobed ovary, a bulblike sperm-storage vessel (seminal receptacle), and a small rosette-shaped gland (the Mehlis gland) surrounding an egg-forming organ (the ootype), which is connected with such ducts as the oviduct from the ovary, the yolk-duct from the yolk glands, and the uterus. The latter leads from the ootype to the female genital opening on the surface of the animal's body, and it contains fully formed eggs which are in progressive stages of development as they move toward the genital pore.

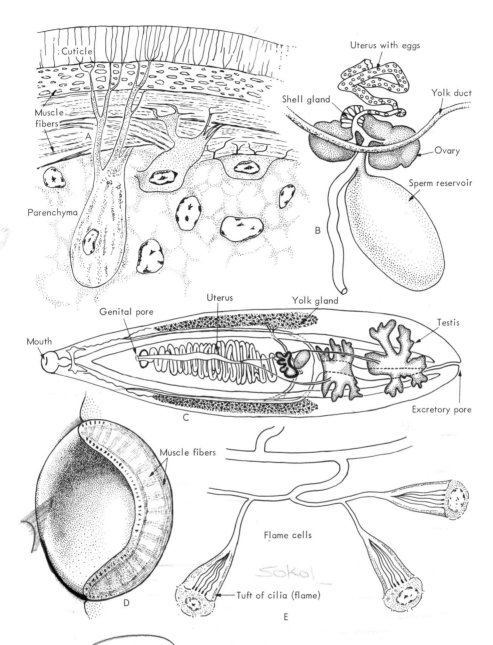

FIG. 8–1.   Trematode anatomy. (A) Section through the body wall of a trematode. (B) Female genital system. (C) Adult trematode. (D) Ventral sucker, showing radial muscles and tangential muscles. (E) Portion of excretory and water regulating system, consisting of several "flame cells" with their tubules.

Several thousands of eggs are usually present in the uterus of a mature specimen of *Clonorchis sinensis*. Figure 8–1 shows the above details of the female system as well as the parts of the male system and other anatomical features. The male reproductive system of *Clonorchis* consists of two greatly branched testes (the name *Clonorchis* means "branched testis") from which sperm cells are transported by sperm ducts to a seminal vesicle in the anterior region of the body near the opening of the female system. In *Clonorchis* and its near relatives there is no elaborate male copulatory organ, and the sperm are presumably ejected into the common genital opening, whence they travel up the uterus to the seminal receptacle, the female sperm-storage organ. (Whether fertilization requires copulation in this group of hermaphroditic animals is unknown. Many flukes have complex male copulatory organs, and copulation with cross-fertilization is believed to occur in these forms.) In addition to this organ, the typical digenetic trematode, such as *Clonorchis*, has an oral sucker, a muscular esophagus, and a U-shaped intestine ending blindly on either side near the posterior end. There is an excretory system consisting of flame cells (Fig. 8–1) whose ducts unite into tubes which empty into an excretory bladder. The bladder opens by a pore at the posterior extremity of the fluke's body. Body musculature consists of longitudinal and circular muscles as well as dorsoventral strands; by the antagonistic action of these muscles the animal can change its shape. *Clonorchis* is covered by a smooth noncellular cuticle; in many flukes this cuticle is spinous. Flukes do not possess a true circulatory system; like all other flatworms, however, they have wandering amoebalike cells capable of phagocytosis, and their tissues are loosely organized to allow some passive movement of fluids throughout the body. The nervous system is a pharyngeal ring from which paired longitudinal nerves arise. Of all the anatomical structures, the parts of the reproductive system are the most prominent—a prominence characteristic of all parasites and not of digenetic trematodes alone.

LIFE CYCLE

The term digenetic means "having two ways of reproduction." The production of eggs by *Clonorchis* is only one reproductive phase of its life cycle (Fig. 8–2). A second phase, as we pointed out in Chapter 7, must occur in molluscs.

FIG. 8–2. Life cycle of *Clonorchis sinensis*. (A) Infective metacercaria in piece of fish. (B) Human host showing site of entry [(1) bile duct] from duodenum to liver. (C) Egg, with (1) characteristic lid or operculum and (2) ciliated larva or miracidium. (D) Snail (*Parafossarulus* sp.) showing sites of development of various larval stages; (1) region where miracidia hatch in snail stomach and penetrate the host, (2) sporocysts, (3) rediae, containing germ masses and developing (4) cercariae. (E) Two views of Cercaria, showing, respectively, keeled tail and penetration glands. (F) Carp, one of many kinds of fish in which metacercariae can encyst and develop to infectivity. [(C) and (D) after Faust and Russell, 1964.]

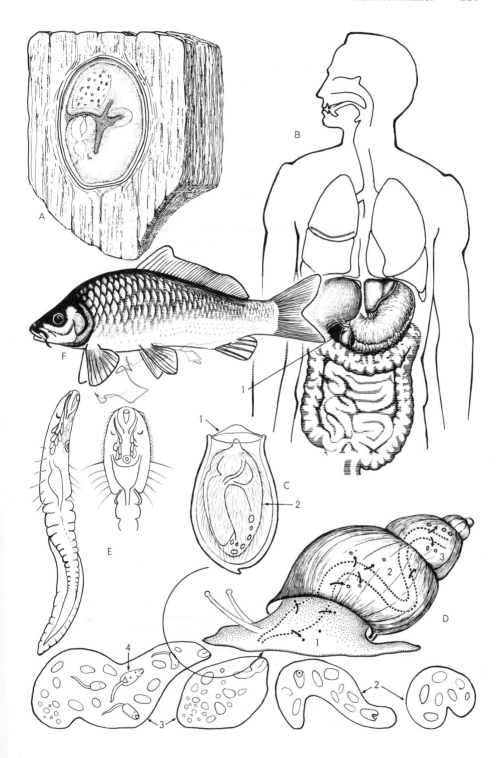

The flask- or bottle-shaped eggs, laid in the bile duct, escape to the host intestine. They are mixed with the intestinal content and are deposited with feces. These eggs contain a ciliated larva but do not hatch unless ingested by a snail. In certain snails of the subfamily Buliminae (family Amnicolidae) the eggs hatch, the miracidia penetrate the intestinal wall of their host, and sporocysts develop. The sporocysts of *Clonorchis* give rise to rediae. A redia is a larva with a primitive and usually rudimentary gut, a feature which distinguishes it from the sporocyst. It also possesses a birth pore through which the cercariae escape. There is continuous production of cercariae from germinal tissue in the redia. The cercariae reach water by escaping from the snail tissues. With a keeled tail they are well equipped for swimming, and they have widely spaced eyes. They also have oral and ventral suckers, rudimentary reproductive tissue (gonad anlagen), a characteristic excretory system, and both cystogenous and histolytic glands. Thus the second, or molluscan host, phase in the reproduction of *Clonorchis* produces a large number of larval trematodes.

Instead of entering the final host directly, however, these larvae penetrate various fishes, which serve as a second intermediate host. Using the secretions of the histolytic glands, the cercariae bore into the skin of fishes such as carp and others of the minnow family; once in the muscles, they encyst. Within the cyst, the cercariae develop further, losing the eyes and growing somewhat larger. After some weeks, the metacercariae are infective for their final host. This infectivity is maintained for an indefinite period.

Any fish-eating carnivore can probably serve as final host. The ingested fish is digested, and the metacercariae are liberated from their cysts. They reach the bile duct of their host and then move into the liver. There the young flukes mature and in about three weeks begin to lay eggs. The longevity of *Clonorchis* and related flukes has not been established for human infections, but the very large number recovered from some cases indicates a life-long accumulation of flukes, with very little mortality on the part of the parasites.

PATHOLOGY

Damage to the host is caused by obstruction of the smaller bile passages and by mechanical erosion of the epithelium of the passages, with consequent thickening and scar tissue formation. If numerous, these local reactions can result in cirrhotic changes, swelling and hardening of the liver, interference with liver function, and generalized weakening of the body due to metabolic failure. In particularly advanced cases involving thousands of flukes, the liver may become so scarred and altered that the portal circulation nearly fails, and parallel veins develop; hence the liver is bypassed entirely. The victim in such cases is pitifully thin, for he is deprived of the essential metabolic controls and reserves mediated by the liver. Death, of course, is the result of such heavy and long-standing infections.

Treatment of clonorchiasis is not very effective. Early cases (which are usually symptomless and seldom receive treatment) may be successfully treated with rather large quantities of Gentian Violet contained in enterically coated capsules. Chloroquine, the antimalarial drug so useful in other diseases, is effective also.

CONTROL

As we have seen, disease caused by liver flukes in man is due to the presence of flukes of the genera *Clonorchis* and *Opisthorchis*, which are acquired by eating uncooked fish containing encysted metacercariae. Infection of the fish depends on the presence of sewage-feeding snails of certain species. Therefore, it appears that clonorchiasis can be controlled or prevented by either of two seemingly simple expedients: cooking of fish before eating or preventing sewage from entering waterways of fisheries. That these expedients are not simple will appear when they are considered in relation to Thailand, the country briefly described at the beginning of this chapter.

Not everywhere is cooked fish preferred to raw fish. In Thailand, as in many other parts of Asia, sliced or diced fish is eaten after being dipped in wine sauce, or spiced, or warmed by placing on mounds of steaming hot rice. Fish so prepared and consumed is considered very good, just as in some parts of the world other delicacies—for instance raw oysters or clams from various seaboards, smoked or pickled fish from the northern lakes of the U.S. or Europe, or sausage prepared from pork—are much esteemed as food. Rare beef, with its risk of tapeworm infection, is widely preferred to well-done. The simple expedient, cooking of fish before eating, would actually require in Thailand a change in diet and a giving up of a favorite kind of food. In a somewhat similar situation the demonstrable danger in cigarette smoking has not, at this writing, appreciably changed the smoking habits of Americans. (I do not know whether totalitarian methods of control might be used effectively in a public health program. Perhaps Communist China may be able to control clonorchiasis by some means not available to a democracy. The value of such a cure seems, to say the least, questionable.) It seems unreasonable to expect the unappreciated risk of clonorchiasis to change the eating habits of the Thai people.

There is an economic factor involved in cooking, moreover, that affects the problem of clonorchiasis control. Thailand produces for fuel each year about 600 million board feet of wood. While this is almost equal to the production of fine woods, such as teak for export, it amounts to less than 30 board feet per individual, which is hardly enough for a good-sized bonfire. Of course, other fuels, such as dried cattle or buffalo dung, are used, but the amounts available are very small in comparison with European or other western standards. An additional need for fuel for cooking certain fish might produce or aggravate a fuel shortage. In lands poorer than Thailand, this would certainly be the case.

Sanitation, then, might appear to be a better method of control than attempts to change the food habits of a people. Two difficulties, however, seem rather great.

The first of these is education. In a literate population such as that of Thailand, which has more than 50% literacy, it is possible to disseminate rather widely facts about politics, agriculture, or health. But to inform is only one part of education; the other phase is to bring about reform and to stimulate people to act more intelligently than heretofore in their own behalf. Merely giving people convincing information on the danger to health resulting from contamination of the canal system and fish-culture ponds with human sewage will not suffice, and change in cultural habits will not necessarily follow. Houses along waterways cannot be equipped with western plumbing, even if the owners desire such things. However, suggestions for the collection and storage of human feces (night soil) seem practical, and it may be possible to persuade those in charge to treat such stored feces with chemicals or to store the material long enough for fermentation products (ammonia, chiefly) to kill the eggs of *Clonorchis* before ingestion by snails can occur. Thus the difficulty of sanitation by education is great but not insuperable.

The second difficulty involved in prevention of clonorchiasis by sanitation is that human sanitation would not affect the behavior of reservoir hosts. (The latter term is appropriate for animals other than man in which a parasite of man can and does complete its life cycle and thus occasionally bypass human hosts.) Dogs and cats serve as reservoirs of clonorchiasis. As long as these fish eaters are part of the environment where suitable snails and fish occur, the intermediate hosts will become infected, and fish will be a source of human infection irrespective of human sanitary practices.

Thus it appears that human clonorchiasis in Thailand is an unsolvable problem; the obvious solution is either very difficult (to change the eating habits of millions) or both impractical and biologically futile (sanitation by control of humen excrement). If the problem cannot be solved in a relatively wealthy, literate, and politically stable country, it seems unlikely that it will be solved elsewhere.

CONCLUSIONS

Clonorchiasis must be considered part of the agenda of public health parasitology. The disease affects millions of people; an undetermined number are affected severely. It can be diagnosed and treated, and its prevention and control, while very difficult, are not impossible. Several kinds of investigation may lead to real progress in the battle to fight this disease. Studies of the actual harm done by clonorchiasis would provide a basis for concern; the reasonable question—"How much does the disease cost?"—should be answered before expensive control efforts are initiated. (Very few nations can afford vast public expenditures for relatively minor health problems.) Further research on the disease itself, especially on therapy and the feasibility of immunization, might lead to simple measures of control by mass treatment. Methods of educating people to protect themselves should be investigated, for the custom of eating raw fish may not be so firm as observers now believe; even firm customs have changed under the pressure of circumstance. Education is essentially a national, indeed a local matter, since it depends on the understanding and acceptance of new ideas and not on the imposition of new patterns by authority. Finally, a broad study of the diseases which share with clonorchiasis the same sewage-dependent epidemiology might show that sanitary control measures which no single disease could justify would nevertheless be justified by their potential effectiveness against a whole group of diseases. In other words, clonorchiasis needs to be put into perspective and evaluated as one cause, among many, of human misery in the countries where it exists.

SUGGESTED READINGS

HARINASUTA, C., and S. VAJRASTHIRA, "Opisthorchiasis in Thailand," *Ann. Trop. Med. Parasitol.*, 54:100–105, 1960.

SADUN, E. H., "Studies on *Opisthorchis viverrini* in Thailand," *Am. J. Hyg.*, 62:81–115, 1955.

STRAUSS, W. G., "Clinical Manifestations of Clonorchiasis—A Controlled Study of 105 Cases," *Am. J. Trop. Med. Hyg.*, 11:625–630, 1962.

YAMAGUCHI, T., K. UCHARA, and M. SHINOTO, "Treatment of *Clonorchis sinensis* with Pankiller (Dithiazanine iodide)," *Jap. J. Parasitol.*, 11:30–88, 1962.

CHAPTER 9

SCHISTOSOMIASIS

INTRODUCTION

Some diseases, like the common cold, become important to mankind through familiarity. Others, like smallpox, cholera, or bubonic plague, impress themselves on us by the ferocity of their assault. As we have seen, malaria has occasionally assaulted through terrible epidemics, but this disease usually has seemed a familiar, widespread, and seemingly unconquerable "force of nature." As a result of the eradication of malaria, which is now in progress, other diseases have been raised by default to positions of importance. Yaws, syphilis, tuberculosis, and the nutritional diseases are now exposed to world attention. Except for the last-named, these diseases are probably not increasing, and the removal of malaria simply points to them as next in line for attack. At least one disease, however, is actually increasing in incidence in spite of strenuous efforts to control it. This is schistosomiasis (or bilharziasis), the group of painful and disabling disorders caused by the blood flukes.

Schistosomiasis occurs wherever three conditions are found. These conditions are (1) pollution of water, (2) use of such water for bathing or irrigation, and (3) the presence of snails in which blood flukes can develop. Pollution of water is almost universal in today's world. Snails suitable for the transmission of schistosomiasis are common. As dams are being built to conserve and direct water for the arid regions, the use of water for irrigation is rapidly increasing. Therefore, more and more people are being exposed to the blood flukes, and the importance of the disease becomes greater annually. In 1963, McMullen cited data indicating that in sixteen African countries, with a combined population of 125 millions, there were 38 million persons infected with schistosomiasis and there was a range of prevalence in individual countries from 0.5% to 70%. It is not possible to compare these figures with early estimates for the same area, because only recently have surveys been made in much of Africa; however, it is interesting to note that Stoll, in 1948, estimated that the world incidence was 114,000,000, that Wright, in 1950, believed the incidence to be about the same, and that Maegraith, in 1958, estimated 11,000,000 victims to be in mainland China alone. Chandler and Read, in 1961, thought Stoll's estimate to be too low; they believed that in lower Egypt 95% of the people were infected. McMullen, in 1963, stated that even after years of extensive attempts at control, 14 million cases remained in Egypt.

The disease is very serious. The victims become progressively weaker and more miserable, as the parasites continue to deposit eggs, which causes the tissues of bladder or bowel (or both, in frequent mixed infections) to become inflamed, thickened, and eventually grossly altered. Liver function may be affected, various other organs, including the brain, may be invaded by the eggs, and general deterioration of health continues while over a period of years the flukes live. Eventually death, often from secondary infection, puts an end to the victim's misery. The economic loss is considerable. Expensive and ambitious water resource developments have had to be abandoned after being begun. Decreases in labor efficiencies of 20–30% have been reported. Based on Wright's 1951 figures for certain Japanese areas, the total annual loss in the world due to schistosomiasis may be as high as 500 million dollars.

The problem of schistosomiasis is very complex. First, there is an ecological dilemma. The world needs food, and the chief source of new agricultural development seems to be certain potentially fertile deserts. Vast irrigation schemes are planned or are being executed; in 1963, for example, Egypt with over 5 million acres under irrigation, expected to add nearly two million more through the use of the Aswan dam. Such projects are necessary, but under present conditions they involve the increase of schistosomiasis. The decision has not yet been made (nor perhaps the problem fully recognized) whether to pursue agricultural aims at the cost of increasing the misery caused by disease or to postpone such plans until schistosomiasis becomes more controllable; the latter approach would tend to increase the amount of malnutrition in an overpopulated world. We may hope that this paradoxical problem will be solved by the cooperative approach recommended by McMullen, who suggested that engineers and public health scientists must consult each other in matters involving water resources and blood-fluke disease.

Second, there are many technical difficulties in the treatment and control of schistosomiasis. The snail hosts in various parts of the world have different habitats, different habits, different population dynamics, as well as different behavioral and physiological responses. Knowledge of all these facts is essential for control. Reservoir hosts are a serious part of the blood fluke problem in the Far East: *Schistosoma japonicum* is harbored by various animals besides man. People, especially those of the underdeveloped countries where new water projects are being pushed most energetically, are often ignorant of the life cycle of the blood flukes and of ways of avoiding infection. The cultural pattern of pollution and water use is age-old and not easily broken by education. Treatment, although progress is continually being made, is not yet cheap or effective enough for general use. The cost of even pilot programs of control is very great—usually too great to be borne by the governments concerned. And if money were available everywhere, it would not be possible to carry on major programs of prevention and control because of the shortage of trained personnel. Thus it appears that the complexity of schistosomiasis may defeat the efforts of men and governments to solve this problem.

The remainder of this chapter concerns much that is known, however, about this very difficult world-health problem. When the reader has finished the following discussions—the parasites, their life cycles, their effect upon the host, the immune responses of the latter, and the details of control through destruction of snail populations—he should recognize that there is at least a fair prospect for reduction in the incidence of schistosomiasis.

THE PARASITES

The schistosomes of man belong to several species, each of which consists of several geographically or otherwise differentiated strains. The recovery of eggs from 5000-year-old Egyptian mummies shows that schistosomiasis has been a human disease since almost prehistoric times; the recent reports of newly discovered strains infective of man, plus the ability of the cercariae of species normally parasitic in birds or wild mammals to penetrate human skin, illustrate the continuing versatility and opportunism of these parasites.

The chief species are *Schistosoma mansoni, S. haematobium*, and *S. japonicum*. The first two species share the same range in the Middle East and in much of equatorial and southern Africa. *S. mansoni* occurs in parts of South America and the West Indies, while *S. japonicum* is common in the Far East. (Fig. 9–1). *S. mansoni* lays eggs with lateral spines and causes lesions chiefly in the walls of the lower intestine and rectum of its host. *S. haematobium* lays eggs with terminal spines, and deposits these in the venules of the walls of the lower bowel and the urinary bladder. *S. japonicum* lays eggs without spines (a small lateral lump or protuberance is present instead) and affects the walls of the upper intestine through branches of the superior mesenteric veins; this fluke lays eggs at a much higher rate (up to 3,500 per day) than the other two worms (about 300 per day). The pathologies caused by these species depend on the above peculiarities in egg laying. *S. mansoni* and *S. japonicum*, which cause inflammation, thickening, and necrosis in the wall of the intestine, are said to be "intestinal" forms, while *S. haematobium*, which damages the bladder, ureters, and urethra, is said to be "urinary." In *S.*

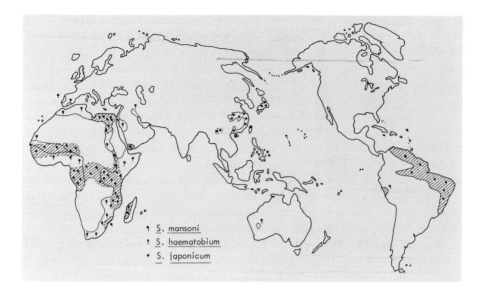

FIG. 9–1. World distribution of *Schistosoma mansoni, S. haematobium,* and *S. japonicum.* The symbols refer to the characteristic eggs of the three species: lateral-spined for *S. mansoni,* terminal-spined for *S. haematobium,* and spineless for *S. japonicum. Schistosoma mansoni* areas are shaded, since *S. haematobium* coexists with *S. mansoni* in Africa. [After Faust and Russell, 1964.]

mansoni and *S. haematobium* infection, the disease is relatively slow to develop, because of the relatively slow rate of egg laying, while *S. japonicum* produces a severe infection which reaches serious proportions in months instead of the years required for the progress of the other two. There are other blood flukes of man. *S. intercalatum* and *S. rodhaini,* as well as *S. mattheei* and *S. bovis,* have been reported from Africa; the former two are normally parasites of sheep, and the latter parasites of cattle. The fact that *S. japonicum* infects many vertebrate hosts besides man suggests that man himself may serve as reservoir or accidental host for perhaps other schistosomes. In Formosa the endemic strain of *S. japonicum* chiefly parasitizes lower mammals.

Various blood flukes are found in cattle, buffaloes, and other domestic animals, as well as wild mammals and birds. *Schistosoma bovis* (Fig. 10–2) is the most widespread of the species of *Schistosoma* affecting animals. Other species are *S. mattheei, S. spindale, S. indicum,* and *S. nasale.* The last-named inhabits vessels of the nasal septum of cattle and causes a "snoring" disease. Several species of the related genus *Ornithobilharzia* parasitize cattle and even elephants.

Although with the exception noted above most of the latter schistosomes cannot complete their life cycle by means of a human host, their cercariae (especially of bird schistosomes) can penetrate human skin to cause the "swimmers' itch," reported from many parts of the world. Victims, sensitized (see Chapter 27) by repeated exposures, develop a severe rash. Apparently the larvae of a large number of species of blood fluke can cause this condition.

LIFE CYCLE

The life cycle of *Schistosoma mansoni* (Fig. 9–2) will serve as an example of blood-fluke life cycles.

Adult flukes occupy those veins of the portal system which drain the colon. It is noteworthy to mention that the schistosomes are unique among trematodes because they are dioecious, i.e., bisexual. The male flukes are much larger than the females, and their bodies are curved into the form of half-cylinders. Within the concavity thus formed the female is held, and copulation takes place presumably quite often. Without leaving their mates, the female worms lay their eggs in the smallest veins they can reach. The lateral-spined eggs lodge within the lumen of the venules. The worms live for years, and although, as mentioned earlier, the rate of egg-laying may be only 300 eggs per day, the number produced by one female can eventually become very large.

The eggs, when laid, remain passively in the blood vessels, while the embryos develop into ciliated miracidia. The latter produce histolytic enzymes which diffuse from the shell of the egg and soften or weaken the host tissues. Partly from this secretory activity of the parasite and partly, it is believed, from pressure of the host vessels surrounding the eggs, the eggs move more deeply into the intestinal wall. The lesions thus formed trap most of the eggs until, in a few weeks, they die. A few escape into the intestine through ulceration caused by the parasites, and these eggs with their developed miracidia may then reach soil or water via the host feces.

FIG. 9–2. Life cycle and pathogenesis of *Schistosoma mansoni*. (A) Usual source of infection, penetration of skin by cercaria in snail-infested water. (B) Pathways of metacercariae in human host after penetration of skin and circulatory system. After briefly stopping in the lungs (1), the schistosomules are carried to the portal system within the liver (2) from which they later enter the vesical and mesenteric veins (3), their permanent home. (C) Pseudotubercle (fibrous tissue infiltrated with pha-gocytes) from the wall of the intestine, showing several dead and shrunken eggs in its center. This is a secondary or immune reaction to the presence of eggs. [After Belding, 1942.] (D) Female adult (1) partially held in the gynecophoral canal (groove) of the male (2). (E,F) Egg, with fully developed miracidium, showing penetration glands (1) and flame cells (2). [After Faust and Russell, 1964.] (G) Snails of the family Planorbidae. (H) Second generation sporocyst containing germinal masses (1), developing cercariae (2), and fully developed and emerging cercariae (3). [After Faust and Russell, 1964.] (I) Cercaria in "floating" position, showing two kinds of penetration glands and forked-tail, as well as flame-cell pattern. The excretory bladder is at the base of the tail.

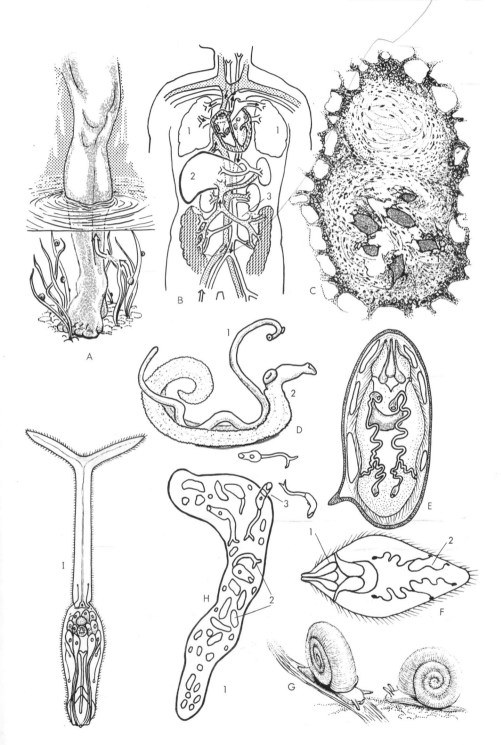

When in water, the eggs hatch promptly. (This fact makes possible a rather simple method of diagnosis; the miracidia hatching from eggs in saline-washed, sedimented feces swim about and may be seen with the unaided eye as bright, motile specks when such a sample is illuminated.) The ciliated larvae swim actively until their reserve of stored energy is exhausted (usually in less than 24 hours) or until they reach a suitable snail host. There is some question about the mechanisms of attraction or recognition which may aid the larval trematodes in their "search." Some workers assert that the presence of a suitable snail merely excites more rapid swimming on the part of the miracidium and thereby increases the latter's chances of touching the snail. Others believe that once an essential proximity is achieved by chance, the miracidium can move directly toward its host along some chemical gradient. Once in contact with a snail (of the family Planorbidae, in the case of *S. mansoni*, but only certain species, or even strains, of snail within this family), the miracidium discharges histolytic enzymes from special penetration glands and bores into the snail's tissue.

The successful miracidium now loses its ciliated epidermis, absorbs food from the host through its own surface (it has no mouth or intestine), and grows rapidly into a saclike sporocyst. The latter, about 1 mm long after 2 weeks, produces numerous daughter sporocysts which escape the parent cyst through the ruptured wall and grow to a length of 1.5 mm. These begin to produce cercariae about 4 to 6 weeks after the snail becomes infected. These fork-tailed larvae leave the snail during the morning hours in bursts, or puffs; they swim about, with frequent periods of rest, for two or three days until exhausted or successful in finding a final host.

The behavior of the cercariae requires special attention. They swim upward rapidly and then become motionless with the forked tail extended. In this position they slowly sink but are delayed by the viscosity of the water and presumably by their own low specific gravity. In alternately rising and sinking, they may be carried far from the habitat of the infected snail. The latter remains for its entire life a source of new cercariae, and total progeny of a single miracidium has been estimated to be greater than 200,000. Thus one may visualize the snails of a polluted stream or drainage ditch as almost inexhaustible sources of infective, fork-tailed cercariae, which, as plankton, are carried far from their point of origin. The distance they travel depends upon the speed of currents and upon the cercarial life span. Facts like the above are useful in planning to prevent the spread of schistosomiasis.

Entry into the final host usually occurs with the mechanical assistance of a drying film of water and the chemical assistance of cercarial enzymes. Although cercariae can enter the human body through the lining of the throat and esophagus (drinking cercaria-infected water is unwise), they usually force their way through the intact skin either while the skin is submerged or while it is drying. Laboratory animals such as hamsters are easily infected by exposing their shaven abdomens to cercariae in dishes of water for several minutes; hence it is assumed that human skin is doubtless penetrated while under water. However, the evidence from "swimmer's itch" make it clear that repeated wettings and dryings (as would happen during agricultural labor in rice paddies or during recreational bathing, for instance) are highly efficient ways to acquire schistosomiasis. Swimmer's itch can be rather easily prevented by wiping dry immediately after swimming in water where the cercariae of schistosomes abound; but allowing such water to dry slowly on the skin results in the painful rash mentioned earlier. It is rather clear that the surface tension of the water film gives mechanical support to the larvae as they work their way in. However, their penetration also depends upon the action of mucous secretions, which aid in adhesion, and of histolytic enzymes, which are formed in penetration glands; such enzymes are capable of dissolving epidermal cells and making a hole into which the cercaria can wiggle. The cercaria leaves its tail outside.

Once under the epidermis the cercaria, without its tail (more properly called a "schistosomule," or miniature adult), migrates sometimes for half a day or more in the dermal layer of the skin. There it enters a small vein and is carried to the heart and thence through general circulation. By the sixth day it is found collected in the portal system of the liver. Within the small liver veins, the males and females copulate; then they move into the hepatic portal vein and into the mesenteric veins which supply the portal. Blood in this system, rich in nutrients, permits growth of the males and females to maturity. (Interestingly, the females cannot develop to maturity unless males are present. Therefore a successful transmission, or completion of the cycle, depends upon the penetration of the final host by cercariae of both sexes.) About forty days after cercarial penetration, eggs begin to be deposited in the venules of the large intestine.

Transmission having occurred through a series of necessary events, the life cycle of *S. mansoni* is thus completed. First, in water contaminated with the feces of a victim of the disease, the eggs hatch. Second, the miracidium successfully penetrates a snail of a suitable variety, and two generations of sporocysts develop in the snail; the result is continuous liberation of cercariae into the water over months or years. Third, cercariae of both sexes, eventually penetrating a second human host, migrate via the blood to the lungs and then to the liver and mesenteric veins, where copulation occurs and egg-laying starts.

PATHOLOGY

The symptoms and injury caused by schistosomes can be described in terms of pathological changes in the host tissues.

When the cercariae first penetrate the host, there may be a prickling, burning sensation at the points of entry. This is due to mechanical injury, no doubt, but since the above sensation is not felt by all victims, allergic reactions may be involved; some individuals are sensitive, or become sensitive, to the invasion. Invasion by "foreign" schistosome cercariae often evokes such a response.

Following invasion of the skin, with a slight, sometimes unnoticed, effect on the tissues, there is little damage until the parasites become temporarily arrested in the lungs. Their presence in large numbers causes local irritation reflected in bronchial cough. During the so-called incubation period, there may be widespread allergic response: generalized rash, fever, eosinophilia, etc. A characteristic diarrhea occurs. These symptoms coincide with the growth of the worms, the release of metabolic waste products by the parasite, and the breakdown of cells and tissues in the affected organs.

When the worms mature and start to lay eggs, the primary pathology consists of destruction of small blood vessels by the migrating eggs as they pass toward the lumen of the affected organs. Thus in *S. mansoni* or *S. japonicum* infections, blood and mucus may appear in the feces; while in *S. haematobium* disease, there may be blood in the urine. Some pain occurs. In histological preparations a relatively small number of eggs can be found in the walls of the intestine or bladder, and the invaded tissues are not yet greatly changed from the normal condition.

After some time, however, local sensitization occurs, and wherever eggs lodge inflammation results, with swelling of the tissues, invasion by leucocytes, and growth of scar tissue. Thus the affected organs lose their flexibility, the blood vessels draining them become blocked, and various necrotic changes occur. In *mansoni* and *japonicum* disease the portal circulation is so damaged that the liver shrinks and becomes fibrous; the spleen and the whole abdomen swell, and the rest of the body, deprived of a functioning liver, becomes emaciated. In *haematobium* disease the damage is mainly to the wall of the bladder. Since venous flow from that organ does not join the portal system, eggs may be carried directly to the heart. From the heart, they are distributed widely throughout the body. The eggs of the other schistosomes may also get into the general circulation. Such disseminated eggs set up local inflammatory changes in lungs, spinal cord, brain, or other organs.

IMMUNITY

All three schistosomiases are progressive disorders, the effect of which becomes serious in proportion to the numbers of eggs deposited. Yet the gravity of later symptoms cannot be explained merely in terms of the number of eggs deposited; immunity, probably in the harmful sense of allergy, plays an important role. This theory of the way in which schistosomes cause disease is borne out by three kinds of facts. First, the above-described shift in tissue reaction from an initial lack of response other than simple repair to a secondary allergic inflammatory reaction wherever eggs may lodge shows that immune changes occur. Second, the infection rates in regions where the disease is prevalent show an interesting shift with age. This fact, mentioned earlier in connection with amoebiasis (Chapter 6), needs further explanation. From age one to approximately age fifteen, there is a simple rise in frequency of infected individuals; the rise reflects the opportunity to acquire the infection. That the latter explanation is so is shown by data mentioned below, where the incidence curve among the young breaks sharply at a time when transmission has been stopped by control measures. Then from age fifteen to age forty, there is, in the complete absence of control, a reduced rise in incidence of disease. After middle age there is no increase of frequency at all. Therefore it is believed that many individuals must acquire light infections which render them immune. Otherwise, as Chandler points out, entire populations in Egypt, for example, would be destroyed. The study of laboratory animals has confirmed the role of immunity in schistosomiasis, a role which, while harmful in its immediate effect upon the victim, may be useful in protecting against increased or new infection.

Some details of research on immunity in schistosomiasis will be given here to illustrate the complexity of the problem and to show that progress is being made in understanding this important disease. As early as 1916, it was suggested by a Japanese scientist that *S. japonicum* induces immunity in man. This was the first suggestion that a many-celled animal parasite could act in such a way, as all the classic studies in immunity had dealt with bacteria and viruses. Because it is impractical, if not prohibited, to study immunity experimentally in man, many kinds of laboratory animals have been used to date. The simplest experiment was to infect such animals with schistosomes and then to "challenge" the infected animals with more cercariae in order to see whether the expected increase in the number of parasites could occur. The results of many such experiments have been rather puzzling. Some strains of parasite produce immunity in certain kinds of host. Mice can be protected against *S. japonicum* but not against *S. mansoni*. Rats, as well as Rhesus monkeys, can be protected against *S. mansoni*. Inoculation of animals with the Formosan strain (a wild animal strain) of *S. japonicum* gave more protection against the Japanese (human) strain than inoculation with that strain itself. Worm extracts, homogenates, and metabolic products have been tested as immunizing agents but with considerably less success than the living parasites. Apparently the time during which an immunizing agent is present in the body is important (see Chapter 27), and the dead worms, etc., may be altered or destroyed too quickly after injection for the slow process of immunization to occur. A method intermediate between immunization with normal, living worms and injection of products and parts of dead worms was the use of irradiated cercariae—a method inspired by the interesting work with the cattle lungworm, *Dictyocaulus viviparus* (see Chapter 20). When cercariae, irradiated rather heavily with x-rays or gamma radiation, were allowed to enter the skin of mice or monkeys, the cercariae failed to establish a harmful infection but produced significant protective immunity against challenge (second) doses. The challenge worms failed by a larger-than-normal percentage to become established, failed to produce a normal number of eggs, and were stunted in size. In order for a challenge dose to be lethal, it had to consist of many more cercariae than an ordinary lethal dose contains, and the time of death was significantly postponed by inoculation with irradiated larvae. The chief value of such experiments seems to have been to prove beyond any doubt that protective immunity against schistosomiasis actually exists. It is very important now to find out the biochemical and molecular bases for such immunity in order that protective inoculants may be isolated, purified, and perhaps synthesized on a mass scale.

The fact of immunity to schistosomiasis should be a warning to planners of control programs, which are the subject of the remainder of this chapter. It is probable that every exposed population has many members who have been rendered immune to reinfection by a continuing low-grade infection with a few worms. If control greatly reduces the opportunity to acquire such low-grade infection, a population might then become largely susceptible, and any failure of control might result in severe disease for many.

CONTROL

Control of human and veterinary schistosomiasis, made possible by the knowledge now possessed by parasitologists, depends upon both the utilization of that knowledge and its increase by research.

Enough is known of the schistosome life cycle to attack it at four, or possibly more, points.

First, contact between final host and infective cercaria may be prevented either by protective ointments (an expensive expedient inapplicable to the real problem, which is mass protection) or by avoidance, on the part of the host, of cercaria-infested waters. Obviously, this first approach to control is unrealistic, except for the protection of tourists, military personnel, or swimmers, and has little application to agricultural populations of irrigated farmlands.

Second, mass treatment of exposed populations, to rid them of schistosomes or to stop the production of viable eggs, is possible, because a number of drugs are effective. Mass treatment, however, is usually too expensive, for it requires resources in money and trained personnel which are not yet available to the countries where control is most needed. Also, unless mass treatment is combined with methods of preventing reinfection, it is futile.

Third, general sanitary measures to prevent pollution of waters by the urine or feces of infected individuals is a way to reduce the number of infected snails, and eventually results in disappearance of schistosomiasis (except where reservoir hosts occur, an important exception). However, sanitary measures are seldom practical, for they involve rather complex changes in old customs and require educational and administrative techniques not yet in existence in the regions where sanitation is most needed.

Fourth, control of snail populations can directly reduce the number of cercariae in the dangerous waters. At times, this control measure, whether carried out by general molluscicidal treatment or effected by biological or ecological measures such as water-level regulation, snail habitat disturbance, or the encouragement of predators which feed on the important snails, has been highly successful.

In addition to the above four points of attack, two others have been suggested: destruction or inactivation of miracidia between hatching from the egg and penetrating the host, and preventive immunization of man and his domestic animals. These, at the time of this writing, are in experimental stages, but one may hope for success. Thus the above six ways of controlling schistosomiasis are all of some real or potential value.

The effective control of the disease in a particular region depends on properly combining the above measures; the requirements and limitations of the particular problem must also be considered. Moreover, each control problem involves not only a combination of the above attacks upon the parasite life cycle but also recognition and manipulation of other factors, i.e., the political, sociological, and economic realities of the government, people, and resources of each country or region. Unfortunately, these realities are apt to be less well known and less subject to manipulation than the strictly parasitological factors. The examples of actual experience to be described in the remaining discussion will illustrate some of these real difficulties. In general, the control of schistosomiasis can be seen as an attempt to save the fruits of agricultural development. The benefits of irrigation programs intended to increase the amount of food-producing land may not be fully realized unless the spread of schistosomiasis can be halted. The prospects for success depend upon the effective use of known control measures and, to some extent, upon the discovery and development of simpler, cheaper, and still more effective techniques.

EXAMPLES OF CONTROL

In Southern Rhodesia, preliminary studies completed by 1957 indicated that control of snails was practical. Watershed units had been treated first by application of molluscicides to the whole area and then by spot treatment of snail habitats which were found to be still active. This success meant that two important diseases, schistosomiasis in man and fascioliasis (see Chapter 10) in cattle, could be controlled by a single effort. Farmers, through their organizations, were encouraged to participate in a country-wide control program; they responded enthusiastically, and only their lack of training prevented rapid success. While some time, money, and energy were wasted in improper applications of molluscicides and uncoordinated work, eventually by 1961 four areas totaling 4.5 million acres had been brought under substantial control, so that "surveillance" (a term appropriately borrowed from the malaria-eradication vocabulary) was sufficient to keep snail populations low. The first stages of snail control (blanket application of molluscicide and mopping up of residual habitats) had been carried on by the farmers themselves under expert supervision. Surveillance was being maintained at government expense. The effectiveness of the program in stopping the transmission of schistosomiasis was revealed by a survey of children from 1 to 15 years of age. Since there is a well-known correlation between numbers and age of infected individuals based on the obvious relationship between age and exposure, any substantial change in transmission rate would be reflected in age-incidence data. Actually, Southern Rhodesian children aged 11–14 showed no higher incidence than the 10-year-old group; the rate of acquiring schistosomiasis had dropped to zero during the three- to four-year period of control. Evidence indicated that fascioliasis transmission, an important benefit which helped reduce the cost of the program, had also ceased. The cost is said to have been 35 cents per person per year. Thus the Southern Rhodesian program, with a large amount of cooperation, volunteer labor, and a plan based on watershed units, seems to have been successful as well as economical. (Compare Fig. 9–3, a hypothetical program leading to eradication.)

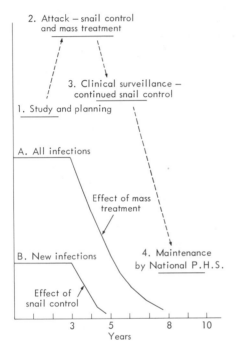

FIG. 9–3.   A hypothetical program of control shows levels of effort at various stages in the program. Note the levels of infection before operation of the program, during the early stages, and at the period when special control measures cease and ordinary public health services become able to cope with the disease. The analogy with malaria eradication is not perfect because of the lack of highly effective chemotherapy for schistosomiasis and the existence of reservoir hosts in the case of S. japonicum and perhaps other schistosomes.

In contrast, some quite expensive programs have proven to be not highly effective. Such a program, depending entirely on government-administered snail control, was initiated on a pilot scale in Egypt in 1954. In the first year, over twelve tons of molluscicide were applied on 5,000 acres of perennially irrigated land. Careful estimates based on the total volume of water, recommended concentrations of chemicals (10 parts per million for 8 hours, three times a year), cost of molluscicide, and cost of labor indicate that snail control by molluscicides in Egypt would cost a minimum $1.35 per irrigated acre per year. Since Egypt has between 6 and 8 million acres of such land, it would require over ten million dollars annually, plus trained personnel, to maintain a complete molluscicide program. Actually, Egypt has tried to control schistosomiasis in a limited way by attacking particular areas. But the results do not seem to have been very successful, for in 1963, nine years after the pilot program, 14 million people in Egypt had schistosomiasis. The above example illustrates the inadequacy of a single approach (in this case snail destruction) to the complex problem of schistosomiasis control.

The literature of schistosomiasis control contains reports of numerous approaches similar to the above. A conspicuously successful campaign against schistosome-bearing snails was waged in the Philippines in the 1950's. There it was possible to combine improvement in land and water utilization with snail control, so that again, as in Southern Rhodesia, secondary benefits helped defray the cost of controlling schistosomiasis. Ecological control of the snail host succeeded not only in eliminating 95% of the snails but also in increasing crop production at the same time.

In summary, available techniques of control have been used under varying conditions and with variable results. Application of molluscicides alone, while very effective in circumscribed areas, is too expensive and too difficult to be very practical in underdeveloped lands with much irrigated acreage. Cooperative endeavors, such as the Rhodesian experiments, seem to be an excellent approach; the concentration of effort in restricted areas (the watershed units), the education of the farmers to do much of the actual work, the shifting of the burden of the program from farmers to government once the surveillance stage has been reached, and eventual mass treatment to reduce existing disease once transmission has been stopped are a series of steps which may be reasonably adopted by countries other than Southern Rhodesia. In the Philippines, the linking of improved agricultural yield with snail control based on water regulation is particularly encouraging, since the cost of the project is borne, essentially, by the value of an obvious economic gain. Yet each schistosomiasis problem is unique. This fact, like the uniqueness of malaria problems, makes it clear that in every project for control there must first be a careful survey of the problem by experts, who must recognize all relevant factors. Some of these factors may present real obstruction to success. Thus it is important that the effort to control schistosomiasis must continue to be directed on a worldwide scale, that the World Health Organization's teams continue to investigate and inform, and that international agencies, perhaps regional in scope, provide funds for pilot projects or initial, expensive stages in control, while laboratory workers in many countries continue to search for new knowledge of the parasites and their various hosts.

SUGGESTED READINGS

HAIRSTON, N. G., and B. C. SANTOS, "Ecological Control of the Snail Host of *Schistosoma japonicum* in the Philippines," Bull. WHO, 25:603–610, 1961.

KAGAN, I. G., D. W. RAIRIGH, and R. W. KAISER, "A Clinical, Parasitologic, and Immunologic Study of Schistosomiasis in 103 Puerto Rican Males Residing in the United States," *Ann. Int. Med.*, 56:457–470, 1962.

MARKOWSKI, S., "The Distribution of the Molluscan Vectors of Schistosomiasis in the Sennar Area of the Sudan, and their Invasion of the Gezira Irrigation System," *Ann. Trop. Med. Parasit.*, 47:375–380, 1953.

McMullen, D. B., "The Modification of Habitats in the Control of Bilharziasis with Special Reference to Water Resource Development," *Bilharziasis Ciba Foundation Symposium in Commemoration of T. M. Bilharz*. Little, Brown, Boston, 1962.

McMullen, D. B. and P. C. Beaver, "Studies on Schistosome Dermatitis. IX. The Life Cycles of Three Dermatitis-Producing Schistosomes from Birds and a Discussion of the Subfamily Bilharziellinae (Trematoda, Schistosomatidae)," *Am. J. Hyg.*, 42:209–301, 1945.

Moore, D. V., R. B. Crandall, and G. W. Hunter, "Studies on Schistosomiasis. XX. Further studies on the Immunogenic Significance of *Schistosoma mansoni* Eggs in Albino Mice when Subjected to Homologous Challenge," *J. Parasit.*, 49:117–120, 1963.

Oliver-Gonzalez, J., "Antiegg Precipitins in the Serum of Humans Infected with *Schistosoma mansoni*," *J. Infect. Dis.*, 95:86–91, 1954.

Pesigan, T. P., *et al.*, "Studies on *Schistosoma japonicum* Infection in the Philippines. I. General Considerations and Epidemiology," *Bull. WHO*, 18:345–355, 1958.

Sadun, E. H., "Immunization in Schistosomiasis by Previous Exposure to Homologous and Heterologous Cercariae by Inoculation of Preparations from Schistosomes and by Exposure to Irradiated Cercariae," *Ann. N.Y. Acad. Sci.*, 113(Art 1):418–439, 1963.

Stirewalt, M. A., "Seminar on Immunity to Parasitic Helminths. IV. Schistosome Infections," *Exptl. Parasitology*, 13:18–44, 1963.

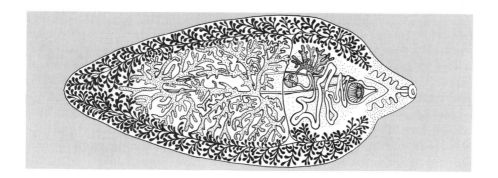

CHAPTER 10

OTHER TREMATODES

In this final chapter on trematodes a zoonosis of man will be discussed, and several trematodes causing economic loss in livestock will be mentioned. The varied and interesting trematode parasites of wild animals will have to be omitted, but the ecological relationships between human and livestock disease and the parasites of other animals will be examined to show the importance of studies in wild life parasitology.

PARAGONIMIASIS

Paragonimiasis, or lung-fluke disease, is common in only a few parts of the world but is a worldwide risk, for the parasites causing it, flukes of the genus *Paragonimus*, have been reported from every continent except Australia (which may not be exempt, for the flukes have been found in New Britain, part of the Australian zoogeographic realm). Only the eating habits of the peoples of the world prevent paragonimiasis from becoming more prevalent, for it is acquired by eating raw freshwater crabs or crayfish. In Korea, parts of Japan, and a few other places raw crabs are used for food, and in these countries lung flukes are important enough to have been the subject of a great deal of study.

133

Over 500 references were cited in a recent review of paragonimiasis (Yokagawa, Cort, and Yokagawa, 1960). Left out of this list were numerous minor reports, case histories, and other articles published in the Japanese language. Lung flukes have been studied since the 1850's by scientists of nearly every nation. A list of hosts, in the order of their discovery, will document the history of this study very briefly and will demonstrate the wide distribution of lung flukes. In 1850, lung flukes of mammals were first observed in a Brazilian otter. Other observations followed: an Indian mongoose, 1859; an Indonesian tiger in a Dutch zoo (the most famous of the early hosts, for its parasite was named for Westerman, the zoo director, and the name *Paragonimus westermani* is a valid name for one of the lung flukes today), 1877; man in Japan and Formosa, 1880; man in China, 1881–3; Japanese dogs, 1890; Japanese pigs, 1892; Japanese cats, 1893; North American cats and dogs, 1894; North American pigs (Ohio), 1898; Venezuelan dogs and pigs, 1918; Indian civet cats, 1926; Chinese rodents, 1935; Japanese laboratory animals (infected by a new species of fluke from crabs), 1939. Human infections outside the orient have been reported in Peru and Ecuador, in the state of Yucatan, Mexico, and in parts of Africa. The species of trematode involved in the above reports have been the subjects of scientific controversy. At present four species are considered to be valid: *Paragonimus westermani* (Kerbert, 1878), Braun, 1899; *P. kellicotti*, Ward, 1908; *P. ohirai*, Miyazaki, 1939; and *P. iloktsuenensis*, Chen, 1940. Details have been accumulated to define these species in terms of the symptoms they cause in man and other animals, details of anatomy of the adult worms, and the kinds of intermediate hosts (molluscan and crustacean) which the above species utilize.

The development of *Paragonimus kellicotti* has been fully described. Snails of the species *Pomatiopsis lapidaria* were used in these studies (Ameel, 1934 *et seq.*). Miracidia, when placed in water near the snail host, gather round the snail closely and move in a crawling fashion upon the snail's body. Soft parts of the snail yield to active penetration by the protrusible anterior tip of the miracidium. Eventually miracidia reach the lymphatic system of the snail, and sporocysts may develop in all parts of that system. After four weeks, the sporocysts begin to produce rediae, of which there are two generations. The mother rediae (about 25 per sporocyst) become lodged around the stomach and digestive glands of the snail. In about 60 to 70 days, well-developed daughter rediae are found within the mother rediae. About 30 daughter rediae are produced by each mother redia. These daughters, potentially about 750 descending from each miracidium, are actually not so numerous. From 11 to 62 daughter rediae have been counted in experimentally infected snails. Each daughter redia has a germinal center where cells proliferate to produce constantly embryos which develop into cercariae. About 25 cercariae are usually present at one time within a fully developed daughter redia, and production of cercariae continues throughout the life of the snail. (One of the differences between the two rather similar species, *P. westermani* and *P. kellicotti*, is the fact that cercariae of the former do not escape readily from the snail, while those of the latter come out in clouds at certain times. This difference is probably significant in the method and rate of infection of the crustacean host.)

Although the infected snail thus becomes a long lasting source of infection for the next host, it is noteworthy that relatively few snails are found infected even in regions where many crabs and mammalian hosts have acquired cercariae and adult flukes, respectively.

Cercariae of the various species of *Paragonimus* are so similar to each other that they cannot easily be distinguished. These cercariae are equipped with short stumpy tails, which are spiny at the tip. The ellipsoidal body, about 0.25 mm long, has sensory hairs along the dorsolateral margin, and the body is covered with small spines. A prominent stylet in the dorsal part of the mouth is equipped with muscles and lies near the openings of the fourteen penetration glands which occupy much of the body. There are two kinds of penetration glands staining differently with intravital dyes. In addition there are glands along the ventrolateral regions which secrete mucus; these glands are believed to aid the cercaria in sticking to the surface of the second intermediate host. The cercariae of *Paragonimus* lack eyes. The many studies of the means by which cercariae enter and become encysted in various crustacea have yielded rather conflicting evidence. Crayfish are usually penetrated by cercariae of *P. kellicotti* at the soft joints of their exoskeleton. They migrate to the thoracic region surrounding the heart. Crabs, the usual hosts of *P. westermani*, may acquire their cercariae by feeding on infected snails. Experiments have shown this to be possible, and the above noted failure of mature cercariae to leave the snail hosts of *P. westermani* suggests some method other than casual penetration as a means of infecting crabs. However, crabs found in nature are usually infected with very small numbers of metacercariae, a fact which casts doubt on the theory of infection by feeding. Obviously, in spite of the many careful studies already made, the life cycle of *Paragonimus* needs further investigation.

Within the tissues of the crustacean host, cercariae, without their tails, form translucent cysts, which grow in size and change considerably in structure during the two to four months required for them to reach an infection stage. A strong, two-layered cyst wall develops, the digestive system becomes fully formed, and the urinary bladder enlarges as it becomes filled with refractile concretions.

Transfer to the mammalian host is accomplished when a raw, infected crustacean is eaten. *P. westermani* and *P. kellicotti* are able to reach maturity in a number of mammals. The former causes human paragonimiasis in the old world; it is suspected that at least some of the few human cases reported in the new world are due to *P. kellicotti*. *P. ohirai* occurs in numerous mammals in Japan. This species, as well as the very similar *P. iloktsuenensis* of the Asian mainland, is believed to be better adapted for development in rodents than in larger mammals such as man; however, human infections may well exist. Like most trematodes, the lung flukes in their adult stage are not very host specific.

Once ingested, the metacercariae excyst. A number of experimental environments have stimulated excystment, which is believed to be an activity primarily of the metacercaria itself. Apparently temperature is a very important factor. The time required for excystment may prevent some species, especially *P. westermani*, from infecting small mammals because of the short emptying time of the latter's intestine. After excystment the metacercariae penetrate the gut wall and enter the body cavity. Actively migrating along the peritoneal surfaces, they reach the diaphragm in from 2 to 8 days and move into the thoracic cavity by way of the connective tissues surrounding aorta and esophagus. About 20 to 30 days after ingestion of cysts (in experimental animals), the young flukes enter the lungs. By this time the parasites, resembling adults in most respects, are much larger than the metacercariae. It is interesting to note that in dogs the young flukes often migrate in pairs and become encysted together at the end of their journey. This contrasts with the human infection, where one fluke per cyst is the usual number. The lung cyst is composed of host connective tissue, which develops in response to active feeding on and mechanical irritation of the lung tissues. The cysts may be found anywhere in the lungs but are usually more numerous in the lower lobes. They are prominent objects (10–20 mm in diameter); the cysts on the surface of the lungs are a dull brown color. The cyst cavities communicate with the bronchi either through necrosis of intervening tissue or by inclusion of air passages in the original mass of cystic tissue. Through these openings, eggs and purulent discharges reach the trachea and, eventually, the environment. In naturally infected animals the cysts may be quite numerous and may not cause much injury or disease. This fact, as well as the prevalence of *Paragonimus* in many mammalian hosts, suggests that the condition is an old, established parasitism in these kinds of hosts but has been rather recently (indeed, sporadically) introduced to man.

Human hosts have little chance of acquiring lung flukes except by eating infected crabs or crayfish. The epidemiology of human paragonimiasis has been thoroughly studied. In China, freshwater crabs are sprinkled with salt and eaten raw during drinking parties; this custom resembles the use of cocktail snacks and hors d'oeuvres in other cultures where, similarly, thirst must be stimulated by salty foods. The salt kills most, but not all, of the metacercariae in infected crabs. Another oriental delicacy is "drunken crab" (crabs which, having been soaked overnight in weak wine, are eaten alive). Sauces of vinegar and spice are used with raw crabmeat in various dishes; these sauces usually do not kill the metacercariae of *Paragonimus* rapidly enough to protect against infection. Infection may be acquired by accident through the preparation of crabs for various well-cooked dishes, since chopping blocks used for preparing the fresh crabs may become contaminated with cysts which then adhere to salad vegetables, hands, or bowls. The juices (blood and body fluids) of crayfish are used as medicine in some parts of Japan and Korea, where infection by this means is possible. Another medical source is the custom of women in the Cameroons, Africa, who eat raw crabs as an aid to fertility. Infection in Buddhist monks in Formosa results from their religious abstention from eating fish and animals; crabs are considered to be neither and hence may be eaten. Thus it appears that there are many ways for man to acquire lung flukes.

The disease resulting is both dangerous and difficult to cure. Two stages in the development of human cases of paragonimiasis can be distinguished. At first, migration of the growing flukes through the abdominal cavity toward the lungs causes lesions along the path. Frequently the flukes get lost and wander sometimes into the brain by way of the foramina, through which pass the cerebral veins and arteries or the cranial nerves. Cerebral paragonimiasis is a very grave condition; its symptoms are epilepsy, meningitis, or other neuropathologies. The chief hazards in the early stages of paragonimiasis are such abnormal behavior of the parasite. Entry into the lungs occurs after the parasites are nearly grown. Pulmonary cysts form, the worm continues its feeding activity, and eggs are laid. Irritation of the lungs, coughing and expectoration of viscous, egg-filled (sometimes bloody material) pain, and weakness occur. Pulmonary tuberculosis may be suspected. Inflammation and pleural adhesions follow. The disease is characteristically chronic and long lasting; cases with egg-positive sputum and no reinfection have been reported to last more than ten years. Just as in schistosomiasis, fluke eggs may get into the circulation and may be carried to various parts of the body, including the brain. Cerebral paragonimiasis can result from local tissue reactions to these eggs. Lung flukes are believed to cause unusually severe disease in children, although diagnostic data are not complete on this point.

Treatment of paragonimiasis cannot be carried out very successfully. Emetine, used by injection as it is in combating systemic amoebiasis, is moderately effective in stopping egg-laying and eventually (after 12 to 33 injections) in killing the worms. Emetine combined with the drug prontosil, which stimulates cellular reactions, is more efficacious than emetine alone. Combinations of emetine with other drugs have shown promising results. Chloroquine, an antimalarial drug, has also been used rather successfully. But no method of treatment yet discovered is suitable for widespread, unsupervised use. Mass treatment is impractical.

Control in the ordinary sense is also impractical. Several approaches to control are usually available in a parasite-caused disease: (1) mass treatment to prevent the spread of eggs; (2) destruction or reduction in number of intermediate hosts; (3) control or destruction of reservoir hosts; and (4) protection by hygienic or other measures against human infection. In paragonimiasis we have seen that treatment is relatively difficult and ineffective. If another snail-borne disease were involved, the molluscan hosts might be controlled as part of a general snail control program, but snails of swift-flowing mountain streams, the crab-crayfish habitats, do not harbor other flukes of medical importance. The crab host is a valuable article of food. While, in Korea, laws prohibiting the catching of crabs may have reduced the incidence of paragonimiasis, the nutritional needs of protein-starved people were surely aggravated. Reservoir hosts of lung flukes are various crab-eating mammals. Control of these hosts would free certain areas of *Paragonimus*, but large areas would remain unaffected, since the so-called reservoirs are actually the normal hosts of these flukes and account for the worldwide distribution of *Paragonimus*. Protection of the population by changing their food habits remains a possible control measure. Educational programs, with illustrated posters, have apparently worked in Korea and Formosa, where the level of human paragonimiasis dropped significantly in certain districts after the institution of such programs. It is not known whether education, which can graphically present the risks of eating raw crabs, can also inform the people of the risk of contaminating food and utensils with cysts released during the preparation of crabs for cooking. But the nature of paragonimiasis forces such educational methods; for this disease, like some forms of leishmaniasis, for instance, is a zoonosis. It is well established in natural cycles, yet not all these are even known, and man is exposed to the disease through occasional and perhaps peculiar food habits. By changing his food habits, man can remove himself from the ecological path of the lung flukes.

FASCIOLIASIS AND LIVER FLUKES

Fascioliasis is a disease of the liver of cattle and other grazing animals, in which large trematodes invasively destroy the hepatic tissue. In the English-speaking world, the disease has long been known as "liver rot." In 1944, Olsen reported the results of an 11-year survey of fascioliasis in cattle in the U.S. Since the livers of 1,400,000 cattle and 60,500 calves were condemned as unfit for human use because of fascioliasis, there was during that period an annual loss of more than one million pounds of liver. At a retail cost of 80 cents per pound, this loss represented nearly one million dollars. Of course, the loss may properly be estimated in other ways to yield other figures, but the order of magnitude remains large. Olsen also reported that parasitized animals gain one half as much in feed lots (where cattle are brought in from pasture to fatten for market) as do nonparasitized cattle, the result is in an annual loss of 1½ million pounds of dressed weight. The value of this meat may be estimated at between half a million and one million dollars. In dairy cattle parasitized by *Fasciola*, decrease in milk production averaged 16%.

Fasciola hepatica inhabits the bile passages of the livers of sheep, cattle, men, and other animals. This large trematode (20–30 mm by 5–8 mm) feeds on blood in the tissues of the liver, and in its feeding and wandering injures the tissues. If enough worms are present, severe damage to the liver results, and the animal may sicken, or the liver itself be rendered unfit for use when the animal is slaughtered. *Fasciola* lays eggs which reach the intestines via the bile duct and are scattered with the feces. After a period of embryonic development, the miracidia emerge while the eggs are submerged in water, and swim until they die of exhaustion or reach a suitable snail. They penetrate the latter and develop into sporocysts, which give rise to rediae; the rediae release cercariae to the environment. The cercariae swim until they come to rest and encyst on vegetation. Grazing animals ingest the cysts with their food. The cysts break open in the duodenum; the metacercariae penetrate the intestinal wall and migrate to the liver, within which they reach maturity. (See Fig. 10–1 for life cycle of liver fluke.) Treatment with hexachloroethane has been 90% effective; emetine has been recommended for cases of fascioliasis in man. Related trematodes (*Fascioloides magna* of deer in North America and *Fasciola gigantica* in parts of Africa and Asia) cause similar damage.

Fasciolopsis buski (Lankester, 1857) Odhner, 1902, causes intestinal and toxic symptoms in man in eastern Asia and the southwest Pacific. Perhaps more than 10 million persons may be infected. This fluke has a life cycle very similar to that of *Fasciola hepatica* and is, like the latter, a large trematode (20–75 mm by

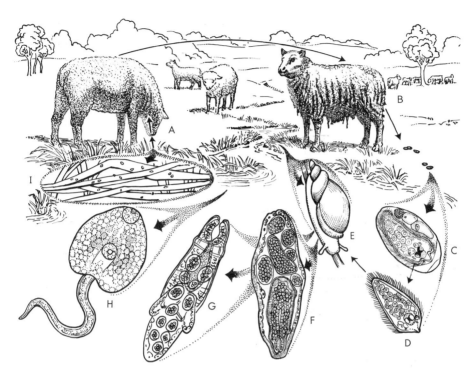

FIG. 10–1. Life cycle of *Fasciola hepatica*. (A) Healthy sheep grazing on cyst-contaminated grass emerging from water. (B) Sheep, in poor condition, voiding feces with
eggs of *F. hepatica*, which reach water and hatch (C,D). (E) Lymnaeid snail, host for
the larval stages [(F) sporocyst and (G) redia]. (H) Cercaria, having escaped from
the redia and the snail, comes to rest on grass, where it encysts (I). After necessary
incubation, it is ingested by a sheep or other vertebrate host. [After Lapage, 1962.]

8–20 mm). Unlike *F. hepatica*, *Fasciolopsis buski* inhabits the duodenum. Its eggs
pass out with feces, the miracidia hatch and swim about, and snails of several
genera can serve as intermediate hosts. After a sporocyst generation and two
generations of rediae, cercariae, which encyst on the surfaces of aquatic vegetation,
are produced. Seed pods of the water caltrop and bulbs of the water chestnut, as
well as other aquatic roots and shoots, are peeled with the teeth when eaten, and
metacercarial cysts are ingested during this process. Symptoms of *Fasciolopsis*
infection are toxic diarrhea, pain in the abdomen, and generalized toxemia. While
the worms irritate the mucosa, continuously obstruct the duodenal passage, and
interfere grossly with intestinal function, the chief clinical harm seems to be the
allergic and toxemic reactions of the host. Death may occur. Treatment with
hexylresorcinol is effective. Prevention can result from snail control, pollution
control, or treatment of vegetables before eating.

FIG. 10–2. Some trematodes of domestic animals. (A) *Fasciola hepatica*, a liver fluke of sheep and other ruminants, as well as rabbits. (B) *Dicrocoelium dendriticum*, a liver fluke of ruminants, called the "lancet fluke." (C) *Clonorchis sinensis*, liver fluke of piscivorous mammals; it was discussed in Chapter 8. (D) *Schistosoma bovis*, male (the larger) and female (the smaller), a blood fluke of cattle (see Chapter 9). (E) *Prosthogonimus macrorchis*, the oviduct fluke of fowl. (F) *Cotylophoron cotylophorum*, an amphistome fluke living in the rumen of cattle. (G) *Paragonimus westermani*, one of the species of lung-fluke infecting various carnivores including furbearers. This fluke is a human parasite where raw crabs can be a source of infection. [After Lapage, 1962.] Scales beside figures represent actual size of the animals.

Control of fascioliasis and similar conditions depends on treatment of human victims and treatment or destruction of infected animals, as well as prevention of pollution of ponds, damp pasture land, etc. Snail destruction by slow release of copper sulfate or any of a number of other molluscicides is helpful. Control of *Fasciola hepatica* is complicated wherever wild rabbits or deer occur, since these serve as reservoir hosts.

Other liver flukes of stock are *Dicrocoelium dendriticum* (see Fig. 10–2), the lancet fluke of sheep and cattle, and *Opisthorchis felineus*, a liver fluke of carnivores and man. The latter fluke is discussed in the chapter on clonorchiasis.

Dicrocoelium, like *Fasciola*, feeds on blood in the liver tissue; its eggs escape by the bile passages. Its first intermediate host is a land snail, in which sporocysts produce daughter sporocysts which constantly liberate cercariae. The latter, in hundreds, secrete slime which eventually surrounds them as they are released from the snail. The slime balls containing cercariae are then ingested by ants, in the bodies of which the cercariae develop into metacercariae. The behavior of cercariae within the ants is remarkable. At least one cercaria encysts near the ant's brain; the ant is then unable to open its jaws when the ambient temperature is low. Thus infected ants are trapped on blades of grass which they happen to be cutting at nightfall; there they must remain, clamped by their jaws, until the sun warms them the next morning. Early grazing by sheep results in ingestion of infected ants.

In the Far East, a related dicrocoeliid, *Eurytrema pancreaticum*, infects the bile passages of cattle and other animals, and the pancreatic ducts of pigs.

AMPHISTOMIASIS

Amphistomes are rather large trematodes (see Fig. 10–2) with suckers at each end (an oral sucker and a larger posterior sucker.) The genus *Paramphistomum* parasitizes sheep, cattle, and other grazing animals; it causes paraphistomiasis, a condition resulting from the migration of young worms and mechanical irritation to the mucosa of the rumen by mature worms. Eggs pass from the host with feces,

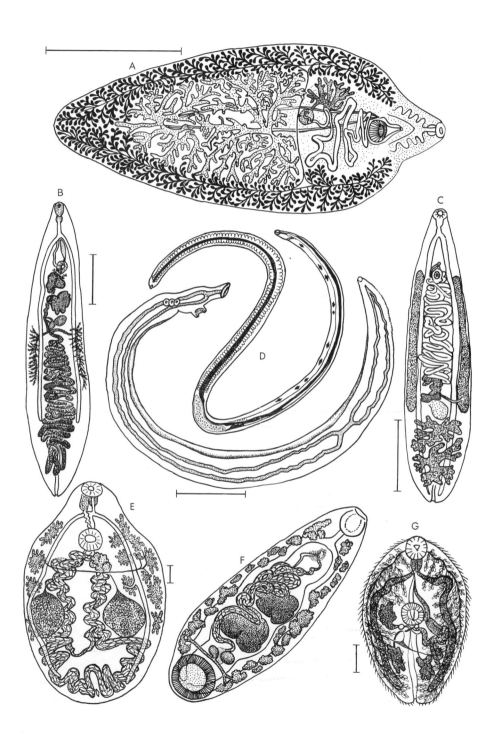

and hatch in the water of wet pasture land. The miracidia penetrate snails, where sporocysts and rediae occur; the latter produce about 25 cercariae each. These escape from the snail and encyst on vegetation, where they remain viable for over three months. Ingestion of the cysts by cattle permits the metacercariae to excyst in the duodenum and to migrate from there to the rumen. Treatment by carbon tetrachloride and prevention by molluscicides spread upon wet pastures are recommended.

PROSTHOGONIMIASIS

The oviduct fluke of poultry is *Prosthogonimus macrorchis* (See Fig. 10–2) of chickens, ducks, and other birds. The flukes, often very numerous, irritate and inflame the mucosa of the bursa of Fabricius and the oviducts. Egg formation is interfered with or prevented, generalized secondary infection of the ducts and peritoneum may occur, and death sometimes results. Carbon tetrachloride is effective in treating moderate infections. The cycle of the parasite involves ingestion of the eggs by small, bottom-feeding snails, in which sporocysts produce cercariae and omit the redia stage. The cercariae are drawn into the breathing chambers of dragonfly naiads and penetrate the tissues. They encyst in the abdominal muscles of the insect and may remain there for almost a year while the naiad moults and grows to the preadult stage. When the naiads emerge from water along the shores of ponds or lakes, they are easy to catch and eat, and are ingested by chickens. Of course, ducks eat them at any time during the aquatic phase of their life. The mature dragonfly effectively spreads prosthogonimiasis in its wide-ranging flight; on damp mornings birds catch and eat adult dragonflies. This disease is worldwide but is focal in distribution because of the requirements of the life cycle. Control consists of treatment of infected birds and prevention of their access to dragonflies or naiads especially during times of emergence of the latter. Since various wild birds serve as reservoir hosts, eradication is probably not possible.

SALMON FLUKE POISONING

A small trematode which utilizes the salmon as its second intermediate host, *Troglotrema salmincola* (Chapin, 1926) Witenberg, 1932, is a dangerous parasite of dogs and related mammals on the northwest coast of North America and in eastern Siberia. While the tiny fluke itself (about 1 mm long) causes no injury in the host intestine, it often carries a bacterialike organism, *Neorickettsia helminthica*, which causes a frequently fatal disease of dogs, coyotes, foxes, and conceivably (although not yet reported) of man. The symptoms, after an incubation period of about one week, begin with fever and sensory disturbance. Purulent discharges and gastrointestinal symptoms follow. After about five days of such symptoms, the temperature drops and death may occur. Recovery produces immunity. This disease is apparently unique, for the transmission of rickettsias usually involves arthropod vectors.

CONCLUSIONS

This concludes the series of chapters on trematodes. In retrospect, schistosomiasis is seen as an extremely important disease which requires further research and practical measures for its abatement. Clonorchiasis, less severe, less obvious, and restricted to those who eat raw fish, is a condition which education and sanitation have some chance of controlling, although cultural factors and the prevalence of reservoir hosts are discouraging. Paragonimiasis, a subject of long and careful study, is a zoonosis which is still not completely known; abstention from raw or drunken crabs and care in preparing crabs for cooking can prevent rather simply human lung-fluke infections. Other trematodes of man and his domestic animals have also been mentioned in this chapter. The omission of any discussion of a vast literature of the trematodes of wild life is intentional but regretted. Further discussion of the biology of trematodes will be included in chapters on host-parasite relationships.

SUGGESTED READINGS

FAUST, E. C., and P. F. RUSSELL, *Craig and Faust's Clinical Parasitology*, 7th ed. Lea and Febiger, Philadelphia, 1964.

LaPAGE, G., *Mönnig's Veterinary Helminthology and Entomology*, 5th ed. Williams and Wilkins, Baltimore, 1962.

REINHARD, E. G., "Landmarks of Parasitology. I. The Discovery of the Life Cycle of the Liver Fluke," *Exptl. Parasitology*, 6:208–232, 1957.

YOKAGAWA, S., W. W. CORT, and M. YOKAGAWA, "*Paragonimus* and Paragonimiasis," *Exptl. Parasitology*, 10:81–205, 1960.

CHAPTER 11

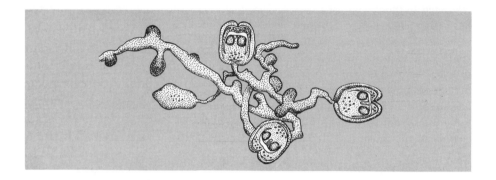

CESTODES

INTRODUCTION

In a text as brief as this one, the inclusion of a chapter on cestodes needs justification. First, cestodes are the "most parasitic" of parasites, at least among the metazoa, and thus demonstrate more emphatically than do other groups of animals the biological features of parasitism. Second, they can reveal something of both their own and their hosts' evolution, a valuable contribution to evolutionary theory if studied and understood. Third, they have excited the curiosity, if not the respect and admiration, of man from the earliest times, and such somewhat morbid fascination with a subject by the public is itself justification enough for a teacher to include it in his course or text. Fourth, there is no better place in this book to consider a group of parasites from a broad, or biological, point of view; such a point of view is important indeed to the future of parasitology and public health, however academic it may seem to the uninformed outsider. Finally, cestodes affect man both directly (by causing disease) and indirectly (by causing losses among man's domestic animals).

The general features of cestodes were described in Chapter 7 (see Fig. 7–5). The scolex, or attachment organ (head), the neck, or growth zone, and the strobila, or chain of segments, are parts of a reproductive unit; they constitute the whole

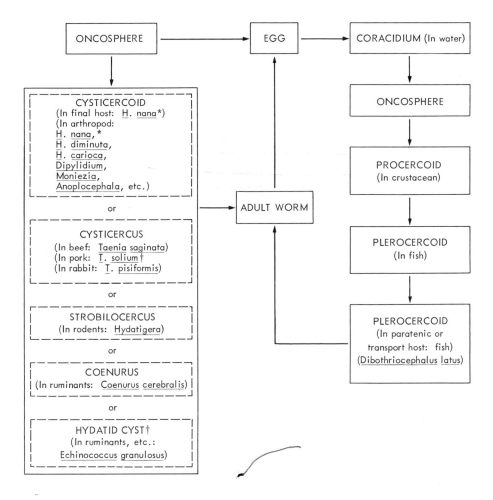

*Hymenolepis nana can develop directly from egg to adult in the final host, or it may utilize insects for its cysticercoid.
† These larval worms are found in man, an accidental host, terminating the life cycle.

FIG. 11–1. Life cycles of some common cestodes. The egg contains a six-hooked embryo (oncosphere), the infective larva. This, if surrounded by a ciliated layer of cells, is called a coracidium, characteristic larva of the fish tapeworm of man. Oncospheres may develop to later stages in vertebrates (in some cases) or in invertebrates; in the latter case the oncospheres often require a second intermediate host in order to reach an infective stage capable of developing in the final host, a vertebrate. The cycle illustrated at right, requiring both an invertebrate and a vertebrate intermediate host, is characteristic of many cestodes of fishes and is believed to be a "primitive" life cycle. The direct life cycle of H. nana, one of two cycles of which this worm is capable, is believed to be a "recent" evolutionary development.

tapeworm, which lives in the intestine (rarely in the bile duct) of its host and proliferates reproductive subunits, i.e., the egg-producing segments. Since the life cycles of cestodes were not discussed in the general introduction to flatworms, they will be discussed here, before we proceed to the main business of this chapter.

LIFE CYCLES

Cestode life cycles show great variety. With the well-known exception of *Hymenolepis nana,* a tapeworm of rodents and man (see below), nearly all cestodes require an intermediate host to complete their life cycles. This host may belong to any of several phyla. It must be a host which ingests eggs or embryos of tapeworms, and it must serve potentially as food for the final host. Sometimes two or more intermediate hosts form successive habitats for a tapeworm. Figure 11–1 illustrates diagrammatically some tapeworm-host relationships.

The larval stages of tapeworms have been classified as follows. The six-hooked embryo which occurs in the egg is called an oncosphere (spherical swelling) (Figs. 11–2A and 11–3D). It bears hooks in three pairs: one median and two lateral. These hooks are frequently distinguishable, the two median hooks are somewhat longer and more slender than the lateral four, and one member of each lateral pair is sometimes stouter than the other. To these hooks are attached muscles within the embryo, and when an embryo is activated (by heat, bile salts, mechanical pressure, for example), the hooks move rhythmically in such a way that the embryo digs into any substance soft enough to be penetrated. The oncosphere thus penetrates the intestinal wall of its host and enters the host's tissues. Then the oncosphere grows larger, and, depending on the group of tapeworms involved, several different lines of development may begin. A kind of cyst may form.

FIG. 11–2.   Cestode eggs and larvae. (A1) Egg and (A2) hatched ciliated larva (coracidium) of *Spirometra,* a pseudophyllidean tapeworm closely related to the fish tapeworm of man. Note that the egg is operculate. (B) Two examples of the infective stage of a pseudophyllidean, called plerocercoid larvae ("spargana" if found in human tissue). (C) Intermediate stage between the coracidium (A2) and the plerocercoid (B), called a "procercoid." Note the taillike appendage and the suckerlike structure opposite the tail. (D) A proliferating example of the solid kind of infective larva called cysticercoid. This is the larva of *Hymenolepis cantaniana,* a parasite of poultry. The larva develops in beetles. (E) Diagram of part of the bladderlike larva called "hydatid cyst." Shown are several developing "brood capsules" on the inner surface of the cyst wall, as well as three scolices greatly enlarged (end view, side view with rostellum extruded, and longitudinal section with rostellum folded back below the suckers). (F) The elongated bladder worm called a "strobilocercus," the larva of the cat taeniid, *Hydatigera taeniaeformis.* The larva has been removed from its cyst, a connective tissue capsule which encloses it during its residence in the liver of an intermediate host, the rat. [(A)–(D) After Wardle and McLeod, 1952; (E) after Chandler and Read, 1961.]

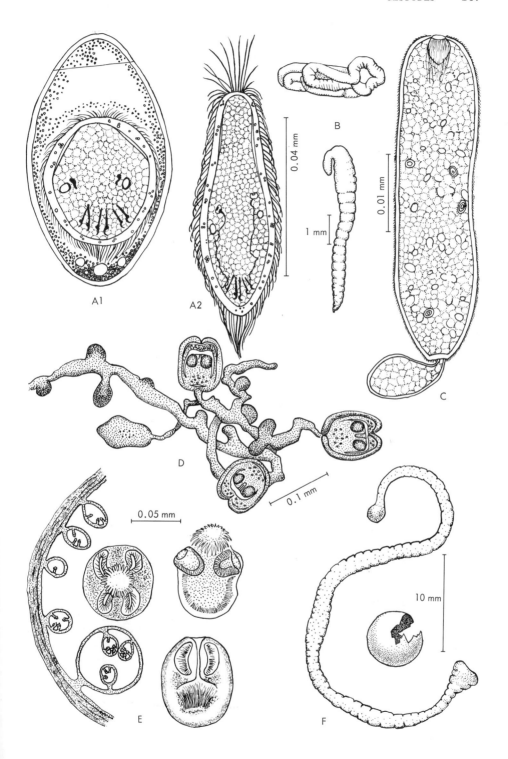

A1

A2

B

C

D

E

F

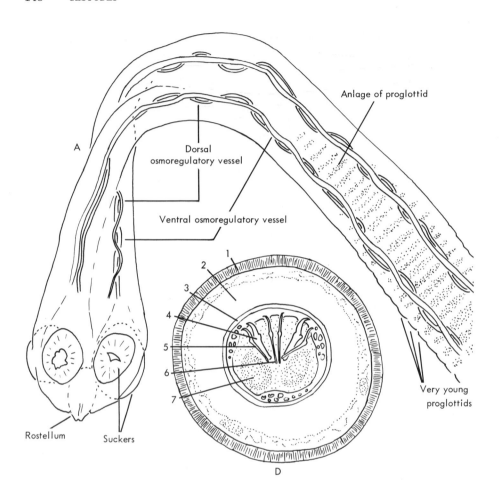

Anlage of proglottid

Dorsal
osmoregulatory vessel

Ventral osmoregulatory vessel

Very young
proglottids

Rostellum   Suckers

D

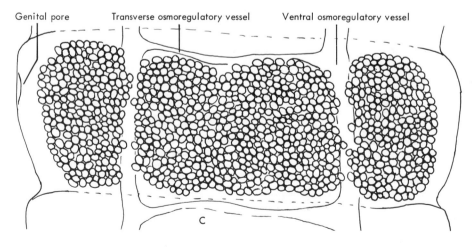

Genital pore   Transverse osmoregulatory vessel   Ventral osmoregulatory vessel

C

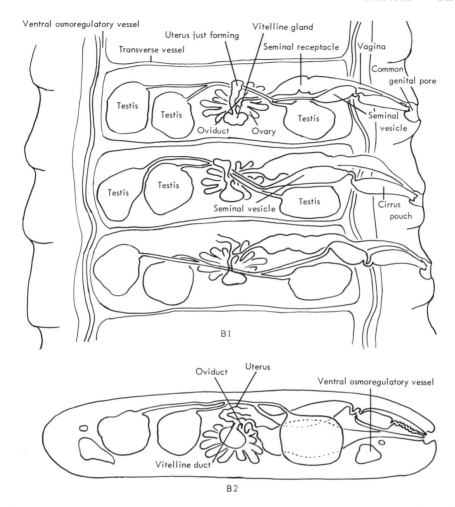

FIG. 11–3. Morphology of *Hymenolepis diminuta,* the rat tapeworm. (A) Scolex and neck, showing rudimentary unarmed rostellum, two suckers, face view (concealing the other two suckers), and the beginning of segmentation or proglottid formation as evidenced by transverse masses of cells. ×100. (B1) Dorsal view of mature region, showing characteristic serial repetition of structure. ×100. (B2) Proglottid as seen from the posterior end. ×100. (C) Gravid proglottid, showing a single layer (one of three or four) consisting of about 1000 eggs. The uterus obscures all other structures except the swollen ventral and transverse vessels and the surface of the genital pore. ×100. (D) Egg consisting of (1) shell or embryophore, (2) jelly-filled cavity, (3) oncospheral (embryonic) membrane, enclosing oncosphere. The latter shows hooklet dimorphism, characteristic of many if not all the cestodes. (4) The median member of each lateral pair of hooks is at least twice as thick as (5) the lateral member. (6) The hooks of the middle pair are both longer and more slender than the other hooks. Part of the glandular (penetration?) tissue of the oncosphere is shown as (7) two granular masses. [Drawn from life at a magnification of about 750×.]

This cyst may be partly filled with fluid and may contain at one end the scolex or head of the developing tapeworm. Usually this scolex, with its four suckers and perhaps a protrusible organ (the rostellum), is invaginated or turned "outside in." Such a larva is called a bladder-worm, or cysticercus. A similar larval cyst, yet solid and without fluid, is called a cysticercoid. (See Fig. 11–2D.) Some kinds of tapeworms in the same order (Cyclophyllidea) as those which form the above larvae have complicated larval cysts and varying numbers of scoleces contained within the cysts. Such are the "coenurus" and "hydatid" cyst (Fig. 11–2E) discussed below in relation to gid disease of sheep and hydatid disease of man, respectively. Other cestodes, especially the members of the Pseudophyllidea, have wormlike larvae. Since these are cestodes of fish or fish-eating animals, one would expect the larvae to develop in fish as they do. The well-known "broad tapeworm," *Dibothriocephalus* or *Diphyllobothrium*, utilizes first small crustaceans, various species of copepod, in which the six-hooked embryo gives rise to a wormlike procercoid (Fig. 11–2C). The copepod is ingested by a fish, in which the procercoid develops further to become finally a plerocercoid larva (Fig. 11–2B). The larva resembles the scolex of the adult cestode. As each fish is eaten in turn by a larger fish, this larva may utilize a whole series of host fish, and the cestode remains unencysted in the muscle tissue. Some few cyclophyllideans as well as pseudophyllideans have proliferating larvae—cystlike or wormlike forms which bud and branch to become many larvae in the process. Larvae of each of the kinds mentioned above cannot develop further until ingested by an appropriate final host.

Adult tapeworms develop directly from infective larvae. Usually only the scolex and a portion of the neck region survive the action of digestion; the cyst membranes and other larval tissues disintegrate. As noted above, the neck region grows and forms segments or proglottids which then mature. However, segmentation does not universally accompany growth; in certain pseudophyllideans the adult worm may be no larger than the larva, and transfer of the larva to a final host stimulates proglottid formation without further elongation. Egg-laying completes the cycle. Some tapeworms (the Pseudophyllidea) have in each proglottid a uterine pore through which eggs are continually discharged. Others give rise to detached "gravid" segments (see Fig. 11–3C), which pass out of the host with the feces so that eggs are subsequently scattered by pollution factors such as coprophagous animals, water, or wind. The gravid segments of some tapeworms move actively and leave the host independently of the fecal matter.

MORPHOLOGY

The details of anatomy of a typical adult cestode are well illustrated by *Hymenolepis diminuta*, the rat tapeworm, which occasionally parasitizes man and has been the subject of much research.

Worms of this species, measuring about 200–300 mm by 3–4 mm, are intermediate in size between such giant worms as *Taenia* and *Dibothriocephalus* of man and certain minute forms (less than 1 mm long) found in insectivores.

They possess a scolex with four circular suckers and a hookless rostellum (Fig. 11–3A). The scolex is a complex organ capable of holding onto the intestinal mucosa or releasing its hold and taking another grip. The suckers, like those of trematodes, are cuplike objects, the walls of which have two sets of muscle fibers: (1) a circular set, which is parallel with the rim of the cup, and (2) a radial set, which passes from the lumen of the cup to its outer surface. These two sets of fibers are antagonistic. When they contract, the circular ones tend to deepen and narrow the opening of the sucker; and the radial ones, having an opposite effect, widen the aperture and flatten the concavity. Each sucker is anchored to the musculature of the neck by several retractors and protractors so that the sucker is actually used as an adhesive foot. The rostellum of *Hymenolepis diminuta*, being unarmed, may have some secretory function but probably is not adhesive. Within the scolex is a ring of nervous tissue, with tracts leading to each of the four suckers and with two large trunks running posteriorly into the neck and throughout the strobila. The scolex, formed in the larval stage, has been shown to be capable of only a small degree of repair when experimentally cut or injured. It is essentially a fully differentiated part of the tapeworm in which growth does not occur.

Back of the scolex a region of perpetual growth is found. Sections of this region, properly stained, show numerous stages of division among the closely packed nuclei of the neck. This region (Fig. 11–3A) is the source of nearly all the cells that make up the main body or strobila. In the neck region, in addition to the mass of proliferating tissue, there are circular and diagonal muscles as well as longitudinal muscles arranged in bundles; the muscles are parallel to the two nerve cords and parallel also to two pairs of fluid-filled vessels which extend from the scolex to the end of the worm. The fluid is believed to be excess water resulting from osmotic absorption through the worm's cuticle. These vessels have been called "osmoregulatory," because the "osmotic" fluid contained therein is constantly being removed from the worm's tissues by the action of flame cells (see Fig. 8–1), the same cells which are found throughout the flatworm phylum (Platyhelminthes) as well as in some other groups of invertebrates. Muscles, nerves, and osmoregulatory system are continuous parts of the strobila and permit the strobila to crawl, undulate, expand, and contract in a coordinated way. Nothing is known about the regulation of cestode behavior, however, since "tapeworm watching" is not a popular avocation.

Another element of cestode structure which is present in every part—scolex, neck, and strobila—is a tissue usually called parenchyma, a spongy network of cells or syncytial elements (nuclei embedded in a continuous mass of cytoplasm) in which muscle, nerve, canals, etc., are found. This tissue resembles an epithelium where it underlies the covering of the worm; the covering itself (formerly called a "cuticula") is a complex extension of cell surfaces and corresponds quite closely to the brush border or microvilli of the intestinal epithelial cells of higher animals. Absorption of nutrients (specifically amino acids, fatty acids, glucose, etc.) takes place through the covering. Presumably each region of the worm acquires its own nourishment directly through its external surface and underlying epithelioid cells. Characteristic "chalk bodies" are found in the parenchyma of most cestodes. These consist of concentric layers of phosphates and carbonates of calcium, and their function is unknown.

The chief function of the cestode strobila is reproduction. Each segment as it matures increases in size and complexity. Genital primordia, groups of nuclei descended from the proliferating nuclei of the neck, can be seen in *Hymenolepis diminuta* about 20 or 30 segments back from the scolex. In each segment at this level there is a small central mass of nuclei which contains three other isolated masses in the parenchyma and a group of nuclei extending inward from the margin on one side. Respectively, these are the beginnings of the ovary-uterus-yolk complex, the three testes, and the genital ducts. As the segments age further, the male system reaches maturity first. Testes grow larger, and active groups of spermatocytes undergo meiotic divisions; hence masses of threadlike sperm enter narrow ducts leading from the testes to a seminal vesicle or sperm storage organ. This vesicle lies next to a muscular organ known as the cirrus pouch, a sac which contains both the convoluted eversible copulatory tube or cirrus and an expanded portion of the cirrus wherein sperm are stored. (The cirrus pouch is called an external seminal vesicle to distinguish it from the vesicle mentioned above.) Such a complete male reproductive system (Fig. 11–3B) is capable of inseminating a fully developed female system. The latter system, however, matures later than the male; it is functional at a level of about twenty or thirty segments farther down the strobila. The female system (Fig. 11–3B) includes a vagina. The vagina opens on the margin of the segment into the common genital atrium, which also receives the male opening. The vagina leads to a seminal receptacle, where sperm from copulation are stored. Connected with the receptacle and vagina is an egg-forming center, into which pass ripe egg cells from the ovary, yolk granules from the compact vitelline gland, and the secretions of a group of glandular cells known as the Mehlis gland. Sperm, packaged with egg cell and yolk, are enclosed within each shell, and the complete egg is then discharged into a constantly forming saclike uterus. As the segment further ages, 2000–3000 eggs form, and the expanded uterus fills the parenchyma, crowding and obliterating the senescent male organs as well as the exhausted ovary and yolk gland (see Fig. 11–3C). A single strobila produces eggs at such a rate that an average of more than 100,000 are discharged with the feces of its host in 24 hours. Segments filled with eggs are called gravid segments. They break off at the end of the strobila in short chains of 10–50 segments at a time, and may either rupture in the host's intestine or be deposited more or less intact within the fecal pellets. A specimen of *Hymenolepis diminuta* may live for a year or longer and may constantly produce eggs at the rate given above.

Such fecundity is one of the commonest kinds of adaptation to parasitism. It implies tremendous loss of intermediate stages and a high risk of failure of transmission from one host to the next.

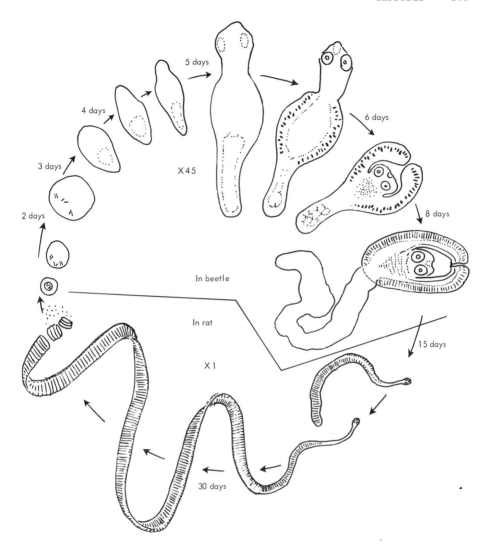

5 days

4 days

3 days

2 days

X45

6 days

8 days

In beetle

15 days

In rat

X1

30 days

FIG. 11–4. Life cycle of *Hymenolepis diminuta*. The complete cycle requires a minimum of about 30 days for completion. Eggs are ingested by a host such as the flour beetle, *Tribolium confusum*. Then, after the periods of time indicated, the oncosphere, having invaded the beetle's hemocoele, grows and metamorphoses into a cysticercoid, or solid-tailed larva. In the process the scolex forms and is withdrawn into a cavity, where it remains protected until ingested by a rat or other final host. At least 8 days are required for infectivity to be reached. When ingested by a rat, the cysticercoid extrudes its scolex, loses its tail, and grows rapidly into an adult worm. The latter, about 20 days after ingestion (or about 30 days after the egg had been eaten by a beetle) begins to lose gravid segments (see Fig. 11–3), and eggs reach the environment, where beetles may find them. [After Voge and Heyneman, 1957, and others.]

LIFE CYCLE OF *HYMENOLEPIS DIMINUTA*

The intermediate host of the rat tapeworm is any of various grain-eating insects, including two or more kinds of flour beetles as well as cockroaches. Egg-bearing rat feces contaminate stored grain and are eaten by the young or adults of insects. When the tapeworm eggs hatch, they liberate six-hooked embryos (oncospheres), which penetrate the intestinal wall of the host and reach the blood-filled body cavity (haemocoele) of the insect. There the embryos grow and within about two weeks become transformed into larvae, each of which has a complete scolex and neck attached to and withdrawn into a fleshy cystlike covering. The whole structure is called a cysticercoid. When rats eat insects containing fully developed cysticercoids, the latter become active. The outer covering is digested away and the scolex and neck are everted. The suckers hold onto the intestinal lining, and the neck cells proliferate. Within three weeks a strobila has been formed, with gravid segments beginning to break off and pass out of the host. Figure 11–4 illustrates the above cycle.

PHYSIOLOGY OF *HYMENOLEPIS DIMINUTA*

A considerable amount of information is available on the physiology of cestodes. The above species has been analyzed biochemically, and estimates of its chief source of energy (stored glycogen) and its protein content have been obtained. *In vitro* studies have shown this worm to be capable of existing under either aerobic or anaerobic conditions. It can utilize glucose from its environment. Recently (Schiller, 1965) this tapeworm has been successfully reared *in vitro* from cysticercoid to adult by a technique which promises to permit critically detailed biochemical studies. On the basis of earlier studies this worm probably absorbs not only glucose but also amino acids and vitamins either from the intestinal fluids in which it lives or by direct contact with the host's cells. Of course it possesses internal enzymatic systems for releasing energy and building and destroying proteins as well as other such mechanisms necessary for its life. The chief nonreproductive use of energy by cestodes is their struggle to remain in place during peristaltic movements of their habitat. Otherwise, nearly all their activity is reproductive, as we have seen.

The rat tapeworm may be considered typical of tapeworms in general. All tapeworms are quite dependent upon their hosts for very specific nutritional requirements. Their transmission to new host individuals depends upon particular food habits of the host and also upon food habits and opportunities of the intermediate host. The risks in such dependency necessitate very high fecundity, which, of course, the sheltered and well-fed existence of the adult tapeworm makes possible.

Within this established pattern many variations exist. The more obvious of these variations are the readily observable differences in form, size, and structure, and serve to distinguish the many taxonomic groups in which cestode species are arranged. It is not mere coincidence, however, that the classification of cestodes on morphological grounds fits very well an altogether different basis of arrangement, that derived from host relationships. This idea will be explored following a few examples of morphological uniformity combined with host specificity.

HOST SPECIFICITY AT THE ORDER LEVEL

The Cyclophyllidea, the order of cestodes to which the rat tapeworm belongs, was briefly described in Chapter 7 and may be further characterized as follows:

cestodes with a scolex of four cup-like suckers plus a rostellum (often the latter and sometimes the former are armed with hooks);

the strobila apolytic (shedding segments);

usually protandrous (with the male system developing earlier than the female); the internal organs of a mature segment including a compact and separate yolk gland (except in a family lacking distinct yolk glands of any kind);

the oncosphere enclosed within three membranes, the outer of which may be lost and the middle of which may be variously sculptured or thickened;

larval stages consisting of bladder worms (cysticerci) or similar forms (cysticercoids) with solid cystlike appendage;

the single intermediate host usually an arthropod, but other invertebrates also utilized, while in possibly two cases no intermediate host is required;

parasites of warm-blooded animals with one exception, i.e., a family parasitic in amphibia.

Another important order of cestodes, the Pseudophyllidea, is characterized thus:

cestodes with a scolex bearing two lateral grooves, never suckers, and without a rostellum;

the strobila anapolytic (not shedding segments);

the internal organs of a mature segment with diffuse, many-follicled yolk glands;

the oncosphere surrounded by a ciliated epithelium, and the egg-shell similar to that of Digenea, which with few exceptions is operculate (with a cap or lid);

two post-embryonic larval stages, the first a minute procercoid (in an arthropod host) and the second a relatively large wormlike plerocercoid (in the tissues of a vertebrate host, usually the prey of the final host); final hosts are predatory fish and piscivorous birds, reptiles and mammals.

Morphologically, other orders of cestodes are similarly distinct and are similarly restricted to particular groups of final hosts. Thus it appears that the basis for classifying the tapeworms in their separate orders is partly descriptive of the worms themselves and partly related to their hosts.

HOST SPECIFICITY IN FAMILIES

A similarly high degree of host specificity is shown by families, genera, and species of cestode. Thus the family Anoplocephalidae (subfamily Anoplocephalinae) are parasites of grazing animals. Within the family there is a genus *Moniezia* restricted to ruminants, a genus *Cittotaenia* restricted to lagomorphs (rabbits, hares, etc.), and a genus *Anoplocephala* restricted to equines. Another example is the large family *Hymenolepididae*, which contains perhaps 800 species. Nearly every one of these species has been reported from only one species of bird or mammal; the rat tapeworm, the mouse-man tapeworm *Hymenolepis nana*, and the mouse bile-duct tapeworm *Hymenolepis microstoma* show some ability to infect several kinds of hosts, but these cosmopolitan species are probably exceptional in this respect. Apparently the taxonomic arrangement of cestodes is a natural one and conforms to the phenomenon of host specificity.

As will be explained in Chapter 26, host specificity may be both ecological and physiological. The specificity of cestodes is probably of the latter type, since each species of host provides a nearly unique environment in terms of temperature, pH, combinations of nutrients, and biochemistry of the host mucous membranes which are usually in intimate contact with the parasite. Special factors such as rate of flow of material through an intestine (emptying time) may be critical for the establishment of larvae "hatching" from cysts. Initiation of hatching (or activation) of larval cysts may also depend on special conditions. Of minor importance, perhaps, are differences in motility, pressures, and peristaltic forces in the gut of various hosts; presumably some of the morphological variations in cestodes— shapes of suckers and rostellum, muscularity of neck and strobila, size and shape of the worms, for example—play a role in host specificity. The above facts, illustrating the similarity or convergence of categories based on host relationships are essentially evolutionary in implication. Host specificity in cestodes is a good example of what has been called "adaptive radiation," the tendency of living forms to occupy every available niche. Each species of host represents an ecological niche, or small environmental specialty, for exploitation by some cestode. Each successful cestode demonstrates adaptation to its particular niche. The fact that adaptive radiation seems unusually complete in the cestodes, an almost perfect correspondence of parasite to host, species by species, requires some explanation.

POPULATION STRUCTURE

Perhaps that explanation is to be found in cestode population structure. Like nearly all parasites, cestodes live isolated lives. The chance of cross-fertilization between any two cestodes depends on their being in the same host at the same time; yet cestodes are seldom found in large numbers in a single host, and often occur individually. Being able to fertilize themselves, moreover, cestodes do not require mates and may get along without cross-fertilization even when other cestodes are present. They are thus isolated both by the accident of limited numbers and by the fact of their hermaphroditic nature. It is well known that self-fertilization, or close inbreeding, results in genetic purity or homozygosity of organisms where this occurs. Therefore cestodes must be considered largely homozygous. (The question of whether a tapeworm is an individual or a population is immaterial here, since a genetically uniform population is, practically speaking, an individual; its response to environmental stress is individual, not statistical, and the survival or extinction of a homozygous population depends on the same factors which affect individuals within it.) Selection, as all plant and animal breeders know, is most effective when it affects homozygous stock. Therefore selection among cestodes may be rigorous, and that is probably the reason for the phenomenon of physiological host specificity.

VARIATION AMONG CESTODES

Nevertheless, all variations among cestodes are not necessarily adaptive. Although the physiological features of a tapeworm must be suited to its environment and the striking ingenuities of each life-cycle must be linked ecologically with the host, some variation in internal structure or even external form may occur. In any survey of forms of the scolex, for instance, one is more impressed by the bizarre diversity of this organ than by any correspondence of form with function. In the single family Hymenolepidae, there are forms, like the rat tapeworm, with relatively small scolex and unarmed rostellum; there are many forms with a large rostellum, bearing hooks of varied numbers and shapes; and there are forms in which the scolex is insignificant indeed and has been replaced in function by the much twisted anterior part of the strobila, or in these species, the "pseudoscolex." In the strange cestodes of the orders Tetraphyllidea, Trypanorhyncha, etc., found in sharks and rays, the scolex is a very ornate object; it often bears muscular leaflike or petal-like bothridia and, in addition, eversible proboscislike organs capable of actually penetrating host tissue. It is a curious paradox that several of those species with scolices and equipped with tremendous holding power are nonetheless hyper-apolytic, for they lose their segments even before they become gravid. Perhaps the extraordinary scolices of the cestodes of sharks and rays have real adaptive value, but at present this would be hard to demonstrate. The same question arises regard-ing variations in the sexual apparatus of cestodes. All cestodes have highly ef-fective egg-producing machinery, but the variations in this fundamental structure are quite large. Some cestodes lack entirely a vagina and are inseminated by puncture with a cuticularized and spinose cirrus. The uterus also is apt to be quite varied. The rat tapeworm's saclike cavity in the parenchyma, which gradually fills with eggs, is the counterpart of reticular or branching uteri in other forms. The uterus of the great taeniids is like a tree with a central trunk and lateral branches; the uterus of the common cat-dog tapeworm, *Dipylidium caninum,* is composed of hundreds of separate pockets, each with one to a dozen or more eggs. Other tapeworms in the same order as the above (the Cyclophyllidea) have par-uterine organs, i.e., curious thick-walled chambers into which the fully developed eggs pass from the true uterus. As we know, some tapeworms lay their eggs. The uterus of pseudophyllideans, such as *Dibothriocephalus,* is a short coiled tube which opens on the surface of its segment. Members of another order, the Proteocephala, common tapeworms of freshwater fishes, amphibians, and reptiles, have uteri which dehisce, or break open, along the surface of the segment, spilling eggs into the

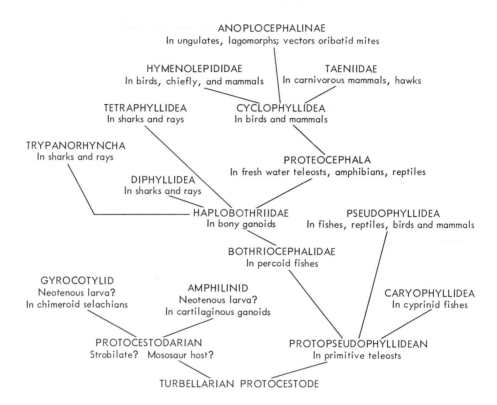

FIG. 11–5. A cestode "family tree." This hypothetical structure implies an origin of cestodes from turbellarian ancestors. Early branching of the stock separated the cestodarian line from the true cestodes. The most primitive of modern cestodes are probably the Pseudophyllidea, and therefore they are placed close to the line of origin of all other modern cestodes. These are thought to have been derived from forms like the present-day Bothriocephalidae, a family of primitive cestodes which combine features of Pseudophyllidea and other modern orders. All the primitive cestodes are parasites of fishes; the more primitive (such as the Caryophyllidea) are found in ancient groups of fishes (such as the Cyprinidae). As the diagram shows, each distinct group of cestodes has its own rather restricted group of hosts. The two families and one subfamily of Cyclophyllidea, a recent order, are chosen to illustrate such host specificity at the family level. As pointed out in the text, host specificity actually extends to the species level in most groups of hosts of cestodes, and an evolutionary diagram such as the above owes much to the fact of host specificity. [After Baer, 1951.]

host intestine. Diversification similar to the above examples has been noted in the arrangement of body musculature, the position of the genital structures, and other details. Such diversity in animals inhabiting a basically similar habitat, the intestine of vertebrates, seems rather extreme when contrasted with the simplicity of the basic cestode pattern defined above (scolex, neck, strobila; great fecundity); perhaps some other explanation is more suitable than the rather facile one of evolutionary adaptation.

All organisms, including tapeworms, mutate spontaneously and with no regard for the helpfulness or harmfulness of mutations. Therefore there must be areas in which selection is inoperative because of the neutral nature of certain mutations. The bizarre morphological variations in the cestodes are perhaps the products of random neutral mutation resulting in the survival and proliferation of changes which in no way affect the basic cestode requirements of security, nutrition, and reproduction.

That is not to say that the morphology of tapeworms lacks internal logic. Each of the major groups, i.e., the orders, shows within itself the relatedness of its members. Basic patterns of structure and life cycle reveal that these orders are the present-day representatives of ancestors who long ago occupied such diverse habitats as ancient sharks and chimney-swifts, marsupials and cetaceans. Students of tapeworm taxonomy are able to predict to some extent the types of cestodes yet to be discovered in various kinds of hosts; and a sense of "surprise" (that evidence of accustomed expectation) follows the announcement of any unusual discovery of which, up to now, there have been few.

PHYLOGENY OF CESTODES

A family tree can be erected for the cestodes. Since fossils of parasites are almost nonexistent, this tree is much less dependable and less accurate than, say, the family tree of mammals, reptiles, or molluscs, but the fact that evolutionary relationships can even be postulated in parasitic worms is noteworthy. Figure 11–5 shows that correspondences among primitive hosts and primitive cestodes are one basis for the tree's structure. Larval characteristics are also utilized, and internal morphological features are not neglected. As will be pointed out in Chapter 29, the pattern by which groups of taxonomically similar cestodes occupy particular groups of hosts is analogous to the pattern which struck Darwin, i.e., the restriction of particular animal and plant taxa to the different geographic realms of the earth.

CONCLUSIONS: THE BIOLOGY OF CESTODES

The above remarks about cestodes were intended to demonstrate the truth of the introductory statement that these worms are the "most parasitic of parasites" and that the study of cestodes leads to better understanding of several principles of parasitology. *Hymenolepis diminuta* was described in detail in order to show (1) the dependency of this animal upon its final and intermediate hosts, (2) its stupendous fecundity, and (3) the fact that its "mating system" leads to homozygous individuals and populations. The results of such homozygosity were then illustrated by recounting some data on host specificity in cestodes. The specificity is so exact that selection must have produced it; each host acts as a screening environment in which only one genotype, essentially homozygous, can survive. The extraordinary variety of some kinds of cestode structures, seemingly a surprising or irrational factor in the selection theory of cestode host-specificity, was explained as being due to random mutation in relatively nonadaptive features. In all species the fundamental features of cestodes are effective holdfast, reduplicative growth, nutritional dependency, and great fecundity; variation which alters these is not permitted. Finally, on the logic of these ideas, it seemed possible to construct a phylogenetic diagram of the cestodes. The plausibility of such a design is evidence that understanding of parasitism can come from the study of cestodes and can justify their inclusion in this book.

CESTODES OF MEDICAL AND ECONOMIC INTEREST; CESTODES OF MAN

The common cestodes of man include five species in the Cyclophyllidea and one in the Pseudophyllidea. The Cyclophyllidean species are *Taenia solium*, (the pork tapeworm), *Taeniarhynchus saginatus* (the beef tapeworm), *Echinococcus granulosus* (the worm causing hydatid cyst), and *Hymenolepis nana* (the small tapeworm of man and mouse). The Pseudophyllidean species is *Dibothriocephalus latus* (the broad or fish tapeworm).

Taenia solium is a large cestode (5–10 M) acquired by eating undercooked or raw pork infected with the bladder worm, *Cysticercus cellulosae*. Its scolex bears a double crown of hooks, and, unlike those of the beef tapeworm, its gravid proglottids, which are passed in the feces, are weak and soft. The eggs, like those of other members of the family Taeniidae, have a thick shell-like embryophore composed of radially arranged prismatic rods cemented together. The eggs are durable and are capable of surviving conditions of heat, cold, and variable humidity. When eaten by hogs, the eggs hatch, the six-hooked embryos penetrate the hog's intestinal wall, and the larvae are carried, probably by blood and lymph vessels, to the muscles and other tissues of the body where the bladder worms develop. In swine, mild infections cause little damage, while heavy infections involving brain, heart, lungs, or eye, as well as the skeletal muscles, generally cause death. The chief importance of *Taenia solium* is the fact that it causes human cysticercosis. Man himself, upon ingesting the eggs of this tapeworm, becomes an "intermediate host," and cysticerci may develop in various organs, including the brain. The consequence may be fatal.

The adult beef tapeworm, *Taeniarhynchus saginatus*, resembles the pork tapeworm in size and structure but differs from it in two ways: The rostellum of the scolex bears no hooks, and the free gravid proglottid actively and vigorously works its way out of the host's rectum, since it is not dependent on defecation by the host. The intermediate hosts, cattle, ingest the eggs with contaminated grass or fodder. It is known that the eggs may pass unhatched through the intestinal tracts of birds, which thus aid in disseminating the eggs widely. As with *T. solium*, the cysticercosis of the intermediate host is usually a mild or symptomless condition. Unlike *T. solium*, the beef tapeworm does not cause human cysticercosis.

Adults of both the pork and the beef tapeworms have been known to cause harm to their host, but there is so much variability in physiological response of the human host to these large tapeworms that no diagnostic symptoms have been recognized. Some reported effects are loss of appetite, excessive appetite, digestive disturbance, dizziness, and occasionally severe nervous symptoms; some persons harboring these worms feel quite well.

Another taeniid, *Echinococcus granulosus*, is a dangerous parasite of man, causing the zoonosis "hydatid cyst." (See Fig. 28–3.) The cyst is the larval stage ordinarily found in sheep and goats which have acquired eggs of the adult tapeworm in grass contaminated by the feces of wolves or dogs. The natural cycle therefore involves canine animals as final hosts and ruminants as intermediates. Man acquires the eggs *via* contaminated food or intimate association with infected dogs. The cysts which develop grow slowly but may become enormous after a number of years; they may reach diameters greater than 15–20 cm. Frequently found in the liver, they may occur in the lung or brain as well as other organs where, by exerting pressure and occupying space, they cause disease or death. Treatment is by surgery; prevention depends on avoiding dogs, and eliminating hydatid disease among sheep and goats. Certain countries, such as Iceland, where hydatid cyst was once common, have successfully reduced the incidence by regulating the disposal of animal carcasses and by requiring the regular worming of dogs.

Two other cestodes of domestic animals occasionally infect man, usually with little harm. *Dipylidium caninum*, very common in dogs and cats throughout the world, normally passes its larval stages in fleas. These animals can be eaten accidentally by children or other persons, and the cysticercoids then survive and become established in the human intestine. As we have seen, *Hymenolepis diminuta* develops in various insects including the tiny flour beetles; accidental ingestion of the latter has resulted in human infections, of which a number have been reported in medical literature. Perhaps the anoplocephaline cestodes of the genus *Inermicapsifer* should be mentioned here also, since these tapeworms of African rodents have occasionally been found in children. Related cestodes, *Bertiella* spp., of old world anthropoid apes have also been reported from man.

Hymenolepis nana, the "dwarf tapeworm," ranges from 7 to 100 mm in length. In heavy infections the worms are quite small, averaging 20–30 mm; apparently, as Read has pointed out, there is a crowding effect that regulates size. The life cycle of this worm is unique in that the eggs are infective to the final host and autoinfection can occur. The eggs shed within the intestine can hatch, the embryos can invade the intestinal villi, and cysticercoids (larvae) can develop within these structures. When the cysticercoids emerge, the scolex becomes attached to the intestine, and an adult worm develops. Thus a true infection is possible, with the parasites increasing in numbers in the same host. However, *H. nana* may develop to the cysticercoid stage in intermediate hosts, which include a number of insects, such as flour beetles (*Tribolium* spp.) and cockroaches. When these insects are ingested, the cysticercoid is freed, and the scolex everts and becomes attached to the host. Rats and mice, as well as other rodents, may serve as alternate or reservoir hosts for the adult worms. Although the rodent "strain" has been called *H. nana* (var. *fraterna*), there is reason to believe that it is infective to man and that the human strain is infective to rodents. *H. nana* causes a severe condition in children, with pain, diarrhea and nervous symptoms. It is the commonest tapeworm of man, having a worldwide incidence of around 1%. Local incidences range up to 25–30%. Chandler believed that rodents were largely responsible for the spread of *H. nana*, in spite of the possibility of direct transmission from person to person by poor sanitary practices; the rather similar incidence in rural and urban areas, he pointed out, argues for the importance of rodent transmission, since crowded and relatively unsanitary urban conditions should otherwise result in higher levels of incidence than the rural.

A last tapeworm of man is *Dibothriocephalus latus*, also called *Diphyllobothrium latum*, the "broad" or "fish" tapeworm. This can become a huge worm, up to 10 m long and 10–20 mm wide. A single worm may have several thousand proglottids. This worm, like others of the order Pseudophyllidea, lays its eggs through the uterine pores of sexually mature proglottids. The eggs must reach water for further changes to take place. The embryos are not developed when the eggs are laid. Then after little more than a week, embryos (called coracidia) with a ciliated covering are formed, and the egg (operculate like a trematode egg) breaks open at the lid or operculum to release the actively swimming embryo. The latter is ingested by a copepod and develops into a procercoid larva (Fig. 11–2c), which remains without further change until the host is ingested by a fish. There the larva reaches the muscle tissue, where it grows into the worm-like plerocercoid stage, which is infective to fish-eating mammals, including man. In some of the regions of the world where fish are an important source of food (chiefly northern United States and Canada and the Baltic region of northern Europe), the fish tapeworm has been very common. Incidence in man reaches 20–100% in some areas, while the corresponding frequency of infective fish may reach 75%. Cats and dogs are important reservoir hosts and in some areas may be the essential link in the cycle, human infection being biologically unimportant. Clinically, human infection with *D. latus* may be very serious. It has been shown by von Bonnsdorf and others that *D. latus* absorbs vitamin B_{12}, the important antianemic factor, much more readily than other tapeworms. (The powdered worms can be used as an effective source of the vitamin.) Thus in some individuals harboring *D. latus*, severe anemia develops, undoubtedly from vitamin B_{12} deficiency caused by the worm. The anemia rate is not high, however, being about 0.05 to 0.10% in *D. latus* infections.

From this brief mention of the cestodes infecting man, it should appear that while certain worms are relatively harmless as well as rare (the beef tapeworm, for instance), others may be quite dangerous (*T. solium* causing human cysticercosis; hydatid cyst; *H. nana* in children). Another, *D. latus*, may be harmless to many of its hosts yet harmful to a few, such as those susceptible to induction of anemia. The cestodes of domestic animals show the same range of effects, but the more harmful species, being also quite common, are quite important.

CESTODES OF DOMESTIC ANIMALS

Monieziasis is a cestode disease of cattle, sheep, and goats caused by *Moniezia* spp. (order Cyclophyllidea, family Anoplocephalidae). This parasite inhabits the small intestine. When present in large numbers the worms may cause obstruction of the intestine; they undoubtedly irritate the intestinal mucosa, and they produce toxic substances which are absorbed by the host. Young animals suffer most from monieziasis. As the animal matures, there is a gradual loss of worms and subsequently immunity to reinfection; however, this immunity may be merely a relatively prompt self-cure or, as Morgan and Hawkins have pointed out, a heightened response to newly acquired worms. Authorities disagree as to the importance of monieziasis. American workers have found little evidence that the cestodes, at least in moderate numbers, have any effect on health or condition of lambs. Yet Russian workers consider the disease important and report numerous primary and secondary effects of monieziasis in heifers and lambs. The differences in monieziasis may be due to the fact that the average worm burden is lighter in one agricultural area than in the other. Ruminants acquire monieziasis by ingesting oribatid mites (Chapter 22), in which minute cysticercoids of the cestode have developed. As Stunkard first pointed out, the life cycle of *Moniezia* appears to be characteristic of a number of related cestodes; *Moniezia* and *Thysanosoma* in sheep, cattle, and goats, *Anoplocephala* in horses, and *Cittotaenia* in rabbits utilize mites as intermediate hosts. Treatment consists of anthelmintic "drenching" with any of several reasonably effective taeniacides, including copper sulfate. Prevention is almost impossible. Infected mites are found on upstanding blades of grass during moist or rainy weather, and young animals should not be put to pasture during such times.

The tapeworms of horses are closely related to the above tapeworms of ruminants. They belong to three species: *Anoplocephala magna, A. perfoliata,* and *Paranoplocephala mamillana.* The first two are relatively large worms, i.e., up to 80 cm (*A. magna*) or 8 cm (*A. perfoliata*) in length with corresponding widths of 50 mm and 20 mm, respectively. *P. mamillana* is a small worm, approximately 50 mm by 4–6 mm. All of these species require oribatid mites as intermediate hosts. Like *Moniezia* (see above) in sheep or lambs, the large tapeworms of the horse may cause mechanical obstruction of the intestine and severe damage to the host, or they may cause chronic intestinal malfunction due to irritation and scarring of the intestinal mucosa. Standard taeniacidal drugs, such as Kamala, oil of male fern, and oil of turpentine, may be used to remove the worms. Reinfection, of course, is frequent, since the infected mites are present everywhere in the pasture. Removal of the worms from horses, followed by transfer of the horses to new pastures, should reduce the incidence and severity of horse tapeworm disease.

The cestodes of fowl are relatively numerous but are believed not to be very important. In geese and ducks, *Drepanidotaenia lanceolata* causes nutritional disturbance, neurological symptoms, and occasionally even blockage of the intestine. Copepods of the genus *Cyclops* are the intermediate host; since these crustacea abound in the stagnant, polluted water of duck ponds, large numbers of cysticercoids are ingested, and both incidence and worm burden may become very high. Sick birds may be treated with taeniacides. Mass treatment of geese after fall removal to pens and treatment before placing geese on ponds in the spring should reduce the incidence and seriousness of drepanidotaeniasis. In chickens, some 46 species of cestodes have been found. Of these, several are thought to be pathogenic. *Davainea proglottina*, a small cestode (less than 5 mm long) consisting of very few proglottids, can cause heavily infected birds to lose condition and weight. Treatment has been ineffective, although many drugs have been tested. Since slugs (shell-less gastropods) are the intermediate host, some protection and control might be achieved by control of slugs; but until this becomes practical, disposal of chicken feces seems the only available method of control. *Raillietina* species (*cesticillus, echinobothrida,* and *tetragona*) are rather large worms which cause loss of weight (or relative failure in young birds to gain weight). They utilize a large number of genera of beetles as intermediate hosts. *Raillietina* infestation may be treated with anthelmintics, and large portions of the worm may be removed; but the scolex resists most attempts to remove it, and the strobila soon regenerates. Control and prevention depend chiefly on disposal of the host feces to avoid infecting dung beetles. Further study of the effects of fowl cestode infection will be needed before the losses caused by it can be estimated. It is not clear how valuable the efforts at prevention and control of these infections may be, or whether the cost of anthelmintic therapy against fowl cestodes can be justified.

Cestodes of domestic dogs and cats include several parasites of veterinary importance. Since some of these are also of importance in human health, they have been treated elsewhere in this chapter (see *Echinococcus, Dibothriocephalus*). *Dipylidium caninum*, the dog and cat tapeworm which utilizes the flea as intermediate host, can become very numerous and can almost fill the intestine. In heavy infections, intestinal symptoms occur, although in light infections the only symptom is scratching of the anal region, where the emerging ripe proglottids cause irritation. Human infections with *Dipylidium* have been reported and are probably common in children. Treatment of this and other tapeworms of domestic carnivores is apt to succeed at least to the extent of reducing the number of worms. Other cestodes of dogs and cats are *Taenia pisiformis*, the bladder worms of which occur in rabbits, and which cause symptoms similar to those caused by *Dipylidium; Taenia* (or *Hydatigera*) *taeniaeformis*, the elongated bladder worm (strobilocercus) of which is found in spherical cysts in the livers of rats; and *Taenia hydatigena, Multiceps multiceps* and *M. serialis*, as well as *Echinococcus granulosus*, all in this group being important as parasites of cattle, sheep, or goats, which serve as intermediate hosts.

The last three species cause a peculiar type of cysticercosis in their intermediate hosts. *Taenia hydatigena* forms variable bladder worms which, after a course of development in the liver, migrate to the mesenteries and become pear-shaped hyaline cysts. *Cysticercus tenuicollis* is the name given this bladder worm. *Multiceps*

multiceps causes "gid disease" of sheep and goats; its larva, a coenurus (a bladder-like cyst bearing several invaginated scoleces), is often found in the brain, where it exerts pressure and produces profound nervous symptoms. *M. serialis*, like *M. multiceps*, has cysts of the coenurus type, but these may proliferate by budding; the intermediate host is the rabbit, not a ruminant. As mentioned above *Echinococcus* causes the serious human disease "hydatid cyst." In ruminants, symptoms of disease seldom occur, because the short life span of these animals does not allow the scolex-filled cysts to grow to an inconvenient or harmful size. Since the cysts never involve host muscle tissue, little economic loss is occasioned by hydatid cysts in sheep and cattle. Cysticerci of the large taenias, however, may make "measly" beef or pork unsuitable for sale. Prevention of such losses due to larval tapeworms can best be achieved by treating regularly with anthelmintics all dogs in the involved area, by improving methods of disposal of human feces, and by preventing the feeding of infected carcasses or scraps to dogs, such methods being dependent on the particular tapeworm involved.

SUMMARY

In review, the cestodes are seen in three aspects: as unique members of the animal kingdom, as minor parasites of man, and as significant agents of animal disease. In the first aspect, they are paradoxical colonies of sexually separate units, united by the same nervous, muscular, and vascular system; they have fantastically high reproductive rates, which compensate for a precarious dependence upon one or more intermediate hosts; and their larval forms are varied and ingenious, being bladderlike, cystlike, or wormlike in different groups and utilizing as intermediate hosts such different animals as ants, copepods, and cattle. The high degree of host specificity of cestodes in their final host can perhaps be explained by the unique population structure of these worms. The relative homozygosity of cestodes makes them subject to rigorous selection and highly perfected environmental adaptation. Such specificity is useful to the biologist in working out the phylogeny of cestodes, which necessarily corresponds to the evolutionary histories of the hosts. In the second aspect, as parasites of man, cestodes affect his intestine—mildly or severely, sometimes producing anemia—or his muscles and other tissue, as in cysticercosis and hydatid cyst. In the third aspect, as agricultural pests, cestodes cause epidemics among geese and ducks, real but unmeasured losses in chickens, injury to lambs and colts, and occasional loss in cattle and swine through condemnation of meat for cysticercosis.

While attention has been focused in the past on the medical and economic aspects of cestodes, these matters are almost trivial; and perhaps study of the biology of these interesting animals should be emphasized in the future, for their peculiarities as "most parasitic" of parasites are useful and instructive exaggerations of parasitic adaptation in general.

SUGGESTED READINGS

BAER, J. G., *The Ecology of Animal Parasites*. University of Illinois Press, Urbana, Ill., 1951, pp. 136–155.

CHENG, T. C., *The Biology of Animal Parasites*. W. B. Saunders, Philadelphia, 1964, pp. 304–353.

ERSHOV, V. S. (ed.), *Parasitology and Parasitic Diseases of Livestock*. State Publ. House for Agric. Lit., Moscow (in Russian), and Israel Program Sci. Trans., Jerusalem (in English), 1956.

HYMAN, L. H., *The Invertebrates. Vol. II. Platyhelminthes and Rhynchocoela, the Acoelomate Bilateria*. McGraw-Hill, New York, 1951, pp. 311–417.

JONES, A. W., *et al.*, "Host Relationships of Radiation-Induced Mutant Strains of *Hymenolepis diminuta*," *Ann. N.Y. Acad. Sci.*, 113 Art. 1:343–359, 1963.

LAPAGE, G., *Mönnig's Veterinary Helminthology and Entomology*, 5th ed. Williams and Wilkins, Baltimore, 1962.

READ, C. P., and J. E. SIMMONS, JR., "Biochemistry and Physiology of Tapeworms," *Physiol. Rev.*, 43:263–305, 1963.

SCHILLER, E. L., "A Simplified Method for the *in vitro* Cultivation of the Rat Tapeworm *Hymenolepis diminuta*," *J. Parasit.*, 51:516–518, 1965.

SMYTH, J. D., *The Biology of Cestode Life Cycles*. Commonwealth Bureau of Helminthology, St. Albans, Herts, England, No. 34, 1963.

VOGE, M., and D. HEYNEMAN, "Development of *Hymenolepis nana* and *Hymenolepis diminuta* (Cestoda: Hymenolepididae) in the Intermediate Host *Tribolium confusum*," *Univ. Calif. Publ. Zool.*, 59:549–580, 1957.

WARDLE, R. A., and J. A. McLEOD, *The Zoology of Tapeworms*. University of Minnesota Press, Minneapolis, 1952.

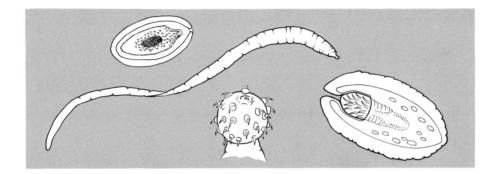

CHAPTER 12

ACANTHOCEPHALA

INTRODUCTION

The thorny-headed worms, which constitute the phylum Acanthocephala, are unique in several ways. First, they are structurally unlike any other animals; yet some of their features, such as their supposedly cylindrical form and their dioecious sexuality, have been used by some writers to link them with the nematodes; and the absence of an intestine as well as certain serological similarities to cestodes have caused other authors to link them with flatworms. The anatomical features of the Acanthocephala taken together describe a form of life different from all others. As will be seen below, this unique group of animals is characterized by a thorny proboscis, a thick body wall interpenetrated with a system of canals and giant nuclei, two sets of muscles with the circular bundles peripheral to the longitudinal, unique reproductive structures (ovarian balls, uterine bell, etc., in the female, and cement glands, etc., in the male), and a pseudocoelomate body cavity. Second, embryological and later development substantiates the above observation since the fertilization and maturation of the eggs, the form of the embryonic membranes and egg shells, the sudden shift from an early spiral cleavage to a syncytial mass,

167

the subsequent metamorphosis from a first-stage to a second stage larva, and the complete absence of any free-living stage in any known species are all unusual features of animal life. Finally, these worms have a peculiar position in the system of classification; they are recognized as a phylum, or major subdivision of the animal kingdom, but are unlike nearly all other phyla in having no major subdivisions (classes) within the group.

This last fact is somewhat puzzling. Every biologist is taught to define taxonomic categories in two ways, either as divisions within more inclusive categories (a phylum is part of a kingdom) or as collections of subordinate groupings (a phylum is made up of classes). Most phyla, including the ones we have discussed so far, consist of more than one class. For instance, four classes comprise the phylum Protozoa, and three classes make up the Platyhelminthes. The classes of vertebrates are well known, and although we may have doubts about the proper classification of certain fossil forms, there is practically no confusion as to whether any backboned animal in existence today belongs to fishes, amphibia, reptiles, birds, or mammals. Most phyla exhibit the same divisibility that the vertebrate subphylum shows. In fact, the phylum Aschelminthes, some members of which we shall discuss in the next section of this book, includes distinctive classes—the rotifers, priapulids, tardigrads, gastrotrichs, etc., in addition to the nematodes and nematomorphs—so different from each other that it seems to be a rather artificial container for possibly unrelated groups. The Acanthocephala were placed in this assemblage until an expert, Van Cleave, in 1948, elevated the then-called class Acanthocephala to the rank of phylum. He recognized the need for making his new category conform to the partial definition that "a phylum is a group of classes" by dividing it into two classes, each of which accommodated two orders. But later writers (e.g. Chandler and Read, 1961) have criticized the division of the phylum Acanthocephala into classes on the grounds that these are merely artificial separations based on very few differences; these authors have not followed Van Cleave's system. It is best at present to consider the Acanthocephala a phylum with but a single class, the class Acanthocephala, the latter being divided into orders according to criteria established by Van Cleave in 1936. As we shall see, these orders are natural groups in the sense that the members of each share both morphological and ecological features, by which the orders can quite easily be set apart. After all, taxonomic categories are man-made conveniences for studying and discussing very diverse and not yet fully catalogued items, the species. It is not surprising that such categories are uneven or inadequate at times. The Acanthocephala are a group of very peculiar animals, and they belong to a peculiar phylum—a phylum with only a single class.

ORDERS OF ACANTHOCEPHALA, A SYNOPSIS

The order Palaeacanthocephala consists mainly of parasites of fishes. A few aquatic birds and mammals harbor members of this order. The proboscis of these worms is dorsoventrally differentiated, for the spines of the ventral recurved surface are larger than the dorsal spines. Spines are present on the trunk or the main part of the body. Uniquely in this order, the nuclei of the hypodermis are fragmented, not limited in size and number. The main lacunar vessels (canals) occur in the lateral hypodermis. There are no special organs of excretion. In the males there are usually six cement glands, the secretions of which seal off or plug the female genital opening immediately after copulation. The eggs are soft and thin-shelled. Aquatic crustacea serve as intermediate hosts.

The order Eoacanthocephala contains parasites of freshwater fishes and turtles. The proboscis hooks are similar dorsoventrally and are radially arranged. Spines may be present on the trunk. The nuclei of the hypodermis are large and very few (only six in some genera). The chief lacunar vessels are dorsal and ventral. There are no excretory organs. The cement glands are syncytial; that is, they consist of a single mass of secretory cytoplasm with a few nuclei embedded in it. A cement reservoir is present. The eggs are thin-shelled. Aquatic crustaceans are intermediate hosts.

Members of the order Archiacanthocephala are generally found in terrestrial animals, including birds but excluding snakes and lizards. These worms have the proboscis thorns arranged symmetrically, either in concentric circles or in long rows. There are no spines on the trunk. The hypodermal nuclei are extremely large and typically few. Dorsal and ventral lacunar vessels (hypodermal canals) are present. Some species possess protonephridia, i.e., bulbed, ciliated excretory organs which empty into the genital canals and which are not found in other orders of Acanthocephala. The cement glands of the males consist of eight separate units. Among the worms found in this order are several Acanthocephala of veterinary or medical importance. Insects which ingest the hard-shelled eggs serve as the usual intermediate hosts.

There are only about 60 recognized genera in the above orders, in about 12 families. New species will probably not be discovered very frequently in the future, because there are relatively few workers interested in this group of parasites. The following discussion of some noteworthy species may make the biology and significance of these parasites somewhat clearer.

SOME IMPORTANT ACANTHOCEPHALA

Among the helminths of swine is the giant thorny-headed worm *Macracanthorhynchus hirudinaceus* (Fig. 12–1). This is a member of the order Archiacanthocephala, defined above. *Macracanthorhynchus hirudinaceus*, like all acanthocephala, is dioecious; the males (5–10 cm long) being much smaller than the females (up to 60 cm). Members of each sex have a retractable proboscis in which are rooted hardened, recurved hooks arranged in five or six rows. There is no intestine; absorption of food occurs through the thin cuticula and thick hypodermis. Embedded in the body wall, in the inner layer of hypodermis, are many giant nuclei, and there is a system of anastomosing canals lying in the base of the hypodermis. The hypodermis is invaginated near the proboscis in the form of two fingerlike flaps or extensions into the body cavity, called lemnisci. The function of the lemnisci is unknown; they merely lie in the body cavity and extend a part of the

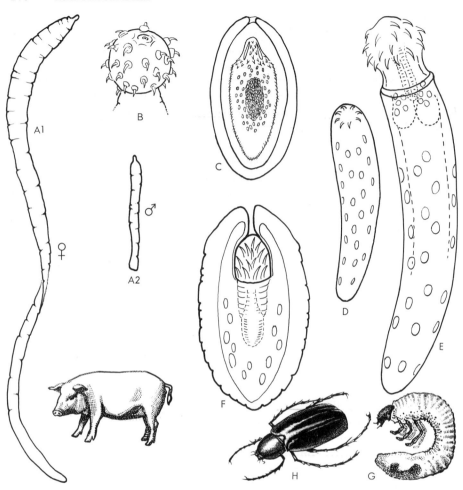

FIG. 12–1.   Stages in the life cycle of *Macracanthorhynchus hirudinaceus*. Adults (A1, female; A2, male) are shown ⅔ life size. The expanded proboscis (B) shows rooted hooks in spiral rows. Development in the hemocele of a grub (G) or beetle (H) includes the acanthor (C) which is shown within the egg membranes, the acanthella (D,E), and the infective cystacanth with retracted proboscis (F).

way from the region of the proboscis sheath to the region of testes or ovarian tissue. Muscles of the body wall are roughly similar to the muscles of nematodes (see Chapter 13) and consist of well-defined groups of fibers associated with more centrally located masses of cytoplasm and nuclei. In the Acanthocephala, however, there are both longitudinally and circularly oriented fibers. Within the body cavity of the male are two sausage-shaped testes suspended by a ligament, and a massive structure, the cement gland, which surrounds the sperm duct. The cement gland opens into a copulatory sac or bursa. In females, the ovary develops at first in a suspensory ligament, but soon breaks up into numerous ovarian balls which lie free within the body cavity and form eggs on their surfaces. The eggs, when fully embryonated, pass through a funnel-shaped egg-laying duct to the exterior. A portion of this duct, the so-called "uterine bell," acts as a sort of filter, permitting only fully developed eggs to pass to the exterior. One female lays up to 680,000 eggs per day and may live nearly 2 years.

The life cycle (Fig. 12–1) involves ingestion of the eggs by the grub or larva of various scarabeid beetles, development of compact, miniature adults in the latter, and ingestion of the infected grubs by swine.

Stages in development are as follows: Within the egg is a spiny acanthor, a larva consisting of a central mass of small nuclei surrounded by a much smaller number of large nuclei. The surface of the acanthor is covered with posterior-directed spines, the larger of which, at the anterior end of the larva, are equipped with muscles by which the spines can be moved. By the movement of its spines the acanthor enters the intestinal wall and hemocoele of the grub which has ingested an egg. The acanthor then loses its body spines and develops a proboscis, lemnisci, and rudimentary sex glands. It is now called an acanthella. This form elongates into a juvenile acanthocephalan. The juvenile, protected by its own cuticle and with its proboscis deeply retracted, remains within the body cavity of its grub or beetle host. Some call this stage a cystacanth. Ingestion of such cystacanth by a suitable vertebrate host (pig or man) results in growth of the juvenile to an adult.

These parasites cause mechanical damage to the intestine, where they remain strongly attached by the armed proboscis. Perforation may occur. Heavy infections (considerably more than 15 parasites per hog) cause nonspecific clinical symptoms, including loss of appetite, restlessness, bloody diarrhea, and emaciation. Treatment is ineffective. Infected lots should be left unused by swine for four years, and infected animals should be slaughtered and not transferred to "clean" farms. *Macracanthorhynchus hirudinaceus* has been known to infect man and no doubt would become a medical problem if man frequently ate scarabeid beetles or their grubs.

The acanthocephalan species *Polymorphus minutus* and *P. magnus* (order Palaeacanthocephala) parasitize domestic as well as wild ducks. These worms are small, the former being about 3–4 mm long in both sexes and the latter being up to 11 mm. Various crustacea, including crayfish and amphipods, serve as intermediate hosts. Although heavy infestations occur and the entire flock of some farms become host to a large number of the acanthocephalans, individual worm burdens are usually light, and little damage is caused. This species, like perhaps most acanthocephala, having very low host specificity in the adult stage, has been reported from more than 40 kinds of wild birds. Thus the parasite is often introduced into clean flocks. But only unusual concentrations of infected birds and intermediate hosts can produce economically harmful disease.

Another acanthocephalan of ducks, geese, and other aquatic birds is *Filicollis anatis*. This worm penetrates the intestinal wall quite deeply, causing emaciation, weakness, and frequently death. Its development in the isopod crustacean, *Asellus* spp., ensures that it can become very numerous in farm ponds, natural lakes, and similar bodies of water, whence feeding ducks can ingest the crustacean host in large numbers. There is no treatment for this (or any other) acanthocephalan.

Other acanthocephalans include species reported from dogs, foxes, and coyotes. The species *Oncicola canis* causes rabieslike symptoms in infected animals. The life cycle is unknown. Dogs probably acquire their parasites from various small herbivores in which encysted stages occur. These herbivores are examples of "transport hosts," because they serve to transmit the parasite from one host to another without being necessary for growth or development of the worm. All acanthocephalans whose life cycles are known are capable of reaching an infective stage in their arthropod host. Obviously, since arthropods are not commonly eaten by carnivores, successful establishment within the latter hosts requires residence in and transport by their prey.

Moniliformis dubius is a species in the archiacanthocephalan order parasitic in rats. While its economic importance is negligible, it is a useful experimental animal, which has been the subject of continuing study. This externally segmented worm is of moderate size, the females being 10 to 30 cm long, and the males 6 to 13 cm. It has unusually large eggs (over 100 microns in length). Intermediate hosts are various insects. The cockroach *Periplaneta americana* is easily infected, and since this insect is also easily maintained in the laboratory, studies of all stages of the life cycle of *M. dubius* can be carried out. Other studies include experiments on the effect of starvation of host on parasite burden, observations on crowding (the inability to superinfect being due, apparently, not to immune response of the host but to space limitations in the rather small gut of the rat), studies on respiration, which reveal that *Moniliformis* carries on primarily anaerobic metabolism, cytological studies which show that there is an X-O sex-determining mechanism, the male being heterogametic, and studies on the effects of ionizing radiation upon development. Although unusual forms such as acanthocephalans may seem to be unlikely sources of information which might be applicable to medical or other practical problems, we should remind ourselves of the contributions to human knowledge which have been made by other strange creatures—*Drosophila*, for instance, or the common molds, *Penicillium* and *Neurospora*.

Another acanthocephalan of laboratory animals is a monkey parasite *Prosthenorchis*. Several species of these have been reported from zoos, and they have been found in the wild in South America, probably their native habitat. These worms are a serious pest. In one modern laboratory they were instrumental in the closing down of an expensive experiment on the tiny marmosets known as tamarins, because these valuable little creatures were often found infected with *Prosthenorchis* when purchased from South American shippers. They sickened and died from peritonitis as a result of perforation of the intestine by the worm's proboscis. This worm develops in cockroaches and doubtless in other insects, the control of which can prevent spread of *Prosthenorchis* in zoos and laboratories.

CONCLUSION

The above introduction to the phylum Acanthocephala may have served two purposes: (1) to describe a group of curious parasitic animals which have puzzled taxonomists for years, and (2) to list a few species of thorny-headed worms which have some impact on veterinary medicine and biological research. Incidentally, the first, or taxonomic matter, explained the nature of a phylum, since the existence of an exceptional phylum like the one-class Acanthocephala points to the rule that ordinary phyla consist of several classes. At the same time, the chief function of the category (phylum) was seen as a separating function to divide the animal kingdom into its natural parts; thus Van Cleave's erection of the phylum under consideration was justified. The economically important acanthocephalans, parasites of swine, anseriform birds, or laboratory simians, need no excuse for inclusion. The scientific importance of *Moniliformis dubius* was emphasized, because the relatively few kinds of laboratory animals have yielded information of great value; the student should be aware of such animals, including the parasitic ones among them.

SUGGESTED READINGS

BAER, J. G., *The Ecology of Animal Parasites*. University of Illinois Press, Urbana, Ill., 1951, pp. 111–116.

VAN CLEAVE, H. J., "Some Host-Parasite Relationships of the Acanthocephala, with Special Reference to the Organs of Attachment," *Exptl. Parasit.*, 1:305–330, 1951.

VAN CLEAVE, H. J., "Expanding Horizons in the Recognition of a Phylum," *J. Parasit.*, 34:1–20, 1948.

VAN CLEAVE, H. J., "The Recognition of a New Order in the Acanthocephala," *J. Parasit.*, 22:202–206, 1936.

WARD, H. L., "The Species of Acanthocephala Described since 1933. II.," *J. Tenn. Acad. Sci.*, 27:131–149, 1952.

WARD, H. L., "The Species of Acanthocephala Described since 1933. I.," *J. Tenn. Acad. Sci.*, 26:282–311, 1951.

YAMAGUTI, S., *Systema Helminthum. Vol. 5. Acanthocephala.* Wiley, New York, 1963.

CHAPTER 13

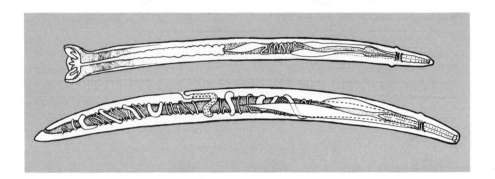

NEMATODES

INTRODUCTION

The nematodes are members of a well-defined class which is one of a number of classes within the phylum Aschelminthes. The phylum includes rotifers, horse-hair worms (Nematomorpha), gastrotrichs, and various obscure groups. The class Nematoda itself is not completely known. Of the more than 10,000 described species, about half are parasitic and half free-living; relatively little attention has been paid the latter, of which perhaps less than one percent of the existing species have been described. The parasitic forms are members of two very important groups: the parasites of plants and those of animals. Man suffers from the direct attack of his own parasitic nematodes in such diseases as the filariases, ascariasis, trichuriasis, hookworm disease, trichinosis, and others. He also suffers severe damage to livestock as well as to important plant crops.

The classification of nematodes is peculiarly difficult because three quite separate kinds of specialists have studied the three major ecological divisions just mentioned. The nematodes parasitic in animals are the best known group; the taxonomic system used for these was started long ago and has been frequently revised. The plant nematodes have also been extensively studied and their classification is well established. It is believed that the free-living nematodes exceed in numbers both the animal- and plant-parasitic species. Yet very few workers have studied these animals, which if better known should form a logical bridge between the systems of classification independently developed for the two groups of parasites. (This situation is analogous to certain problems in paleontology where a few fossils discovered at random appear to be members of quite different groups, while lying undiscovered in the rocks may be a multitude of intermediates, members of the single taxon to which the known forms ultimately belong.) Substantial attempts have been made to classify under one system all nematodes, parasitic and free-living. The success of these attempts will be established by future discoveries, especially in the neglected field of free-living nematodes. Meanwhile, most animal parasitologists tend to ignore the higher categories within the class Nematoda, and for practical purposes they deal with eleven groups which can be rather easily distinguished from each other.

But before describing these groups, it is necessary to present basic information about the morphology of nematodes, since this information forms the language of their classification.

MORPHOLOGY OF NEMATODES

The overall pattern of nematode structure is rather simple. These worms are cylindrical, and they taper at the ends. At mouth and rectum the straight digestive tube is attached but otherwise it is free from the cylindrical body wall. Between body wall and intestine is a fluid filled space, the pseudocoele, within which lie the gonads. The gonads, which are tubular structures and solid at their inner portions, are attached only by their ducts in females and males, respectively, to the openings for oviposition or sperm ejaculation. Muscles, nerves, and excretory ducts lie in the body wall.

The digestive system begins with the mouth or buccal capsule, a highly varied structure usually in the form of a cavity. The buccal capsule is lined with cuticular ridges, thickenings, or rings, which strengthen it. As in hookworms, it may contain teeth or cutting plates. The external opening of the capsule may have lips or papillae surrounding it, or it may be surmounted by a "leaf-crown" as in the strongyles. Separated from the buccal capsule by a slight constriction is the esophagus or pharynx, usually a muscular tube with a cavity shaped like a "T" in cross section. The muscles of the esophagus act on the walls of its lumen to expand the esophagus, producing a pumping or sucking action.

Ingestion is aided in many nematodes by a bulblike valve between the esophagus and the intestine. The nematode intestine appears to be quite simple. It is usually flattened or longitudinally folded when empty, and has nonmuscular walls made of a single layer of digestive-absorptive cells. At the posterior end, the intestine connects with a cuticle-lined rectum, a muscular organ analogous with the esophagus in that its activity forces material to pass through it.

The neuromuscular system of nematodes consists of a circumesophageal ring and dorsal and ventral nerve cords which give rise to special sensory and motor nerves. Longitudinal muscle cells lie just beneath the syncytial layer of nuclei and cytoplasm which underlies the cuticle. Each muscle cell has a parietal portion, composed of a bundle or sheet of contractile fibrils, and a central portion, which is protoplasmic and contains a single nucleus. These tapering cells may be packed together in large numbers forming a continuous sheet of muscle (in polymyarian—"many-muscled" —worms), or they may be distributed in limited number around the body wall (in meromyarian—"separate-muscled"—species). Each muscle cell has a strand of cytoplasm which crosses the body cavity to join one of the nerve cords. The nature of this neuromuscular junction is not well understood. The action of nematode musculature is rather unusual, for antagonistic muscles are absent. The necessary stretching of a muscle after contraction seems to be due to the elasticity of the cuticle, which is compressed on the concave side of a bending worm but springs back to normal length after the compressing force is relaxed.

Parts of the excretory system lie within the so-called lateral lines i.e., two cords of nuclei and cytoplasm continuous with the hypodermal syncytium. Excretory canals in these cords usually open to the surface of the body by a pore ventral to the esophagus. Sometimes bladderlike vesicles called renettes are present. These canals, renettes, and pores are the main parts of an excretory system which is peculiar to nematodes.

Reproduction in the nematodes results from the mating of male with female individuals. The female nematode has a pair of ovaries, solid cords of cells (oogonia) lying free in the pseudocoele. These cords become thicker toward the uteri, which are tubular structures containing eggs. The uteri usually have glandular walls, which probably secrete shell-forming material. The two uteri unite in a common vagina, which may posses a saclike enlargement (the seminal receptacle) where sperm are stored after copulation. The vagina opens to the surface by a vulva, or female genital opening. This is usually at the middle of the body but may be near either end. Egg production starts with the mitotic proliferation of oocytes in the solid, innermost portion of the ovary. Then the oocytes round up and pass through a short oviduct to the uterus, where sperm penetrate their cytoplasm, and egg membranes form. Meiosis, reduction of chromosome number, and fertilization occur in the uterus, and the egg may be laid at various stages of ripening from the one-celled zygote to the shell-enclosed larva, depending on the species. The rate of egg production is very high in parasitic nematodes; the larger hookworm (*Ancylostoma duodenale*) lays up to 30,000 eggs per day. The male genital system is structurally like the female but somewhat more complex. While there is usually only one testis (a solid cord of cells like the ovary of the female), there is a rather large storage place for sperm, the seminal vesicle, connected to the testis by a slender, tubular vas deferens. The seminal vesicle continues posteriad as a muscular ejaculatory duct which opens into the rectum. The latter cavity is properly called a cloaca in the male nematode, since it receives both digestive waste and spermatozoa. Copulation is carried out by the action of two chitinous half-cylinders, pointed at their distal ends, which can be protruded through the cloacal opening. These are the copulatory spicules; they are controlled by muscles and are guided by accessory chitinous structures, most commonly a split cone-shaped gubernaculum and sometimes an ornate telamon. The copulatory spicules form, united, a tube through which spermatozoa are injected into the vulva and vagina of the female worm.

The life cycles of parasitic nematodes are varied and interesting. Detailed descriptions of several nematode life cycles appear in later chapters.

The external features of nematodes form an important element in their classification. The structures of the mouth have already been mentioned. The lips and papillae which surround the mouths of many nematodes are useful in identification. It is possible to cut off the anterior end of a worm, mount it vertically in some viscous or gelatinous medium, and examine it *en face*. Usually, lateral and posterior to the mouth, there are two amphids, gustatory or olfactory organs, in the form of depressions in the cuticle. Somewhat similar caudally located structures called phasmids are present in a few groups of nematodes. These organs lie just posterior to the rectal opening. Their importance seems to rest in the fact that the class Nematoda can be divided into two subclasses based on the presence or absence of phasmids. Among other external features are cuticular ridges (longitudinal or transverse), alae (lateral expansions, usually cervical but in the males sometimes caudal), and cordons (cuticular thickenings forming loops or curves in the head region). The copulatory apparatus of the male nematode is both internal (as described earlier) and external. In some groups of nematodes the males have a conspicuous expansion of the tail region called a bursa. This hand- or scoop-shaped structure is supported by fleshy rays, which form characteristic patterns by which taxa can be defined. Most male nematodes possess innervated perianal papillae, thin-walled elevations of the hypodermal tissues; these are sensory organs which aid in copulation. The pattern formed by the perianal papillae is often diagnostic of species.

TAXONOMY OF NEMATODES

Classification of nematodes depends on information such as the above. When facts about the forms of nematodes are organized into a system, certain groups can be recognized by having similar species and by not including the members of other groups. At present, several such systems have been proposed, all of which have merit. But until experienced parasitologists can agree that one system of classifying nematodes is better (truer, more practical, based on more facts, etc.) than other systems, the writer of a general textbook should be cautious. For the arrangement of nematodes we shall use eleven groups, referred to by some authors as orders, by others, as superfamilies, and by still others, as a mixture of orders, suborders, and superfamilies. We shall use the form (ending in "-oidea") commonly used for most of these names, and shall call these groups superfamilies, which means a collection of families.

There are two superfamilies of nematodes parasitic in vertebrates characterized by the absence of phasmids. These are considered by some authors to be members of the subclass Aphasmidia.

The first superfamily of aphasmid nematodes is the **Trichuroidea**. These worms have a very long, thin esophagus running through a column of large gland cells. The mouth lacks lips. The anterior end of the body is much slenderer than the posterior region. The females have but one ovary; the males have a single copulatory spicule or none at all. This group of nematodes includes two important parasites of man: *Trichuris trichiura* (the whipworm), and *Trichinella spiralis.* Both of these will be discussed later. The genus *Capillaria* contains some parasites of domestic animals, including the laboratory mouse.

The second superfamily of the Aphasmidia is the **Dioctophymoidea**, named for *Dioctophyma*, the genus of the giant kidney worms. All dioctophymoids are large worms. The mouth lacks lips, but is surrounded by 6 to 18 papillae. The esophagus is robust, not filiform as in the trichuroid aphasmids. The females have one ovary, and the males have one spicule and a terminal bursa which lacks supporting rays. The kidney worms infect piscivorous carnivores and usually completely destroy one kidney.

A third superfamily of the Aphasmidia should be mentioned, the **Mermithoidea**. These worms, although not important medically or economically, are commonly found in such insects as grasshoppers, as well as in other invertebrates. These worms are parasitic only as larvae and free-living in the soil or water as adults. Their intestine is not complete, the posterior part being modified for storage of nutrient reserves.

Eight superfamilies of the subclass Phasmidia follow.

The **Rhabditoidea** have two rings of cervical papillae, an inner ring of six and an outer ring of four, six, or ten. An esophageal bulb is present, and the esophagus may be swollen anterior to the circumesophageal nerve ring. Females have one or two ovaries. The copulatory spicules of the males are equal in length, and a gubernaculum is present. Many of these nematodes are free-living, others being parasitic in plants and invertebrates. They are important as agricultural pests, and one species, *Neoaplectana glaseri*, has been used in attempts to control harmful insects.

The **Rhabdiasoidea** lack an esophageal bulb. These are small nematodes, resembling the Rhabditoidea in most respects. (Some authors join these two groups in the same order.) An important rhabdiasoid species is *Strongyloides stercoralis*, discussed under host-parasite relationships (Chapter 26) as an example of life cycle complexity. This species alternates between pathogenic, parasitic habitation in man and other vertebrates, where it multiplies by parthenogenesis, and a free-living existence in the soil, where it reproduces sexually.

The **Oxyuroidea** are small worms. They lack cervical papillae. A definite, valvular esophageal bulb is present. The females have a very long, tapering, pointed tail (hence the common name "pinworms"). Males may have one spicule, or two unequal in length. There is in the males a simple bursa formed of caudal alae. Included in this group of worms are the pinworms, very common inhabitants of the caecum of men, horses, and many other mammals.

The **Ascaridoidea** are large worms. Cervical papillae are present. The esophagus may or may not have a bulb. The males, with two spicules, have ventrally recurved tails, sometimes with alae. The superfamily is named for the family Ascarididae, which contains *Ascaris lumbricoides*, perhaps the commonest as well as the most often noticed of the helminth parasites of man. *Ascaris* and other worms of this group will be discussed in a later chapter.

The **Strongyloidea** are the bursate nematodes. The males of this group have copulatory bursas supported by rays. The mouth structures, especially the buccal capsule, are usually well developed; the esophagus is muscular. There are two spicules in the male. This important group includes several families containing harmful parasites. Members of the hookworm family, Ancylostomidae, have cutting mouth parts within a prominent buccal capsule. Worms of the strongyle family, Strongylidae, have a collar, variously spined or ornamented, surrounding the mouth proper. These small, slender worms are important parasites of stock. The family Trichostrongylidae contains worms with a simple buccal capsule or none at all (exceptional among the superfamilies); they parasitize livestock extensively, and have been reported from man. Lungworms belong to the family Metastrongylidae, containing parasites injurious to cattle, horses, swine, sheep, and goats. These worms have rudimentary buccal capsules and reduced bursas in the males. The family Syngamidae contains the gapeworm of poultry as well as the kidney worm of swine.

The **Spiruroidea** are worms with two lateral lips, sometimes an additional pair of dorsoventral lips and sometimes none. There is a well-developed buccal capsule. The esophagus lacks a bulb, but has two distinct portions: an anterior muscular part and a posterior glandular part. The males, lacking a bursa, have unequal spicules. The tail of the male spirals (cf. the name Spiruroidea), with broad caudal alae supported, like the bursa of the Strongyloidea, by fleshy rays. Spiruroid species are found in many animals, including domestic ones. They occasionally infect man.

The **Dracunculoidea** have a simple mouth, without lips or buccal capsule, but surrounded by a ring of papillae. The esophagus, like that of the spiruroids, has an anterior muscular portion and a posterior glandular portion, and lacks a bulb. In adult females the vulva and vagina atrophy, and the uterus, swollen with embryos without egg membranes, bursts to liberate the embryos in water. The males are much smaller than the females; spicules are equal in length. The guinea worm, discussed in a later chapter, is a member of this group.

The superfamily **Filarioidea** is characterized by long, slender worms, which live in the tissues, blood vessels, or serous cavities of vertebrates. The females usually give birth to embryos which, either in the blood or in the tissues, can be ingested by insects in which larval development occurs. Filarioid worms lack lips and seldom have prominent buccal capsules. The esophagus resembles that of dracunculoids and spiruroids, being two-parted and lacking a bulb. The vulva is near the anterior end. Males are relatively small; their tails are coiled as in the Spiruroidea, but they may lack the caudal alae of the latter. Important filarioid worms occur in man and domestic animals.

The above descriptions of eleven groups of nematodes parasitic in animals is not intended as a system of classification. It is true that the first three groups, lacking phasmids, are readily separable from the other eight. Among the latter, oxyuroids and ascaridoids share several features, differing from each other chiefly in size. The Strongyloidea are such a large and important group that they might reasonably be subdivided further, as some authors do, into orders or suborders. Note our treatment of the families Ancylostomidae, Metastrongylidae, etc. Spiruroids, filarioids, and dracunculoids are similar to each other in several ways; the first two have males with spirally coiled tails, and the last two are exclusively tissue and serous cavity parasites which give birth to larvae, not eggs. But these similarities and differences may not be of deep biological significance, indicating basic relationships or evolutionary ties. Until the class Nematoda is better known, any system of classifying the parasites of animals without reference to the free-living forms or the parasites of plants should be considered merely a listing for convenience, or a sort of key for separating various groups.

In the chapters which follow, certain important nematodes will be discussed. They are not presented in a "taxonomic" arrangement for various reasons, including the artificiality of such an arrangement. Filaroid worms will be discussed first, because of the importance of the diseases they cause and because of the parasitological principles so well illustrated by these diseases and so useful in discussing other parasites. For similar reasons, hookworms, ascarids, and *Trichuris* will follow; these three parasites are aided by poverty and filth and are subject to similar remedies. Then the puzzling or paradoxical nematodiases will be presented: (1) trichinosis, which is prevalent among meat-eaters, that is, among the prosperous or the Eskimos; (2) enterobiasis, the disease that is not a disease, transmitted by unsanitary behavior in the most industrialized, "developed" nations of the world; and (3) dracunculiasis, a water-borne parasite prevalent in desert lands, curable by prehistoric methods, and still a misery for millions. Other nematodes will be grouped together in a chapter which includes some important parasites of livestock.

SUGGESTED READINGS

HYMAN, L. H., *The Invertebrates. Acanthocephala, Aschelminthes and Entoprocta. The Pseudocoelomate Bilateria. Vol. III.* McGraw-Hill, New York, 1951.

CHAPTER 14

FILARIASIS

To the people of the temperate zones of the world, filariasis was once an exotic disease. The early explorers who visited Tahiti were shocked by the sight of grossly swollen limbs, breasts, or genitals which occasionally disfigured the beautiful natives of Polynesia; books on tropical medicine have for many years featured pictures of elephantiasis due to filarial infection with *Wuchereria* or *Brugia*. Some villages in Guatemala have a pitifully high percentage of blind persons and even a higher percentage of people with glistening, irritated skin, a condition due to the presence of millions of microfilariae (first-stage larvae) of *Onchocerca*. In parts of Africa eye worms (*Loa loa*) wander under the skin and occasionally cause considerable pain when they appear on the surface of the eye. Yet conditions like the above, until not many years ago, had not impressed the more favored peoples of the world except as curiosities, subjects for horror or pity, and inspired little rational concern.

Then, during World War II, American military personnel by the tens of thousands invaded the Pacific paradise, where eight out of ten people harbored the larvae of *Wuchereria bancrofti* in their blood, and where mosquitoes day and night transmitted infective larvae to new human hosts. There, in addition to the traumas of displacement, fear of death, and actual attack by the enemy, the men suffered a

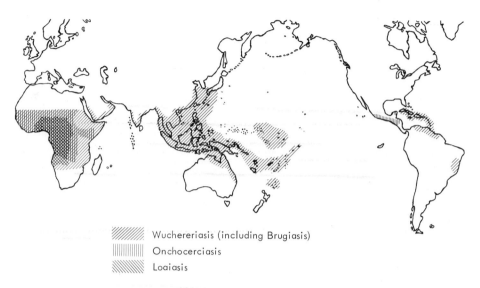

Wuchereriasis (including Brugiasis)
Onchocerciasis
Loaiasis

FIG. 14–1.   Distribution of human filariasis.

severe psychological dread of elephantiasis. Filariasis greatly impressed the American armed forces.

In another part of the world, during the 1940's and 1950's, a great highway was being constructed which eventually was to link North with South America. This highway passes through some of the country most heavily infested with the parasite *Onchocerca volvulus,* the worm which causes "the blinding filariasis." Tourists enjoying the convenience and scenic beauty of the road must pass with some terror through the filarial regions.

In much of Africa, the emergent new nations ask for and receive technical and educational aid in their struggles to become independent and self-sufficient. Among the many obstacles to such success are endemic malaria, amoebiasis, sleeping sickness, and other diseases of the tropics; sharing the blame for obstructing foreign assistance projects is filariasis in at least three forms.

In short, what were once considered exotic diseases have become serious and important health problems in this shrinking world (Fig. 14–1). The history of research on filariasis reflects this transition from the exotic to the near at hand.

HISTORY

The basic work on filariasis as "tropical medicine" was done long ago. Even before the life cycles of the parasites were known, some of the symptoms of the diseases could be cured. Surgery for elephantiasis was, and still is, important. In 1878, Manson discovered the transmission of *Brugia malayi* by mosquitoes. This major discovery in tropical medicine showed the way for Ross and others working on the transmission of malaria. But in spite of the knowledge of the life cycles of the filarial worms, i.e., knowledge acquired through the last part of the 19th century and first part of the 20th, there was only minor progress in control until modern chemistry, combined with the demands of wartime medicine, made the control of filariasis practical. The use of modern insecticides (the well-known chlorinated hydrocarbons and the organic phosphorus compounds) dates from the 1940's; these insecticides can be used to control the mosquitoes that transmit the filariases of Polynesia, the Far East, and central Africa, the black flies that transmit the blinding filariasis of America and the nodule-forming disease of Africa, and, perhaps, the tabanid flies that transmit the African eye worm, *Loa loa*. Further research in the biology of the vectors and the engineering problems of insecticide application should bring greater efficiency in the control of filariasis by insecticides. Improvement in treatment stems from the discovery and testing of the piperazines, especially diethylcarbamazine, called Hetrazan. Although antimony compounds and arsenicals have been used effectively against filariasis, they are relatively toxic and must be given intravenously; they are not so suitable for mass treatment as Hetrazan, which may be given by mouth. Suramin (Bayer 205), an aromatic diamidine, is also useful, but its toxicity is high. There now exist numerous drugs which can kill the larval worms in blood or skin and can sterilize or kill the adults. Further research aimed at improving these drugs is being carried on.

The role of experimental animals should be mentioned here. Many potential antifilarial drugs have been screened by using rats which harbor *Litomosoides carinii*, a murine filariid worm. The piperazine compounds were tested successfully first against this rat parasite. Later clinical tests confirmed the filaricidal properties of piperazine, and the use of Hetrazan in mass treatment is now practical.

Progress in recent years has been encouraging. There seems to be no good reason why systemic (Bancroftian and Malayan) filariasis cannot be soon eradicated in all areas except the truly inaccessible. However, onchocerciasis (cutaneous filariasis) is still a difficult problem, since neither treatment nor vector control has been perfected. And the sequelae of the human filariases—elephantiasis and blindness—are existing medical problems which require more knowledge and skill than we now possess. Also, the filariases of domestic animals are costly and harmful diseases which require much study before they can be controlled. Filariasis then is a disease in transition—from the status of an exotic and relatively unimportant aspect of tropical medicine, inadequately handled both in control and in treatment, to the status of a world public health problem, which has been under study recently and is rapidly being solved.

LIST OF THE FILARIASES

There are actually five or six kinds of human filariasis. The minor filariases of man are loaiasis, dipetalonemiasis, and mansonelliasis. Loaiasis is caused by the African eye worm, *Loa loa*, which lives as an adult in the loose subdermal tissues; it migrates under the skin and occasionally becomes visible on the eye surface.

Prominent swellings, presumably reactions to some antigenic stimulus, appear and disappear on the surface of the body. The blood contains the microfilariae of the parasite, which is transmitted by tabanid flies. Loaiasis is common in western and central Africa. Dipetalonemiasis is a condition (perhaps not a disease) of man in the same geographical area. It produces no symptoms in most persons who harbor adult worms, but there is evidence of allergic response in some cases. The worms are transmitted by midges or gnats of the genus *Culicoides*. Mansonelliasis, an infection found in parts of South America and the West Indies, is due to *Mansonella ozzardi* and is believed to be transmitted by gnats (genus *Culicoides*). It produces no symptoms, unless increased eosinophilia be considered such. Of course, the latter is a symptom of many different parasitisms.

Filariasis of livestock, while not as important as human filariasis, is a significant part of veterinary medicine. Several diseases of horses are caused by filarial worms. Parafilariasis is apparently widespread in the Soviet Union in the forest-steppe regions, where it reaches high infection rates in the warm season. Adult *Parafilaria* in the subcutaneous tissues feed on the walls of the small blood vessels of the skin, causing bloody swellings to appear. The blood from these lesions contains fully embryonated eggs and hatched microfilariae, which must develop to the infective stage in the (unknown) intermediate host. Treatment has not been sufficiently developed but control measures aimed at killing or repelling biting insects have been recommended. Stephanofilariasis of cattle seems to be similar to parafilariasis of horses. *Setaria*, the abdominal worm of cattle and horses, seldom causes symptoms. The adults live embedded in various organs; the microfilariae are found in the circulating blood. In onchocerciasis of horses (and rarely of cattle), adult *Onchocerca* live in the tendons and ligaments of the back and neck region, giving birth to unsheathed microfilariae which migrate (as in human onchocerciasis) to neighboring tissues. Sometimes mild swelling with some hardening of the skin constitutes the symptoms of a self-curing form of the disease. In other cases extensive invasion with dermal hyperplasia and scarring occurs; allergic factors are probably involved here. Onchocerciasis of stock is transmitted by midges or gnats (*Culicoides* spp.) which feed on the serous exudate of wounds inflicted by their cutting mouthparts (see Chapter 24). Microfilariae are ingested with this fluid and develop to infective stages in the midge. Since midges become extremely numerous during warm, wet weather, seasonal outbreaks of onchocerciasis occur at such times. Treatment and control of onchocerciasis are interrelated, since destruction of the microfilariae in the skin of horses makes transmission impossible. Elaeophoriasis of sheep and goats involves the skin of the poll. Invasion of this area by microfilariae produces a dermatitis, with lesions which are aggravated by the host's scratching. The lesions may also occur on the foot used for scratching.

Dirofilariasis (or heartworm disease) of dogs is of considerable importance, not only because it injures many dogs but also because its causative agent, like *Litomosoides* of rats, has been useful in the study of human filariasis. (Observations on the periodicity of microfilariae in the circulating blood of dogs revealed that the lungs serve as a collecting place for these forms when they are not in the peripheral circulation.) *Dirofilaria* adults occupy the right ventricle and pulmonary arteries.

Microfilariae enter the circulating blood and may then be ingested by bloodsucking arthropods. Development can occur in fleas and in a number of species of mosquitoes. Infective larvae escape from the mouth-parts of the arthropod host, entering the final host through the bite or through the intact skin. Adults and microfilariae may persist in dogs for years. Symptoms of heartworm are collapse after exercise from congestion of the pulmonary arteries, with rather prompt recovery. Chronic symptoms include enlarged liver, poor condition, and abnormal heart sounds. Treatment by antifilarial drugs, including the piperazines, has been successful. Treatment of infected animals is probably the best method of controlling dirofilariasis.

The main portion of this chapter will be concerned with the major human filariases, i.e., the systemic disease caused by *Wuchereria bancrofti* and *Brugia malayi* and the cutaneous and ocular disorders caused by *Onchocerca volvulus*. These provide a remarkable contrast in matters of treatment, prevention, and control. After a brief description of each of the diseases, this contrast will be considered, and it should become clear to the reader that even quite similar diseases may not yield to the same treatment and that research may be both rewarding and frustrating.

SYSTEMIC FILARIASIS: BANCROFTIAN AND MALAYAN

Wuchereriasis, or Bancroft's filariasis (as well as the similar disease caused by *Brugia malayi*) may be properly called systemic filariasis. It has several stages recognizable by their symptoms. An early acute stage often characterized by swelling and pain of certain lymph nodes may be followed by a symptomless period during which microfilariae abound in the blood. Then, sometimes years later and sometimes never, the gross symptoms of elephantiasis (enlargement by swelling of an affected limb, genital organ, etc.) appear, eventually resulting in severely crippling deformity. Incidence of infection with *Wuchereria bancrofti* or *Brugia malayi* is usually focal, i.e. restricted to (or highest in) particular villages or communities, where nearly all the inhabitants may have larvae in their blood. The geographic range of these parasites is large; *W. bancrofti* is found in Polynesia as well as most of the tropical world, and *B. malayi* occurs with *W. bancrofti* in much of Southeast Asia. Transmission of these parasites by mosquitoes was established very early (first by Manson, see above). (It is noteworthy that in most of the range, where the chief transmitters are nocturnally active mosquitoes, the parasites appear in the circulating blood only at night; in the islands of Polynesia, however, both the transmitting mosquitoes and the parasites may make contact at any time. This periodicity and the lack of it will be discussed under the more detailed treatment of wuchereriasis.)

WUCHERERIASIS, LIFE CYCLE OF *W. BANCROFTI*

The life cycle of *Wuchereria bancrofti* (Fig. 14–2) is a two-host cycle, involving mosquito and man. The mosquitoes may be any of several species of *Culex, Aedes, Anopheles,* or other genera. These ingest microfilariae which occur in the blood.

FIG. 14–2. Systemic filariasis. A victim of elephantiasis is surrounded by various elements in the life cycle of *Wuchereria bancrofti* or *Brugia malayi*, the similar worms which cause this condition. (A1) Female and (A2) male adults, life size. (B1) Head of female and (B2) tail of male of *W. bancrofti*, showing, respectively, circumoral papillae and spicules with perianal papillae (C) Microfilaria (sheathed larva) with human erythrocytes. (D) Mosquito, with ingested microfilariae entering esophagus, reaching stomach, and penetrating into hemocoele, where short "sausage" forms (first- and second-stage larvae) are seen. Finally an infective (third-stage) larva is seen in the region of the mosquito's mouth parts. (E) The microfilaria and three larval stages, shown at the same magnification, illustrate metamorphosis and growth in the insect vector.

The periodic appearance of microfilariae in the blood of the surface capillaries during the hours from 10 P.M. to 4 A.M. each night, a periodicity which corresponds well with the biting activity of the vector, is probably due to the aggregation of the larvae in the capillaries of the lungs during times of activity of the host followed by the release of the worms to the general circulation during the host's period of sleep. This explanation is true for the similarly periodic microfilariae of dogs, the larvae of *Dirofilaria immitis*. Since that variety of *W. bancrofti* found in Samoa, Tahiti, etc., as well as in the Philippines (where its range overlaps that of the more common, periodic strain), does not show any larval periodicity, it must be assumed that the periodic strain differs genetically from the nonperiodic and that the parasite's periodicity is therefore an evolutionary adaptation to some yet undefined environmental stimulus. Once ingested by a mosquito, the microfilariae penetrate the stomach lining and enter the host's thoracic muscles, where they change form rather radically, first to a shortened, sausage-shaped stage, and then to a rather robust infective larva about 1 to 2 mm by 0.02 to 0.03 mm. The above changes take from 8–10 days to more than two weeks, depending on temperature and relative humidity (80°F and 90% relative humidity are ideal for the parasite). If 100 or more microfilariae are ingested with a drop of blood, the mosquito dies of heavy parasitism; if fewer than 15 larvae per drop are ingested, usually no larva completes its development. In a successfully infected mosquito, one or more larvae will have migrated from the thoracic muscles to the mosquito's proboscis, where the larvae are ready to escape. The stimulus to escape comes from the warm, moist skin of a person being bitten by the mosquito; the infective larvae break through the thin cuticle covering the labial-labellar joint and reach the skin of the human host, entering through breaks in the skin such as mosquito bites. Then until sexual maturity and mating occur and microfilariae appear in the blood, little is known of the whereabouts, rate of growth, or behavior of the adult worms. In some individual hosts over a year has elapsed between the last possibility of infection and the appearance of microfilariae. Shorter rates of maturing have been observed. It is believed that the adult worms may live five years or more. These worms inhabit the lymphatic ducts and glands, usually coiled and entangled with each other in these structures. Most workers believe that the adult worms give rise to both the early (lymphoid) and late (elephantiasis) symptoms of wuchereriasis.

SYMPTOMS AND PATHOGENESIS

The progress of the infection is quite unpredictable. Some people with microfilarious blood show no symptoms; others suffer from mumu and from elephantiasis. Mumu is a local name for the painful swelling of the lymphatics. Although it is not by any means universal, it is quite common among the victims of wuchereriasis in its early stages. It is probably a complex of local allergic responses to the presence of foreign (worm) antigens in the lymph nodes. It has been suggested that these symptoms resemble those caused by various nonparasitic factors, such as mechanical injuries or the invasion of the system by various microorganisms. In this view, mumu may be only one of a number of possible conditions predisposing to elephantiasis, which occurs with some frequency outside the filarious tropics. Elephantiasis is directly caused by blockage of lymph channels from any distal portion of the body. Most workers believe that adult *Wuchereria*, perhaps upon dying, so irritate the lymphatic tissues among which they lie that these tissues become inflamed, fibrotic, and, in effect, nonfunctional. Lymph accumulates in the blocked-off limb, forcing the connective tissues to stretch. The added tension stimulates proliferation of connective tissue fibers and cells, and the result is a steady, often extreme, hypertrophy of the organ involved. This reaction is an example of "positive feedback" in terms of tissue stimulus and response; the pressure of undrained lymph stimulates expansion and fibrous repair of the tissues, while expansion of the tissues permits more lymph to accumulate, and so on.

TREATMENT

Treatments of the two phases of wuchereriasis are quite different. For ordinary infection, revealed by microfilariae in blood samples, Hetrazan (see above) may be taken orally. Such treatment kills microfilariae almost at once and probably sterilizes the adult worms. There is some allergic shock as a consequence of the death of many microfilariae with the liberation of their proteins all at once; some patients experience very severe symptoms upon being treated. As will be seen, the related disease, onchocerciasis, cannot be safely treated with Hetrazan, because nearly every person treated experiences such severe allergic reaction to the death of larvae. Elephantiasis, whether caused by worms or not, is a condition which has progressed too far for any anthelmintic (drug which kills worms) to be useful. Surgery must be employed to remove excess tissue and create new lymph channels. Pressure bandages with accessory administration of cortisone have resulted in marked reduction of the moderately swollen limb. Neither drug treatment nor surgery, however, can be always relied upon to prevent or to cure elephantiasis. Therefore, problems of control and prevention of the infection itself are of great importance.

CONTROL

Until the development of diethylcarbamazine (Hetrazan) for the treatment of wuchereriasis, no method of breaking the transmission phase of the life cycle was possible on a large scale. Mosquito control, although made feasible by the discovery of DDT and other modern insecticides, offered only moderate hope of reducing the incidence of filariasis, because unless all persons with microfilariae in their blood could be somehow cured while the mosquito population was small, an increase of mosquitoes after control would simply reestablish filariasis as a spreading disease. Malariologists had discovered that because of the insecticide resistance often acquired by mosquitoes during mosquito-control programs, long-term suppression of mosquitoes would be difficult and expensive, and might not succeed.

Therefore the mosquito-control experiments begun in the mid 1940's on the island of St. Croix in the West Indies were shifted in 1948 to a program of mass treatment with Hetrazan. The results were a great reduction in cases of filariasis, complete cessation of transmission, and the probability of the eradication of wuchereriasis from that island within the near future. As is true in all diseases caused by parasites, however, local conditions are very important. Combinations of mosquito control with mass treatment will probably work well in most areas where *Wuchereria* or *Brugia* prevails, but the details of insecticide use, drug administration, and the management and conduct of the control program will depend on the kinds of vectors present, the actual incidence of disease, and the economic and technical resources which are available.

ONCHOCERCIASIS

Onchocerciasis, the other major filariasis, instead of affecting the lymphatic tissues and blood (as in wuchereriasis), involves the skin. The worms occur in dermal nodules, where their microfilariae are deposited and from which the latter migrate through the dermal tissues. Sensitization of the skin (an allergic response) results in thickening of the skin, with itching, eczematous symptoms; in some parts of the world (Central America) invasion of the eye tissues causes blindness in 5% of those infected. There is evidence that other factors—allergy, avitaminosis—may contribute to this effect. Onchocerciasis affects people differently in the two regions where it occurs; in Western Africa it causes nodules of the skin of the trunk, with accompanying symptoms, while in Central America nodules occur chiefly on the head and neck, with a much higher rate of ocular involvement than in Africa. Incidence of onchocerciasis is very high in some regions and has reached 100%. Like wuchereriasis, it has focal distribution and varying infection rates. Although in wuchereriasis the reason for such variation is obscure, it is probably related rather closely, in onchocerciasis, to ecological factors of transmission by the vectors (*Simulium* spp.). These factors will be considered below.

THE PARASITE

Onchocerca volvulus, being much more accessible as an adult than the deep tissue inhabiting *Wuchereria* and *Brugia* species, will be described here as an example of a filarioid worm, typical of the nematode superfamily Filarioidea.

The adults are very long and slender, resembling pieces of thread. Females may be 500 mm (two feet) long, while only about half a millimeter thick. Male worms are only one tenth as long. These worms, like all nematodes, have a flexible but tough cuticle which acts as an exoskeleton to which characteristic longitudinal (but not circular) muscles are attached. There is a fluid-filled cavity between the musculocuticular body wall and the simple intestine. Within the body cavity are the genital systems, long convoluted tubes modified at their openings. At this point the tubes are attached to the body wall and in their innermost portion they

FIG. 14–3. *Onchocerca volvulus* and its pathogenesis. (A1) Adult female and (A2) male of *O. volvulus*, 1.3× natural size. (A3) Head of female, showing circumoral papillae, and (A4) tail of male, showing cuticular ridges characteristic of adults of both sexes and spicules and papillae diagnostic of the species. (B) Tissue from a dermal nodule, showing two sections of a female worm: (B1) the region where the uteri contain developing microfilariae in delicate membranes, and (B2) the region of the esophagus (B3) showing the dorsal and ventral nerve cords and the relatively few muscle cells of the body wall; a microfilaria (B4) in the connective tissue of the nodule is shown greatly magnified (B5). (C1) Microfilaria freed from host tissue, and (C2) its head, showing characteristic minute striations. [(A4) After Chandler and Reed, 1961.]

consist of cords of cells which continually multiply and differentiate into sex cells, the eggs and the sperms, respectively. Filarial worms, like practically all nematodes, have separate sexes. In female worms the genital systems open on the body surface by a vulva and vagina in the anterior region of the worm. The embryos with their larval sheath develop in the uterine (middle) portion of the double female duct. The male system opens into the anal region; at this place there are paired, pointed half-cylinders, called "copulatory spicules," which are protruded and inserted into the vagina of the female during copulation; the sperm travel through the lumen to fertilize the female. The sperm complete their development and are stored in the distal portion of the male system, the sperm duct, which gradually narrows in its proximal regions into the testis, where sperm-producing cells continually proliferate. Figure 14–3 shows some of the details of the anatomy and life cycle of *Onchocerca volvulus*.

LIFE CYCLE

The embryos leave the uterus of their mother by way of the vulva. They wander in the skin, occupying both the dermal layers and the deeper layers of the epidermis. From these tissues the larvae may be ingested by the intermediate host and vector, the black flies (*Simulium* spp., see Chapter 24). The latter acquire microfilariae by ingesting blood and tissue fluids from the small lesions they scrape in the skin. Within the flies the microfilariae become infective in 7 days; the flies can ingest more than 100 larvae, but if more than three or four invade the fly's thoracic muscles, the fly dies. Thus each infective fly can transmit no more than two or three worms. Black flies have rather unusual breeding habits (Fig. 14–4); they lay their eggs on wet leaves, stones or debris at the edge of flowing streams. Living in such streams, attached to rocks, etc., by silk threads, the larvae are protected by open-ended cocoons and catch microscopic organic materials by means of

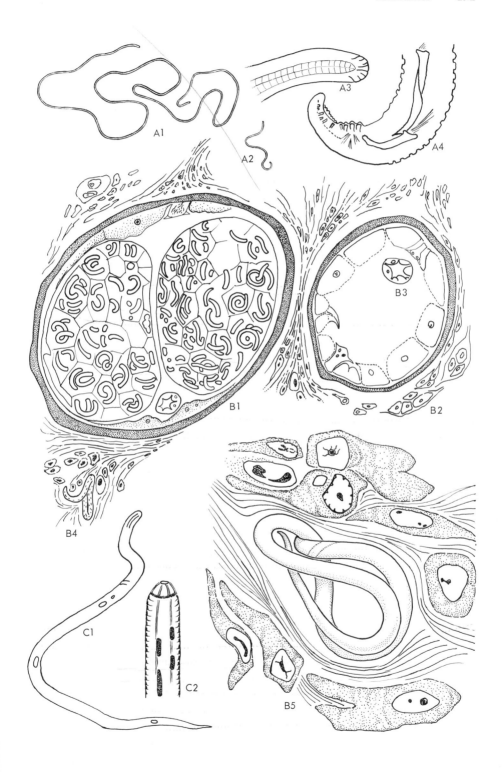

FIG. 14–4. The biology of black flies in Central America. (A) Map of Guatemala and neighboring regions. The mountainous areas in the south are coffee-growing regions where *Simulium*, the black fly, and *Onchocerca* abound. (B) Black flies lay their eggs in streams, where the larvae live attached to rocks or vegetation and form pupae which are protected by cone-shaped membranes. (C) The flight range of black flies from a breeding place is up to 15 miles, as shown by the concentric circles. (D) The flies transmit *Onchocerca volvulus* by biting, in bright light, the exposed parts of the human body. Coffee pickers are exposed in their work.

buccal fans, brushlike appendages which sift food from the water and propel it into the mouth. The adults can fly considerable distances (up to 10 or 12 miles); they are diurnal feeders, requiring fairly bright light. Black flies in tropical regions need no more than four weeks for a complete generation. These facts, it will be pointed out, have a considerable bearing on attempts to control onchocerciasis.

When a black fly with infective *Onchocerca* larvae bites man, the larvae penetrate the lesions caused by the black fly bite, migrate, and grow under the skin until trapped by fibrous tissue, probably locally inflamed. At the point where migration ceases, a nodule forms; it consists of the entrapped worms and dense fibrous connective tissue. Such cysts range in size from about 5 mm in diameter up to several centimeters, and may be observed as definite lumps on scalp, neck, or body. Within the cysts mature male and female worms copulate, and the fertile females then deposit unsheathed microfilariae which escape into surrounding tissues. There is constant migration of these larvae away from the nodules until most of the skin may eventually be invaded. The possibility of transmission begins when larvae first escape from a nodule to the skin. This can occur about two or three months after the victim has been bitten by infected black flies. The adult worms are known to be able to live for at least seven years, during which they continually supply microfilariae to the skin.

SYMPTOMS AND PATHOLOGY

The progress of the disease depends upon the above microfilarial invasion following the establishment of the nodules or cysts (Fig. 14–3B).

The location of the latter is different in the two geographical areas (Central America and Equatorial Africa) where *Onchocerca volvulus* occurs. In America the nodules are chiefly about the head and neck, and number from one to several. Each contains two to four worms. There is probably a relationship between eye involvement, which is quite common in American onchocerciasis, and the above location of nodules. In Africa eye damage is not common, and rashes and thickening of the skin occur upon the trunk, where most of the nodules are found. The

African nodules may be very numerous, and as many as 100 worms may occur in the larger cysts. Lymphatic involvement occurs in African onchocerciasis, but not in American; the difference is considered to be due to the failure of some African worms to form dermal cysts, invading instead, deeper tissues including the lymph nodes.

The microfilariae cause allergic responses as they invade the skin and other tissues. The skin becomes thickened and scaly, and a rash may occur, characteristically localized, as mentioned above. The microfilariae, especially in America, invade the tissues of the eye and cause various degrees of interference with vision. Mild to extreme photosensitivity is noted in up to 25% of the infected persons in parts of Guatemala, for instance. As more and more microfilariae invade the eyes, the cornea, conjunctiva, perhaps the retina, and even the ocular nerve may be damaged or destroyed, causing blindness. The above pathological changes require time; usually about five years of infection must precede serious ocular damage.

TREATMENT

Onchocerciasis is treated by chemotherapy plus surgery. Hetrazan destroys microfilariae in the tissue and stops transmission. Unfortunately, Hetrazan therapy may be quite harmful, since the dead microfilariae often cause severe allergic or anaphylactic response (see Chapter 27). The sudden release of large amounts of worm antigen in the skin, a notable site of antigen-antibody response, and a tissue already sensitized, may cause fever, inflammation, itching, irritation of the eyes, and other unpleasant effects. The symptoms brought on by treatment may constitute a dangerous disease in themselves. Therefore, treatment with drugs must be done carefully, under medical supervision, and is not the relatively simple procedure possible in the related Bancroftian and Malayan filariases. Surgical removal of nodules is easy, since the nodules are visible and accessible. Following the removal of nodules, microfilariae gradually disappear from the skin. A combination of chemotherapy (oral doses of Hetrazan or injections of the aromatic diamidine, Suramin) to destroy the microfilariae with surgery to remove the sources seems practical where clinical facilities are available.

CONTROL

The control of onchocerciasis, although more difficult than that of wuchereriasis and brugiasis, has been successfully carried out in several areas. In parts of Central America a sort of mass treatment by the above-mentioned surgical and chemical techniques has reduced incidence by stopping transmission at the source —the human skin. This method would not work so well in Africa, because there the dermal cysts may be very numerous, and, as mentioned, it is likely that not all adult worms may be found at the surface of the body. Also clinical treatment stations and trained personnel are scarcer in Africa than in America, and the disease itself is far more widely distributed. Large-scale treatment of black fly breeding areas with DDT is very effective. The black fly larvae are highly susceptible to DDT and can be killed by concentrations in water as low as 0.01 part per million. (Their buccal fans probably remove DDT adsorbed to tiny particles and in this way selectively poison the larvae.) It is thus possible to control black flies over large areas by aerial spraying (done in the Congo) or in small areas by applying DDT in controlled amounts to small flowing streams. Altering streams to make them unsuitable as breeding places is also sometimes possible. Eradication of black flies is more difficult than control because of the flight range of 10 to 12 miles. But control may be an effective way to reduce onchocerciasis drastically, for the infection rate in black flies in endemic areas is quite low (about 1%), and any substantial reduction of total fly population should bring the transmission rate to the vanishing point. In spite of promising attempts at control, and the favorable results of systematic surgery upon the nodules of whole populations, the problem of onchocerciasis is far from being solved.

CONCLUSION

Thus it seems that while systemic filariases are susceptible to effective control (perhaps eradication through the drug Hetrazan), nevertheless the other major filariasis, onchocerciasis, is not susceptible to the same methods. The difference between the two kinds of disease is the location of the microfilariae. It is true that killing the microfilariae in the blood causes harmful symptoms, but these effects are not severe enough to obstruct a program of mass treatment and control. Having larvae in the skin is a worse problem, because presumably their death there triggers unusually violent immune reactions. Drugs which are not carefully administered may be fatal; the risks and symptoms of cure, therefore, interfere with mass chemotherapy. There is need for research in both diseases; but the one, systemic filariasis, needs chiefly operational study, while the other, onchocerciasis, requires much fundamental research in biology, ecology, and the medical aspects.

What has just been said about the two major filariases applies to the several other human diseases caused by filariae as well as to the veterinary filariases. All of these are chronic infections. All permit innumerable opportunities for transmission by some biting arthropod. Control, which is the objective of public health planning (as well as animal health planning), can be achieved theoretically by any of several approaches, each aimed at breaking the cycle of transmission. First, the worms in the final host may be killed pharmaceutically. Either the microfilariae may be killed (a temporary victory but an important one in stopping transmission) or the adults themselves may be destroyed. Second, the susceptible host may be protected by insect repellants from contact with the vector. This method is not considered practical as a control measure but may protect individuals, such as tourists, military personnel, or particularly valuable livestock. Third, the vector itself can be attacked. Eradication of vectors is probably not possible, but reduction in numbers is. Probably there is a lower limit of population size—MacDonald's critical density (mentioned in our discussion of malaria)—for every species of parasite, including the filarial worms. Below this level there are not enough infected hosts to provide for transmission. The hosts in question may be "final" hosts, or "intermediate" hosts; actually, enough of both must be infected for transmission rates to keep up with the natural disappearance of the parasites by death of the infected intermediate hosts. Vector control, except in systemic filariasis, is more practical in bringing the parasite level to its critical density than control through treatment. Consistent and continued reduction of vector populations should cause the parasites inevitably to decline and vanish. As further knowledge of the ecology of filariasis is gained, attacks on the vectors will prove more and more practical. Fourth, a combination of attacks should be tried. The principle of combined effort, which in many programs has worked out so well for the eradication of malaria, should be applied to filariasis. The Central American experience is an example of the way in which a combined attack can succeed. In this area the incidence of onchocerciasis was greatly reduced by denodularization of the infected people, careful therapy with Hetrazan, and control of black flies by treatment of breeding places. Finally, control measures will depend on the particular disease to be dealt with, its geography, its economics, its victims, and the resources of knowledge, money, and personnel. Among these resources, knowledge is perhaps the most important. Therefore research in all aspects of filariasis should continue in order that facts may be discovered and made available for the successful control of these diseases.

SUGGESTED READINGS

BEYE, H. K., and J. GURIAN, "The Epidemiology and Control of Filariasis (*Wuchereria bancrofti* and *Brugia malayi*)," Expert Committee on Filariasis, WHO, Geneva, 1961.

BUDDEN, F. H., "The Incidence of Microfilariae in the Eye and of Ocular Lesions in Relation to the Age and Sex of Persons Living in Communities where Onchocerciasis is Endemic," *Trans. R. Soc. Trop. Med. Hyg.*, 57:71–75, 1963.

BURCH, T. A., "The Ecology of Onchocerciasis," *Studies in Disease Ecology*, J. M. May (Ed.). Hafner Publishing Co., New York, 1961.

DAVIES, J. B., R. W. CROSSKEY, M. R. L. JOHNSTON, and M. E. CROSSKEY, "The Control of *Simulium damnosum* at Abuja, Northern Nigeria, 1955–1960," *Bull. WHO*, 27:491–510, 1962.

DUKE, B. O. L., and D. J. B. WIJERS, "Studies on Loaiasis in Monkeys. I. The Relationship between Human and Simian *Loa* in the Rainforest Zone of the British Cameroons," *Ann. Trop. Med. and Parasitol.*, 52:158–195, 1958.

EXPERT COMMITTEE ON FILARIASIS, "*Wuchereria* and *Brugia* Infections," Tech. Rept. Ser. No. 233, WHO, Geneva, 1962.

FAUST, E. C., "Human Infection with Filariae of Nonhuman Hosts," *Bull. WHO*, 27:642–643, 1962.

HAWKING, F., "A Review of Progress in the Chemotherapy and Control of Filariasis since 1955," *Bull. WHO*, 27:551–568, 1962.

JORDAN, P., "A Pilot Scheme to Eradicate Bancroftian Filariasis with Diethylcarbamazine," *Trans. R. Soc. Trop. Med. and Hyg.*, 53:54–60, 1959.

MARCH, H. N., J. LAIGRET, J. F. KESSEL, and B. BAMBRIDGE, "Reduction in the Prevalence of Clinical Filariasis in Tahiti Following Adoption of a Control Program," *Am. J. Trop. Med. Hyg.*, 9: 180–184, 1960.

RAGHAVAN, N. G. S., "Epidemiology of Filariasis in India," *Bull. WHO,* 16:553–579, 1957.

RODGER, F. C., "A Review of Recent Advances in Scientific Knowledge of the Symptomatology, Pathology and Pathogenesis of Onchocercal Infections," *Bull. WHO*, 27:429–448, 1962.

TRENT, S., "Reevaluation of World War II Veterans with Filariasis Acquired in the South Pacific," *Am. J. Trop. Med. Hyg.*, 12:877–887, 1963.

CHAPTER 15

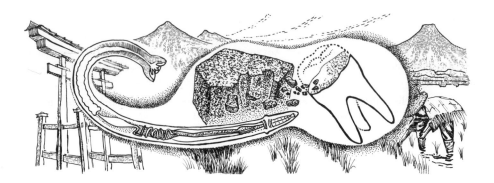

HOOKWORM

INTRODUCTION

Nearly all the parasites of man are probably ancient; the few exceptions, such as *Leishmania tropica* in South America, merely emphasize the fact that the human species has inherited its parasites rather than acquired them. This view of human parasites is part of the general concept of parasitism discussed in Chapters 26 and 29, the view that parasitism is evolutionary and that hosts are isolating factors permitting the independent occurrence of hereditary changes and the accumulation of such changes until the isolated populations become new races and then new species. But the evolutionary successes and failures of the parasites of man have been influenced greatly by a peculiarly human factor, i.e. culture and cultural change. The hookworm is unusually sensitive to man-made, cultural factors. Both the prehistory and the history of hookworm in man show how human parasites are affected by human control of the environment.

HISTORY OF HOOKWORMS

The presence of hookworm in stone-age man may be inferred. Other primates have hookworms rather closely related to those of man, and we may be certain that these simian and anthropoid species of hookworm are descended from ancestral species which once infected the ancestral apes and monkeys. However, the hunting and food-gathering cultures of early man must have prevented heavy parasitism by hookworms, since this parasite depends for its transmission upon frequently polluted soil, which is in contact with human skin. The changing of campsites with the exhaustion of game and food from particular areas, as well as migrations which people of stone-age culture undergo for seasonal water and other necessities, would make a hookworm's survival precarious. Levels of parasitism with hookworm should be quite low in such cultures. However, with the discovery of agriculture and the immediate consequences—a settled village economy, rapid growth of population, and concentration of human excrement in populated areas—hookworms as well as other parasites must have increased greatly in incidence, becoming for the first time a health problem. Among the many "diseases of civilization" which we recognize today, hookworm is surely one of the oldest.

Only in the twentieth century, however did hookworm receive the attention it merited. Recognized by Dubini in the midnineteenth century as the cause of anemia in Italian mine-workers, and studied at the beginning of the twentieth century by Looss, who worked out the basic life cycle of hookworm, the parasite first received public attention because of the efforts of Stiles, an employee of the United States Public Health Service. In 1902 his survey indicated that throughout the southeastern United States many individuals had hookworm. In an official report Stiles stated, "Indications are not entirely lacking that much of the trouble popularly attributed to 'dirt eating,' 'resin chewing,' and even some of the proverbial laziness of the poorer classes of the white population are in reality various manifestations of uncinariasis (hookworm disease)." This cautious statement was interpreted by American journalists to mean that "the germ of laziness" had been discovered and that rather sensational news had a major effect on the public. Hookworm disease became rather well known rather suddenly.

John D. Rockefeller contributed a million dollars toward the eradication of hookworm, and the Rockefeller Sanitary Commission was established for that purpose in 1909. The Commission became a coordinating agency for federal, state, and local public health offices. (The United States as late as 1910 consisted of many semiautonomous political units, i.e., the states, whose sovereignty was weakened by the American Civil War fifty years earlier yet was still strong enough to be a source of jealousy and occasional obstruction when interstate cooperation was needed. The hookworm campaign of the Rockefeller Commission was marked by effective cooperation among the states. Such cooperation was an important step in strengthening the acceptance of the United States as one nation by citizens who still remembered the bitter and bloody war of the 1860's.) In the control measures that followed, public education in the cause and prevention of hookworm disease brought acceptance of such drastic procedures as mass treatment (see below) and improvement in sanitation. In the first year of the project, Stiles gave 246 public lectures on hookworm. At the end of a five-year period, the campaign had reduced the incidence of hookworm very substantially. Both the experts and the public were aware of this success.

FIG. 15–1. World distribution of hookworms. [After Belding, 1942.]

The indirect benefits of the antihookworm campaign in America were much more important than the immediate effects. Appropriation of funds for public health agencies increased by 80% during the five-year period of the campaign against hookworm, as the publicity associated with that effort aroused the interest of the public in other health problems. (Similarly in recent years the launching of the first satellites influenced support for education and research in sciences other than physics and astronomy.) The Rockefeller Foundation was endowed with $100,000,000 for its International Health Board, which has carried on through the years many important investigations and demonstrations throughout the world. An effective use of the Board's money has been to organize "model operations," such as the reduction of hookworm incidence in an Egyptian village from 42% to 12%, a demonstration or model of what the people themselves can do with the aid of their own public health officers. The multiplication of effort by the above principle is now an expected result of most "aid" programs. Even a great Foundation cannot undertake the conquest of a worldwide disease (see Fig. 15–1) on more than a demonstration and training level, and the governments of many countries have carried out, with their own funds and personnel, the programs that were stimulated if not initiated by an international aid organization. In the field of formal education, the success of the hookworm campaign, with the aid of Rockefeller funds, helped lead to the establishment of the School of Hygiene and Public Health of the Johns Hopkins University, a research and teaching institution which has trained many outstanding public health scientists. Without claiming that the hookworm campaign of 1910 was solely responsible for the above and other benefits, it is suggested that the influence of this campaign is evident in the following: (1) support for the U.S. Agricultural Extension Act of 1914 (an educational, nation-wide program of demonstration and assistance to farmers); (2) the concept of technical advice which was basic to Truman's "Point Four" program of assistance to war-ravaged and underdeveloped nations in the 1950's; (3) the basic principle upon which the malaria-eradication program of the United Nations' WHO (see Chapter

2) is founded; and (4) perhaps, the whole pattern of international cooperation emerging today, a pattern which is similar in structure to the interstate antihook-worm campaign which Stiles successfully directed in 1910.

In the light of the above story, we must conclude paradoxically that humanity owes a substantial part of its present progress to the existence of a dangerous para-site, the hookworm. That this paradox is implicit in the nature of culture is a thought which goes beyond parasitology and for the moment will not be pursued further.

HOOKWORM DISEASE

Many readers of this text are probably familiar with the hookworm's life cycle, its mode of transmission, the special conditions (sandy, moist, feces-contaminated soil, contact with human skin) which favor its establishment and spread, and the damage (continual hemorrhage from intestinal lesions) which hookworm causes. The details of the cycle are included in Fig. 15–2 to remind readers who may have forgotten facts learned in general biology courses. The rest of this chapter will be concerned with some special aspects of hookworm disease, including methods and importance of diagnosis, the significance of "worm burden," some principles of immunity and pathology, and an evaluation of the importance of hookworm disease.

DIAGNOSIS

Diagnosis of hookworm serves two useful purposes. First, it reveals the incidence of hookworm infection within a population, indicating whether or not control measures should be initiated. Second, it may reveal the extent of infection in an individual, i.e., the total number of worms inhabiting the patient's intestine. While the first revelation of diagnosis tells something of the risk of human infection, the level of sanitation, and the number of sources of reinfection, the second function of diagnosis is far more important than the first, for, as we shall see later, the harmful effect of hookworm is directly related to the number of worms per host, that is, to the "worm burden." Therefore methods of estimating the latter by fecal egg counts have been devised.

The first effective method of quantitative diagnosis was the Stoll Egg Count, proposed in 1923 and modified somewhat in later years. Stoll suggested that a fecal specimen could provide an estimate of the number of adult hookworms present if certain facts could be ascertained. These are (1) the number of eggs produced per day by a female worm and (2) the number of eggs found in the feces over a given period. The original method is as follows: Three grams of feces are thoroughly mixed, by shaking with glass beads, with enough 0.1 normal sodium hydroxide solution to make 45 cc of mixture. Then 0.15 cc of the suspension is placed on a fecal slide (a wide slide used for such examination), covered with a wide coverglass, and the total number of eggs is counted. This number $\times$ 100 is the number of eggs per gram and is accurate to within 10 to 20 percent. Then this

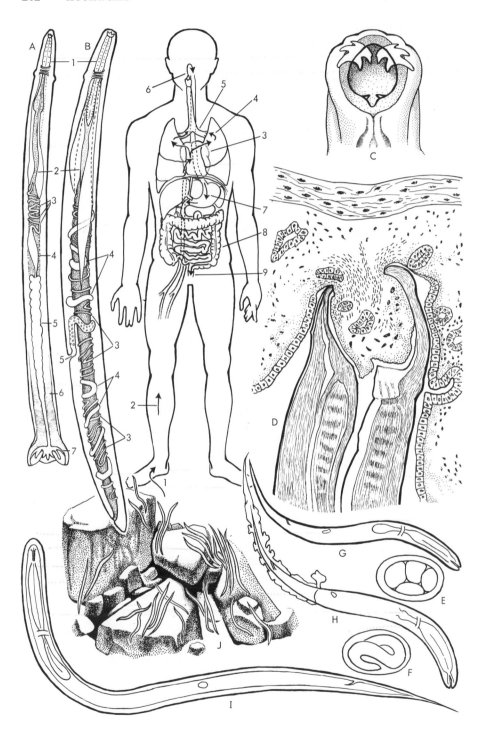

FIG. 15–2. The life cycle of hookworms. A human figure is surrounded by details of the life cycle. (A) Adult male worm shows the following structures: (1) esophagus, (2) salivary gland, (3) testis (convoluted), (4) seminal vesicle, (5) sperm duct and associated glands, (6) transverse muscles used in copulation, (7) copulatory bursa. (B) The female illustrates structures (1) and (2) above, plus (3) ovaries, (4) uterus, and (5) vulva. (C) The buccal capsule of a hookworm, *Ancylostoma duodenale*, is shown enlarged. Note teeth rooted in the walls of the capsule and projecting inward. (D) Section of mucosa of the intestine shows hookworm with part of a villus inside its buccal capsule. First- and second-stage larvae (G, H) are found in the soil, where the egg (E) laid in a four-celled stage, has incubated (F). The third-stage larvae (I), called filariform or strongyliform, awaits a victim among particles of loose moist soil (J). Pathway of the infective larva in the human host is labeled as follows: entry into skin (1) by circulation (2) to heart (3) and to lungs (4), where fourth-stage larva develops and reaches trachea (5) and pharynx (6), and is swallowed (7); in intestine (8) adults live and mate; to complete the cycle, eggs pass out with feces (9).

number is multiplied by the estimated number of grams of feces produced per day in order to get the total number of eggs produced. The latter number is then divided by the average daily egg production of one female worm (9000 for Necator and up to 30,000 for *Ancylostoma*, a much larger worm) to obtain the number of female worms present, and then the latter number is multiplied by 2 to get the total worms of both sexes.

Based on experience and experiment, a rough correlation can be made between the number of eggs per gram of feces and the number of worms. Thus 2000 eggs per gram of ordinary feces means about 60 adult *Ancylostoma*. If the feces are hard, the latter number is too large, and if the feces are diarrheic, the figure is too small, because of the correspondingly smaller and larger volume of feces respectively obtained in these situations. Correction factors based on fecal consistency are used in making estimates.

It should be obvious that the above method (or methods, for numerous variations have been tried) is far from precise. Even if the estimated number of eggs per gram of feces be accurate within 10 or 20 percent, the sample of feces chosen may not be representative of the whole fecal mass. Also, the egg output per worm per day may vary. Still another variable is the uncertainty as to which species of hookworms may be present; it is an uncertainty factor potentially as great as three, because *Ancylostoma* produces three times as many eggs per day as *Necator*. The two species cannot be separated on the basis of eggs in ordinary egg-counting preparations, and the eggs must be presumed to be of the species commonest in the survey area. In spite of the above inherent inaccuracies in egg counting as a basis for estimating worm burdens, these techniques tell whether a surveyed population has, on the average, light infections (about 100–300 worms per host), clinically significant infections (500–1000 worms), or severe infections (over 1000 worms). Such information is essential to public health decisions.

The relationship between worm burden and symptoms is not very simple. It is said that, "other things being equal," the severity of hookworm disease—anemia, stunting of growth, neurological and other symptoms—is directly proportional to the number of worms in the patient. Each worm draws its not negligible share of blood, the amount being about 1 cc per worm per day; this figure is based on experiments with dogs infected with *Ancylostoma caninum*. Thus a worm burden of 500–1000 worms may mean a very severe daily loss of blood from a pint to a quart. Other things being equal, each worm burden should cause a proportionate degree of anemia. But "other things" are seldom equal. Some individuals, indeed some populations, live on the brink of starvation. The loss of iron and protein from such individuals is very serious. Even a few hookworms—perhaps less than the 50 which are considered by most workers to be a symptomless burden—may seriously weaken such poorly nourished individuals. Other individuals and populations are very well nourished, with adequate mineral and protein diets, and such fortunate people can tolerate rather high levels (perhaps 500 or more worms) without obvious symptoms. While the above facts seem to explain rather well the correlation and occasional lack of correlation between worm burden and symptoms, by suggesting that diet and hookworm disease are intimately related, the full meaning of this relationship can be seen only in the light of another factor, i.e. immunity.

IMMUNITY

Immunity to hookworm is caused by the presence of worms in the tissues and, presumably, in the intestinal tract as well. It manifests itself as a protective reaction, resulting in the establishment of fewer worms than would normally be acquired, and in a marked inhibition of egg-laying by established worms. Like most immunities to parasites (see Malaria, Chapter 2, as well as Chapter 27), hookworm immunity is of the transient type called premunition. However, it offers considerable protection against heavy infection and is therefore an important feature in hookworm disease.

The principles of immunity will be discussed collectively in a later chapter. Here they may be sketched very breifly, as they relate to the problem of the relationship between host diet and establishment and effect of hookworms. Proteins and other substances of foreign origin, when present in the tissues of the body, stimulate certain cells to produce antibodies capable of reacting with the foreign protein. This antigen-antibody reaction is often useful; that is, the combination of foreign substance (antigen) with antibody may result in destruction or other change in the antigen, rendering it less harmful. In hookworm disease, the nature of the antigen (salivary secretion of the hookworms?) is unknown, but it stimulates the production of effective antibody. Hookworm numbers and activity are consequently reduced. The production of antibody, however, requires a precursor substance

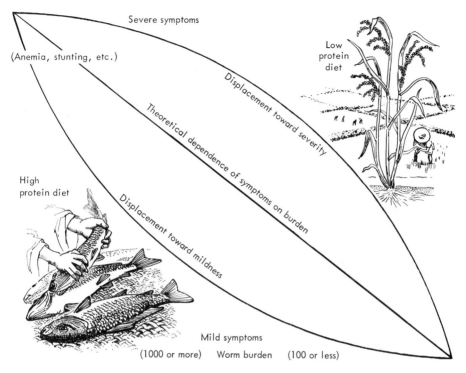

FIG. 15–3. The effects of diet and worm burden in hookworm disease. Assuming some direct relationship between the number of worms present and the severity of anemia and other symptoms, this relationship changes with diet. A high-protein diet, represented by fish, helps restore protein deficiency caused by hookworms, and permits a larger worm burden to exist with relatively mild symptoms. A protein-deficient diet, such as polished rice, makes worm-induced protein deficiency critical and aggravates the symptoms caused by only a few worms. Thus the effect of worms can be shifted away from or toward severe symptoms by the adequacy or inadequacy of diet.

found in the blood, a protein called gamma globulin. This is one of the serum proteins. When it is plentiful, rates of antibody production are high and protection (high level of immunity) occurs. When gamma globulin levels are low, the cells are unable to manufacture antibodies, and immunity cannot protect against new infections or effectively resist current ones. Thus the mechanism of immunity is directly linked with the blood proteins, especially gamma globulin.

From the above it should be clear that loss of blood during an infection with hookworm is more than a loss of nutritional iron and protein, as serious as this loss may be. Also lost is much of the chief defender of the body, gamma globulin, without which antibodies cannot be made. Thus hookworm anemia reduces resistance to entry of new hookworms, which cause further anemia, with further breakdown of immune resistance. The hookworm starts a vicious circle or a self-accelerating chain of reactions which leads to the most serious consequences. Good diet, with enough protein to restore the lost gamma globulin, can break the vicious circle, provided the worm burden is not too high. Figure 15–3 illustrates the interrelationship of diet, worm burden, and symptoms. The influence of immunity may be considered to be exaggerating, or further bending, the curve of response in that diagram.

Before leaving the subject of immunity to hookworm, it is worth mentioning that hereditary or racial differences probably play a part. Members of the Negro race, while capable of acquiring and supporting hookworm infections, generally have light worm burdens and very few symptoms. Since this difference in susceptibility persists in Negroes who have changed their home environment from Africa to parts of the United States, the difference must be genetic. Probably hookworm is a parasite native to Africa, as is the *falciparum* malaria organism. There has been time enough in that geographical area for the hookworm to become adjusted to its human hosts and for the latter to become relatively resistant to hookworm.

SYMPTOMS

The well-known symptoms of hookworm disease are probably all attributable to loss of blood (except, of course, the rash at points of entry into the skin, and involvement of the lungs during larval development and migration). Thus delayed puberty, mental retardation, and physical stunting are the effects of physiological starvation of the growing child, i.e., starvation for the oxygen-carrying cells and the protein-carrying serum of lost blood. Apathy, sallowness, and emaciation are symptoms of hookworm among the older. Women in pregnancy, with increased demands for blood minerals and nutrients, are particularly harmed by hookworms. They may experience difficulty during labor, and their babies are frequently still-born. Susceptibility to infection with bacteria and other pathogens is increased by hookworm through the removal of the serum gamma globulin.

Some unusual, indeed bizarre, symptoms occur in hookworm disease. One of these is a "reverse iris-dilation reflex," in which the pupil of the eye, instead of contracting in the presence of strong light, dilates. No complete explanation of this peculiar phenomenon exists, but a reasonable theory is that failure of the iris to receive enough oxygen damages this structure's ability to respond to light. Another strange symptom is geophagia, or "dirt-eating." In hookworm patients (and in some other persons, also) a craving for earthy, gritty substances may develop. Usually the craving is indulged secretly. Clay, earth, and hard substances like bricks or mortar are usually preferred. Needless to say, the habit is harmful, particularly to the teeth, which may literally be ground down. Some geophagic people cook earth, presumably to dry it to the proper consistency, and store a supply in some convenient place from which a spoonful or two may be taken whenever desired. Possibly geophagia may be an "instinctive" attempt to replace minerals, especially iron, lost by hemorrhage. It is not known whether dirt-eaters have a particular craving for iron-bearing soil.

HOOKWORMS OF VETERINARY IMPORTANCE

As in human hookworm disease, hookworms of domestic animals cause effects proportionate to their numbers and to the condition of the host. Hookworms of the genera *Uncinaria* and *Ancylostoma* parasitize many species of carnivore, including foxes, dogs, and cats. Transmission is by infective larvae which actively penetrate the skin, reach the intestine via the lungs, and are swallowed after undergoing some development. Infection probably occurs also by ingestion of larvae with food or water. Since the larvae require moisture and hatch in feces, from which they migrate, improvement of rearing conditions to eliminate excess moisture and frequent removal of feces should help control hookworm disease. The worms can be eliminated from the host by various anthelmintics, including tetrachlorethylene and carbon tetrachloride.

Hookworms affecting sheep and goats belong to the species *Bunostomum trigonocephalum*. Young sheep are particularly affected. Infection rates may reach 100%, and the worm burden may be 5–6 thousand worms. Heavily infected animals lose weight, develop severe anemia, and show retardation in growth. Death may occur. Treatment by phenothiazine is effective. Seasonal changes in infection rates, noted in the USSR and elsewhere, are due to changes in moisture content of pasture soils, as well as changes in temperature. Mild summer temperatures and heavy rainfall increase the risk of heavy infection. Since the larvae of hookworms are much less resistant to environmental factors than are strongyle larvae (see Chapter 20), hookworm control by rotation of pastures is relatively effective.

Sheep and goats in India and Africa may be infected with an extremely virulent hookworm, *Gaigeria pachyscelis*. This worm can kill its host if only a few worms are present (15–20). In such infections, death comes rapidly from loss of blood.

The zebu of India and Southeast Asia, as well as other cattle throughout the world, suffer from *Bunostomum phlebotomum*. Stabled cattle suffer from the itching caused by penetrating larvae and from loss of blood and other effects of adult worms in the intestine. Another hookworm of cattle, apparently restricted to the zebu and ox in India and Sumatra, is *Agriostomum vrybergi*. Little is known about this worm.

Swine are affected by hookworms of the genus *Globocephalus,* which occurs throughout the world. It is common, however, only in parts of tropical Asia and the Pacific. Little is known of its importance, but any hookworm can be presumed to be potentially dangerous.

A troublesome medical aspect of some veterinary hookworms is "cutaneous larva migrans", a creeping eruption caused by the larva of various hookworms of animals which cannot complete their development in man but can enter human skin. *Ancylostoma braziliense* and *A. caninum,* common hookworms of dogs, often cause creeping eruption. Sandy beaches and other moist, loose soils where dogs defecate and people congregate for work or play are sources of infective larvae. This condition may be compared with the much more serious but less easily diagnosed "visceral larva migrans," caused by the ascarids of dogs and cats (Chapter 16), or with "swimmers' itch" caused by cercarial migration of nonhuman schistosomes (Chapter 9).

IMPORTANCE

A disease like hookworm, which affects the body so profoundly, is closely related to dietary insufficiency as a cause of stunted growth and damaged intelligence, and flourishes where insanitation and poverty prevail, is bound to affect vast numbers of people and to be considered an important disease. Yet attempts to evaluate its actual importance have not been very successful for the following reasons. First, the mere presence of hookworm does not imply the existence of disease. When it was discovered that 80% of the people of Bengal (at the time a region with almost 50,000,000 people) harbored hookworm, the problem seemed insurmountable; but egg counts then revealed that less than 17 percent of these people had more than 200 worms, that the average number per person was only 20, and that worm burdens of 400 or more were almost entirely absent from the population. There was no problem, because there was almost no disease. By contrast, many of the families in southeastern United States during the early stages (1902–10) of the antihookworm campaign were found to be heavily infested (a thousand or more worms per person); this was a real problem, and it was solved by well-justified expenditures of time and money. Conditions of soil, personal habits, poverty, and race combined to make certain groups in the U.S. highly susceptible to heavy infections. Thus while the pattern of distribution of hookworm follows a broad band around the globe from North Temperate to South Temperate zones, in much of that region hookworm has little importance because of improved sanitation, the wearing of shoes, the absence of porous, moist, sandy soil, or the prevalence in the region of members of the resistant Negro race. Within the hookworm regions pockets of disease occur wherever conditions are conducive to soil pollution and contact with such soil. The tea agriculture in Ceylon has provided the elements of such a problem, since the soil is a sandy loam, tea culture requires numerous pickers in the fields, and the tea plants themselves provide shade and protection to billions of infective larvae migrating from the feces of barefoot or sandal-wearing laborers. Until detailed information from all parts of the hookworm belt is made available, facilitating the separation of the figures for incidence from the much more significant ones for average worm burden, the real importance of hookworm cannot be arrived at.

CONCLUSIONS

The hookworm problem, like most of the problems presented in this book, has its unsolved puzzles, yet it illustrates well the way in which a public health problem can be solved.

Hookworm was quickly brought under reasonable control in the United States. (As Stoll pointed out in 1962, however, hookworm remains a problem even in the United States.) There were many factors that made this relative success possible. The parasite had to be recognized as important, and before that could be done, diagnostic procedures had to be invented. In order to make any attack practical, the life cycle of the parasite had to be discovered so that education in prevention of transmission could be given to the exposed public. The prerequisite techniques and knowledge were supplied by European scientists. Impetus and interest in attacking hookworm were supplied by the unusual cooperation of a dedicated scientist and a sensation-seeking journalist. A wealthy philanthropist then made funds available for attacking hookworm as a national rather than a local problem. But the Foundation set up by Rockefeller wisely undertook the task to educate, organize, and demonstrate rather than to dictate procedures or pay for the program. Thus eventually, local, state, and national agencies took over the duty of controlling hookworm, releasing the Foundation for other projects. Hookworm incidence in the United States was reduced, with far-reaching consequences for international public health. The irony of this course of events is that fifty years after the Rockefeller Board's successful intervention, hookworm remains a serious problem to many of the world's peoples.

Hookworm disease still affects many millions of people, causing the same terrible waste that Stiles called attention to long ago. Children fail to grow, and their minds are stunted. Adults seem lazy, lacking energy to work. Stillbirths occur and women die in labor because of hookworm. Poverty, ignorance, malnutrition and lack of sanitation reinforce each other in a vicious circle. Diet rich in protein is unavailable to hookworm victims who need it most because of constant hemorrhage, and the lack of a protein source for gamma globulin prevents immunity from developing and allows unnaturally heavy infestations of worms to be built up. Certain strange symptoms of hookworm, i.e., abnormal cravings and neurological disorders, are still untreated and unexplained.

Thus the hookworm problem today is for much of the world what it was in the United States 50 years ago. It surely will yield to the same approach that was tried there and then. That is, some organized approach is needed to find out exactly where hookworm disease is important, to see that victims are treated, and to educate the people in the rudimentary sanitation which will stop transmission and reinfection. All the basic knowledge needed is available, just as it has been for over 50 years. Major obstacles, now as then, are the poverty and ignorance of the victims, the apathy of government at all levels, and the pressing competition of other diseases or other problems for solution. Perhaps the attention of world leaders (or some attention at any rate) can be directed to this problem, which has had such a clear demonstration of its solvability and has contributed so much to the solution of many other problems of world health.

SUGGESTED READINGS

ACKERT, J. E., "Some Influences of the American Hookworm," *Amer. Midland Nat.*, 47:749–769, 1952.

BEAVER, P. C., "Visceral and Cutaneous Larva Migrans," *Public Health Reports*, 74:328–332, 1959.

CHANG, K., W. K. TONG, H. T. CHIN, and C. H. LI, "Studies on Hookworm Disease in Szechuan Province, West China," *Am. J. Hyg.*, Monograph Series No. 19, 1949.

FARID, Z. and A. MIALE, JR., "Treatment of Hookworm Infection in Egypt with Bephenium Hydroxynaphthoate, and the Relationship between Iron Deficiency Anemia and Intensity of Infection," *Am. J. Trop. Med. Hyg.*, 11:497–504, 1962.

LAPAGE, G., *Mönnig's Veterinary Helminthology and Entomology*, 5th ed. Williams and Wilkins, Baltimore, 1962, pp. 226–238.

STOLL, N. R., "On Endemic Hookworm, Where Do We Stand Today?" *Exptl. Parasit.*, 12:241–252, 1962.

UCHIDA, A., *et. al.*, "Field Trials on Mass Treatment of Hookworm Infection with Bephenium Hydroxynaphthoate," *Jap. J. Parasit.*, 11:59–67, 1962.

WAKS, J., "The Hookworm Problem in the Greater Buenos Aires Area with Particular Reference to the Therapeutic Efficacy of Trichlorofenol Piperazine," *Am. J. Trop. Med. and Hyg.*, 12:56–59, 1963.

WEINER, D., "Larva migrans," *Vet. Med.*, 55(8):38–51, 1960.

YOSHIDA, Y., *et al.*, "Experimental Studies on the Infection Modes of *Ancylostoma duodenale* and *Necator americanus* in the Definitive Host (nineteen volunteers)," *Jap. J. Parasit.*, 7:704–714, 1958.

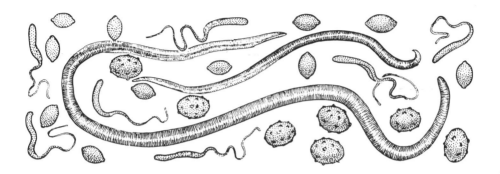

CHAPTER 16

ASCARIASIS AND TRICHURIASIS

INTRODUCTION

Ascaris lumbricoides is believed to be the most prevalent of all the parasites of man. *Trichuris trichiura* is one of the most harmful parasites of children in the moist and crowded parts of the tropics. These two parasites, while members of widely different groups of nematodes, have enough similarities of a practical nature to justify their being discussed together. For example, ascariasis occurs wherever trichuriasis occurs. Both parasites require no intermediate host, and both depend on fecal contamination of the environment for their establishment. Both have been known to cause extremely severe symptoms under certain conditions, yet each, in small numbers, may create little inconvenience or discomfort. Both, needless to say, are part of the unhappy world of the poor, the crowded, the unsanitary. Yet there are important differences between ascariasis and trichuriasis. *Ascaris*, living as a commensal in the small intestine, usually causes harm indirectly by stealing food, while *Trichuris*, living in the colon, is a "predator" much like the hookworms,

211

feeding on the mucosa in which the threadlike anterior region is embedded. The eggs of *Ascaris* are among the most durable of biological units, being capable of survival in strong chemical solutions and remaining infective for many years in the soil. On the other hand, the eggs of *Trichuris* do not resist even moderate desiccation or extremes of heat and cold, a fact which explains why the range of *Ascaris*, while covering that of *Trichuris*, greatly exceeds the latter. *Ascaris* is readily removed by well-tolerated drugs, while *Trichuris*, until quite recently, was among the most difficult of parasites to expel. Finally, *Ascaris*, while very numerous, has attracted little medical attention; the less numerous *Trichuris* has received somewhat more. In order to make clearer the foregoing general comparisons, let us now consider each disease in its turn. Then, at the end of the chapter, we may be able to perceive an interesting relationship between these parasites as they affect man and as man can affect them in the hopefully rational future.

ASCARIASIS: HISTORY AND INCIDENCE

Ascariasis was probably the first of all parasitisms to be recognized by man. The parasite is large, durable, and very common. The Greeks called it by the name it has today, a name more or less synonymous with helminth, or "intestinal worm." Uneducated people in North America less than a century ago knew *Ascaris* and referred to these pale worms so often seen in children's feces as "guardian angels"—a curiously inappropriate extension of the primitive belief in God's providence. Stoll has stated that *Ascaris* infections are the most numerous helminthiases in "this wormy world," estimating the number of human hosts at over 500 millions (see Fig. 16–1). It is interesting to speculate that *Enterobius*, the human pinworm (see Chapter 18) or the hookworms of man (see Chapter 15) may compete with *Ascaris* for the honor of being first. Since enterobiasis is a condition of civilization and industrialization, we may expect its incidence to increase at the expense of ascariasis (which is chiefly agricultural in background) during the coming decades of change in human affairs. Hookworms, like *Ascaris*, are largely agricultural and will probably decrease in incidence with the spread of industrialization. Of course, as Harrison Brown has pointed out, mankind may reproduce to the point of no escape from an agrarian bare-subsistence economy. In the latter case, *Ascaris* will thrive fantastically, celebrating the demise of civilization.

MORPHOLOGY AND LIFE CYCLE

The female *Ascaris* (Fig. 16–2) is large; it ranges from 20 to 35 cm in length and is as thick as a pencil (3 to 6 mm). Males are smaller than the females and have a ventrally curved posterior end. The female has the usual straight intestinal tube of the nematodes (see Chapter 13); it opens at the anterior in a small mouth surrounded by three lips. The genital pore, or vulva, is in the posterior part of the anterior one third of the body. Connected with the vulva by a short vagina are the

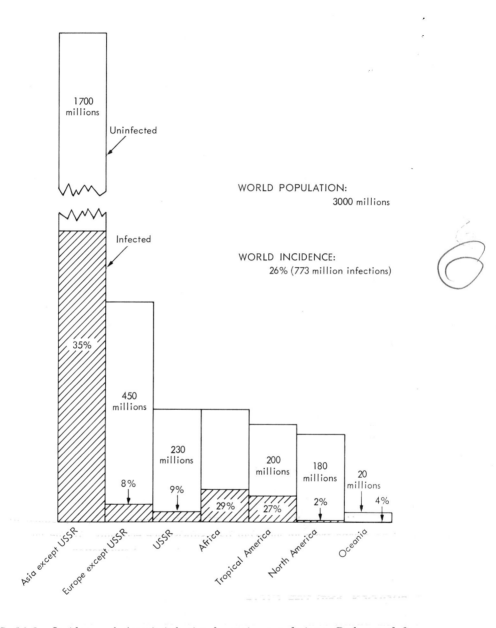

FIG. 16–1.   Incidence of *Ascaris* infection by major populations. Both actual figures (based upon estimates quoted by Faust and Russell, 1964) and percentages of total populations are given. Thus Asia, exclusive of the U.S.S.R., has both the highest percentage, 35%, and the largest number of infections (over 500 millions). North America has the lowest percentage, about 2%. The correlation of incidence with economic deprivation is probably reflected in all the data.

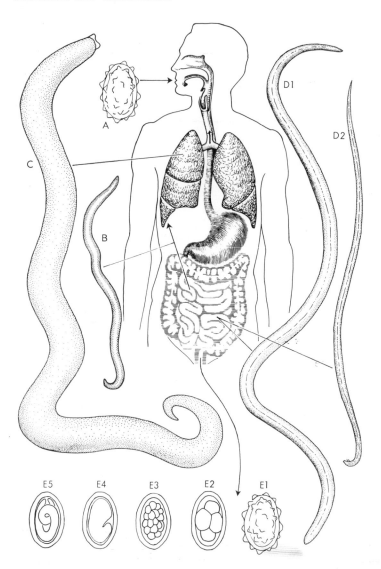

FIG. 16–2. Life cycle of *Ascaris lumbricoides*. After ingestion, an infective egg (A) hatches in the small intestine into a migratory larva (B), which reaches the lungs via the circulation. In the lungs, the larva molts a second and a third time (having already molted once in the egg), grows larger (C), drawn here at the same scale as eggs and infective larva, and enters the trachea and pharynx to be reswallowed. In the intestine such larvae grow to adulthood (D1, D2) and mate. The females (D1) lay eggs (E1) which pass out with feces and require 9 to 13 days under optimum conditions to reach infectivity. Embryonic development follows the sequence (E2–E5). Eggs are drawn 250×, as are larval stages. Adults are shown approximately ½ natural size. [After Belding, 1942, and others.]

paired uteri, which narrow gradually as they coil and over a long distance of more than twice the entire body length change into the delicate ovaries. The egg-producing system may contain at one time 27 million eggs, of which about one percent (200,000) are laid daily. During copulation, the male injects amoeboid spermatozoa into the vagina of the female. The sperm move up the twin uteri to the region where oocytes are released from the ovarian part of the female system. (Fertilization and cleavage of an *Ascaris*—a species from the horse, incidentally—were one of the earliest studies in the history of cytology, and that species is still used for instruction.) The egg of *Ascaris* (see Fig. 16–2) reaches the fecal mass in a stage of early clearage. After about two weeks in the soil or feces, the egg becomes fully embryonated with a larva which has moulted once inside the egg. This larva is infective to man. It rests within the protective membranes of its covering until it reaches a human intestine. *In vitro* experiments have shown that hatching of *Ascaris* eggs in the intestine is a two-step process. First, there is stimulation by environmental factors, the chief of which seems to be a reducing agent. Stimulation causes the infective larva to release "hatching fluid," the action of which is the second step in the hatching process. This fluid contains several enzymes which attack the different layers of the egg shell from the inside out and eventually free the larva. The larva immediately escapes from the intestine by boring into the mucous membranes. The larvae are carried by the blood, first to the liver and the capillaries of the portal circulation and then to the lungs, where they usually become lodged in the walls of the air sacs. Growing there to a length of 1 to 2 mm (about ten times their length at hatching), they break into the alveolar cavities and move actively into bronchioles, bronchi, and trachea. Upon reaching the pharynx, they are swallowed. Their second visit to the intestine is prolonged for their lifetime of about one year. As the worm grows during the first two months in the intestine, it moults once more (about four weeks after ingestion of the eggs). Adult males and females mate, and eggs usually begin to appear in the host's feces about eighty days after exposure to infection. At the estimated rate of 200,000 eggs per day for a period of about 10 months (the average length of egg-laying), some sixty million eggs will have been produced by one female. While this amounts to a volume of only about 5 cc, the number of worms in many infections may run into the hundreds to account for annual production of perhaps a liter or more of eggs. The result is a substantial theft of high quality protein from the host's diet. Obviously, such a large number of eggs can "seed" the environment rather thoroughly if the feces of human hosts are indiscriminately deposited.

HUMAN AND PORCINE STRAINS

An interesting fact about *Ascaris lumbricoides* is the existence of two strains: human and porcine (see Fig. 16–3). Although the worms of both strains are almost identical in appearance, chemical composition, and details of the life cycle, they differ in the important respect that neither strain can easily infect the host of the other. Experimental attempts to inoculate people with the eggs of swine ascaris

FIG. 16–3. Effects of *Ascaris lumbricoides* on swine. (A) Liver, showing gross mottling due to lesions caused by migrating *Ascaris* larvae. (B) Portion of small intestine, unopened, packed with adult *Ascaris*. (C) The same gut split to reveal worms. (D) Two young hogs from the same litter; one was reared free from *Ascaris*, and the other was heavily infected. [After Lapage, 1962.]

have succeeded only a few times, and worms of the human strain cannot reach maturity in swine. The existence of two physiologically isolated strains of parasite in animals so frequently contiguous as man and hogs is hard to explain in evolutionary terms. Perhaps the distinctness of these strains is not so complete as available evidence indicates. The porcine strain should not be neglected in public health planning, for larval migration of many ascarids occurs in abnormal hosts and, as we shall see, may cause serious harm to man.

ASCARIDS OF DOMESTIC ANIMALS

Parascaris equorum is a dangerous parasite of horses throughout the world. The invading larvae, passing from intestine to lungs as in other ascarid cycles, cause mechanical damage and also introduce bacterial or viral agents of disease into the system. Later, the worms in the lungs cause pneumonia with fever. Finally, adult worms in the intestine may cause perforation, damage to the liver or bile ducts, and general symptoms of toxemia probably from metabolic waste products of the worms. Foals are the chief sufferers from parascariasis. Carbon tetrachloride or carbon bisulfide effectively removes the worms. Protection of young colts from feces-contaminated stalls or pastures is important. Such protection is difficult, however, since egg-contaminated soil may remain infective for months or years.

The ascarids of dogs, cats, and other carnivores are of little direct importance, although heavy infections of young or even new-born puppies may cause death or disease. The eggs of these worms can hatch in both normal final hosts and in various "intermediate" hosts. In the latter (various rodents), the larvae migrate without reaching the lungs or intestine. When children ingest the eggs of dog or cat ascarids (*Toxocara canis, Toxocara cati,* or *Toxoscaris leonina*), the migrating larvae ("visceral larva migrans") may cause symptoms of hepatitis, with eosinophilia and lung involvement (Fig. 16–4). Beaver (1956) has called attention to the seriousness of this aspect of ascariasis. Control of dog and cat ascarids depends upon regularly "worming" these domestic animals.

Ascaridia galli, the intestinal ascarid of fowl, and *Heterakis gallinae,* the cecal worm, are very common parasites. The eggs of each must be ingested; the larvae hatch in the gizzard and duodenum, briefly inhabit the mucosal tissues of the intestine, and then reach maturity in their respective sites. Neither worm, in average numbers, produces symptoms. However, as was mentioned in chapter 5, *Heterakis gallinae* is a means of dissemination and transmission of *Histomonas,* an important protozoan parasite of poultry.

SYMPTOMS AND PATHOLOGY OF HUMAN ASCARIASIS

Pathogenesis of *Ascaris* infections has three important aspects: (1) larval migration, (2) toxemia due to reactions of the host to metabolic products of the worms, and (3) direct injury to tissues and organs caused by the robust, active adults.

Larval migration by only a few worms at a time is unnoticed. There are small lesions in the lungs where the larvae break into the alveoli, but these affect only a tiny part of the lung tissue. However, if many larvae are migrating, not only the lungs but also other organs are affected, for apparently some of the larvae may pass through the pulmonary capillaries to the left heart and from there be distributed widely. They may come to rest in lymph nodes, thyroid, spleen, or even the central nervous system. In these places severe tissue reactions to the migrating worms occur to block the worms but also to produce symptoms characteristic of the place of injury.

Toxemia, a general systemic poisoning, has been noted in many cases of ascariasis of long duration. Apparently the waste products of these large worms can be absorbed directly through the wall of the intestine. Some individuals become sensitized to these wastes, which act as poisons. All sorts of symptoms ensue. Among those observed have been epileptiform seizures, urticaria, bronchial asthma,

FIG. 16–4. Visceral larva migrans. The dog ascarid (A), *Toxocara canis*, lays eggs (A1) which, when ingested by man, hatch into larvae capable of migrating to various tissues and of causing local reaction (B). Winglike structures ("alae") in the neck region and characteristic papillae in the caudal region of the male, are diagnostic of the dog ascarid. The cat ascarid (C), *Toxocara cati*, is quite similar to the dog ascarid in both appearance and life cycle. (D) The life cycle is completed in the cat after ingestion of eggs (C1) or larvae migrating in rodents. The eggs hatch; larvae migrate to the lungs, are coughed up and reswallowed, and reach maturity in the intestine. [(B) After Faust and Russell, 1964; (A) and (C) after Faust, 1949.]

photophobia, retinitis, meningitis, and even hematuria. These symptoms, of course, are not common; the great majority of persons harboring *Ascaris* do not exhibit them.

Direct injury by the adult worm is also relatively uncommon, but the effects may be extremely severe. Large masses of worms may become so entangled as to block the intestine, causing gangrene and death. Occasionally adult ascarids wander. They have been known to enter such places as the bile duct (causing jaundice), the appendix (causing acute appendicitis), or the pancreas (causing pancreatic hemorrhage). *Ascaris* adults can perforate the intestine to bring about peritonitis. *Ascaris* sometimes leaves the host by the nose or mouth, or via the anus, causing more alarm than injury. One of the dangerous aspects of chemotherapy is the risk that drugs may irritate the worms without killing or paralyzing them but only stimulating them thus to wander. Where therapy for other intestinal parasites is planned, caution should be exercised, and if a question of *Ascaris* exists, a suitable ascaricidal drug should be used prior to or combined with the other treatment.

TREATMENT

Recommended treatments for ascariasis give a rather wide choice. The earliest ascaricides used in modern medicine were very toxic. Santonin and oil of chenopodium were standard remedies until Ascaridol (related to the latter drug but much less toxic) and tetrachlorethylene came into use in the 1940's. These two drugs may be given together for elimination of hookworm and ascaris; however, care must be taken to prevent absorption of too much of these drugs through the intestine. Hexylresorcinol crystoids have also been used successfully for the past 20 or 30 years. At present the choice of drug is probably piperazine citrate (Antepar). This drug, related to Hetrazan (see Chapter 14), is suited to pediatric use; it is well tolerated, palatable to little children, and highly effective if administered orally in several daily doses estimated on the basis of the child's weight. Thiabendazole, a recently tested anthelmintic, is probably no better than Antepar for use against *Ascaris* alone, but because of its effectiveness against other worms, especially *Trichuris*, it may supersede Antepar in regions where both *Ascaris* and *Trichuris* are a problem. It is believed that the above treatments do not affect the larval migratory phase of ascariasis. Also, it is obvious that surgical procedures are necessary to repair the gross physical trauma sometimes caused by *Ascaris* in its adult wandering.

CONTROL

Control of ascariasis is one of the most difficult problems in public health. It is well known that young children have to be taught to restrict their defecation to proper times and places. In a crowded tropical or subtropical community, families are usually too large (from almost every point of view), and children are of necessity neglected. Their habits of defecation differ little from those of dogs and cats, and the soil of the yards where children play can be a rich mixture of *Ascaris* eggs, *Trichuris* eggs, hookworm and strongyloides larvae, amoebic cysts, and, of course, many enteric bacteria. It is not impossible to transform such a community into a sanitary place, and it is possible to train children to use toilets or latrines. Unfortunately, even if these very drastic changes can be made in human behavior, infective ascarid eggs by the billions will remain in the soil, and reinfection year after year can be expected to occur. Perhaps the best prospect for control of ascariasis lies in the general hope for improvement in living standards throughout the world. That improvement, in the opinion of most experts, is more a necessity than a dream; like the "dream" of world peace, it is the realistic alternative to a nightmare.

TRICHURIASIS

Trichuriasis, a second disease which flourishes in poverty, crowding, and filth, is, like ascariasis, cosmopolitan and highly prevalent. The discovery of *Trichuris* eggs in a mummified Inca child proves that nearly 500 years ago this parasite was present in the Americas. Today, as no doubt always during the history of man, *Trichuris* is found wherever conditions permit its survival. The estimated number of *Trichuris* victims is about 400 million, which is fewer than the number of *Ascaris* infections but still more than any other helminth except, perhaps, *Enterobius*, or the hookworms. Not being alarmingly large, as is *Ascaris* or certain tapeworms, and not causing spectacular symptoms like the schistosomes or the filarioid nematodes, *Trichuris* is a much underrated danger. Its toll of childhood health, vigor, growth, and joy is unmeasured but has been said to be second only to the harm resulting from childhood malnutrition in areas where each condition (too frequently) occurs.

MORPHOLOGY AND LIFE CYCLE

Members of the genus *Trichuris*, misnamed long ago (it is translated "hair-tail," although the threadlike, delicate part of this worm is actually the head), are known as the whipworms. There are species related to *T. trichiura*, the human whipworm, found in many other animals, including dogs, hogs, and various primates. Distant relatives of the trichurids are *Trichinella spiralis* (see Chapter 17) and the species of *Capillaria*, found in various carnivores. All of these worms (superfamily Trichuroidea) are characterized by an esophagus in the form of a slender tube running through a single column of large cells. In *Trichuris* (Fig. 16–5) the esophageal region of the body is long and hairlike, while the posterior region, where the sex glands and related organs are found, is greatly swollen and resembles the handle of a whip. The females are 35 to 50 mm in length; the males are slightly smaller. The female has a single ovary. The sac-like uterus connects with the vulva by a narrow, sinuous duct. The vulva lies in the anterior part of the swollen portion of the worm's body. The males possess a curved posterior end, with a spiny copulatory sheath surrounding the single, pointed spicule. The eggs are barrel-shaped, with a sort of plug in each end. No other helminth of man has similar eggs. The worms live in the large intestine, chiefly in the caecum, but in heavy infections they invade even the rectum. They penetrate the mucosa with their threadlike anterior portion as they become very firmly affixed ("woven" into the host tissues) and feed on tissue juices and blood. They live for many years.

Eggs are in the single-celled stage when laid. They require an incubation period of about 3 weeks in the soil before becoming fully embryonated and infective. Infection is direct. Unlike *Ascaris*, *Trichuris* larvae do not leave the intestine. The newly hatched larvae are found in the crypts of the small intestine, from which they penetrate the soft tissues of the mucosa. As the worms grow, they move toward the large intestine, and in about ten days begin to invade the region of the caecum. It takes three months for the worms to become mature and begin laying eggs. Adults live for years. The average number of eggs laid per day per female worm has been estimated by various workers to be between 1000 and 45,000. Modern experts think the number is in the neighborhood of 6,000, an estimate based in part on critical studies of the dog whipworm, *T. vulpis*. Persistence of eggs in soil or water is variable. Moderate drying inhibits embryonation; heat from the sun's rays, freezing temperatures, decaying matter such as stored feces, various strong chemicals, and prolonged drying kill the eggs and embryos. The implications of these facts for control will be considered after a discussion of the pathogenesis of whipworms.

SYMPTOMS AND PATHOLOGY

As in ascariasis, whipworm disease shows two kinds of symptoms: those due to toxic effects and those due to trauma. The former effects are probably allergic in nature (see Chapter 27). It is not surprising, therefore, that the symptoms are many and varied, for it is well known that individuals react differently to antigens

FIG. 16–5. Life cycle of *Trichuris trichiura*. (A) Female adult and (B) male adult, 10×. (C) Adults with anterior portions embedded in the mucosa of a part of the colon. (D) Eggs, magnified 525×, as they appear in a fresh fecal smear. (E) Eggs showing larval development, which requires about two weeks and favorable conditions. (F) Larva newly hatched, as it might appear in the human small intestine. Crowded, tropical slums provide ideal conditions for *Trichuris*.

and that states of nutrition, age, and general health are quite important determiners of immune reactions. Some of the symptoms attributed to toxic products of *Trichuris* are nervousness, exaggerated reflexes, insomnia, loss of appetite, and moderate eosinophilia. Of course it is difficult if not impossible to sort out symptoms by attributing some to chemical causes and others to physical irritation. It is a truism nowadays that the organism and the environment are essentially one, and it should be equally obvious that the parasite and its host are ecologically bound together in a unity. The irritation of nerve endings by the probing mouths of whipworms must cause release of the chemical products of neurosecretion; likewise the feeding worms must secrete histolytic enzymes into the host tissues. Physical trauma and chemical reactions are inextricably linked. The physical trauma, however, is relatively easy to observe and imagine. Each worm penetrates the colonic lining and leaves its thicker hind-body hanging free. The worms are not entirely quiet. While they do not shift their place of attachment, so far as we know, they can be seen to make muscular movements. No doubt the mouth actively seeks new sources of tissue fluid as old sources become blocked by local tissue response. In moderate infections, various symptoms of damage appear, such as periodic abdominal pain, chronic constipation, indigestion, loss of weight, neurotoxic symptoms, perianal pruritus, and nausea or vomiting. In heavy infections bloody diarrhea, severe abdominal pain, tenesmus (straining at defecation) and loss of weight are observed. The skin is dry, and there is a greatly reduced erythrocyte count, with 20–30 percent reduction in hemoglobin value. Occasionally the rectum is lined with whipworms, and may prolapse (extrude). A rough, but not exact, correlation exists between egg count and the severity of symptoms. An egg count of 30,000 per gram of feces indicates the presence of two or three hundred worms. Such counts usually relate to obvious symptoms. Occasionally, however, larger counts occur in symptomless individuals, and smaller counts may come from persons showing typical whipworm symptoms. This variation is reminiscent of that seen in hookworm disease. In the latter, as was pointed out in Chapter 15, nutrition plays a major part in the reaction of the host to the presence of worms. While this is probably true in trichuriasis also, the picture is not very clear. What part immunity plays in trichuriasis is not known. Because the larvae of *Trichuris* fail to migrate throughout the system, the usual antigen-antibody process that accom-

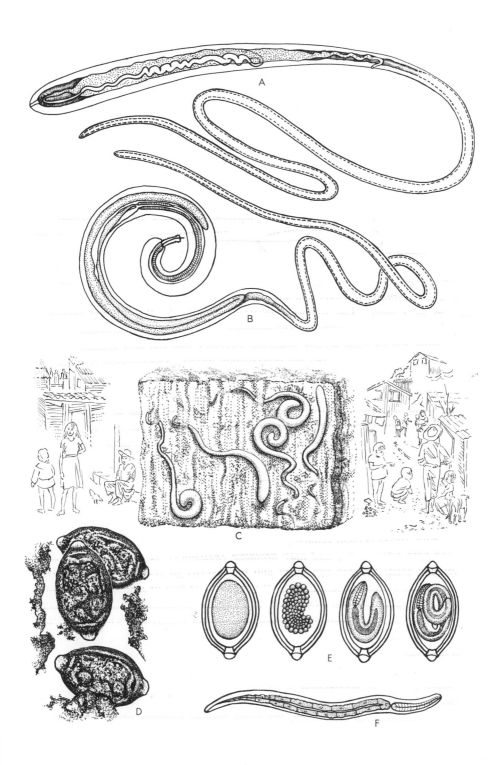

panies hookworm and ascaris infections does not occur. Therefore, this immune reaction to larval stages, a reaction which undoubtedly protects the well-nourished (immunologically active) host from heavy hookworm infection, is unavailable in trichuriasis.

TREATMENT

Treatment of trichuriasis was once very difficult. The location of the worms, far from the point of entry of alimentary drugs, makes it necessary to use such large doses that ordinary anthelmintics, such as hexylresorcinol, tetrachlorethylene, or oil of chenopodium, had to be given in dangerously high quantities so that they would not be diluted to ineffectiveness by the time they reached the caecum. Retention enemas with hexylresorcinol solution were effective yet not easy to administer because of the severe irritation this chemical causes if spilled on the skin. Now, fortunately, dithiazanine iodide can be taken orally for a week with good effect. In South and Central America local plants, *Ficus elabrata* and *F. doliaria*, produce a latex which is very effective in curing trichuriasis. This latex must be used fresh, as its anthelmintic property, a proteolysin, deteriorates unless refrigerated. Thiabendazole, recently tested against a number of helminths, may yet prove to be the best drug.

CONTROL

Control of trichuriasis depends on the same factors as control of ascariasis, hookworm disease, and amoebiasis, that is, proper disposal of human excrement. Often it appears that a particular disease must be controlled by a particular method, and the difficulty of the method may discourage attempts at control. Yet we should always bear in mind the possibility of multiple benefits from a single control program. Thus general improvement in sanitation would reduce the incidence and severity of several kinds of infection. Even such drastic programs as the reeducation of whole communities to improved sanitary practice might be justified by the gains which would be realized in general health, gains such as freedom from hookworm, *Ascaris*, *Trichuris*, *Entamoeba*, to say nothing of enteric bacteria and other filth-born pathogens. *Trichuris*, incidentally, like hookworm, would be one of the first parasites to be conquered by a sanitary program, for its eggs have a life span in the soil perhaps measured merely in weeks (Fig. 16–5). Provided suitable drugs for mass treatment are available, the simultaneous treatment of trichuriasis victims during a program of sanitation should make possible the eradication of this disease over the managed area.

CONCLUSION

In summary, ascariasis and trichuriasis are two very common diseases, which are similar in some respects and dissimilar in others (see Fig. 16–6): they can be successfully controlled, perhaps eradicated, if attention is given them.

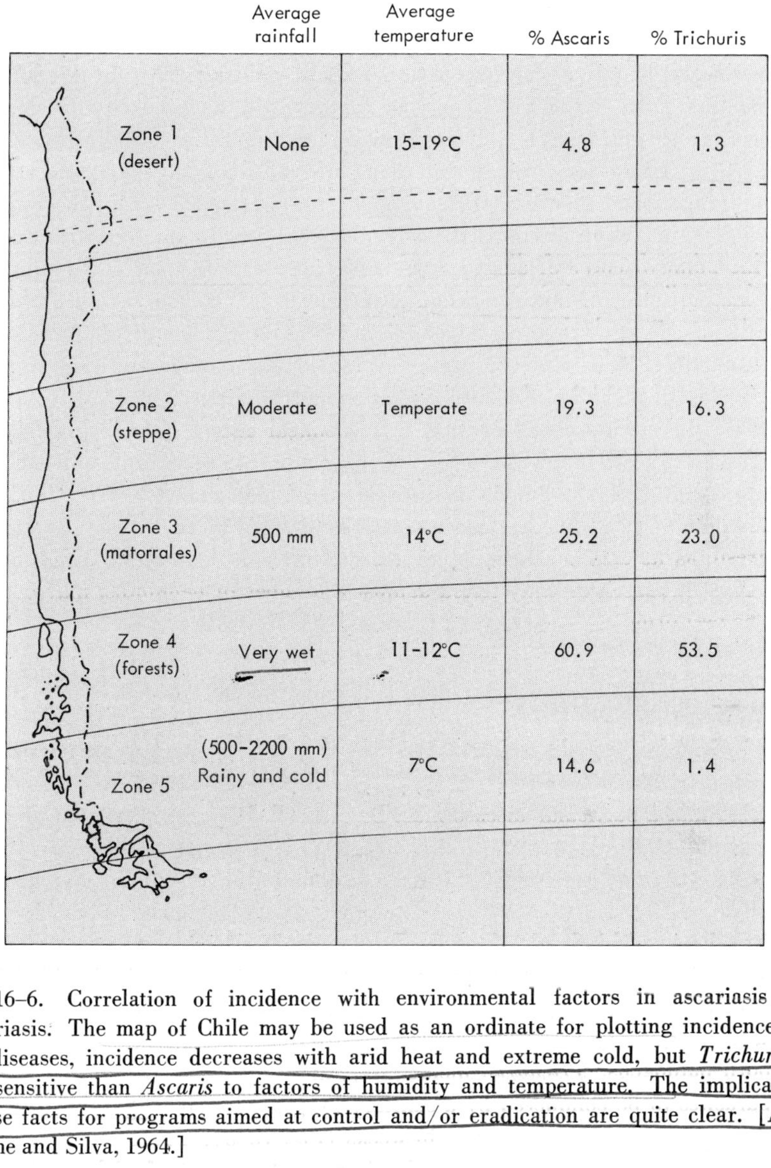

	Average rainfall	Average temperature	% Ascaris	% Trichuris
Zone 1 (desert)	None	15–19°C	4.8	1.3
Zone 2 (steppe)	Moderate	Temperate	19.3	16.3
Zone 3 (matorrales)	500 mm	14°C	25.2	23.0
Zone 4 (forests)	Very wet	11–12°C	60.9	53.5
Zone 5	(500–2200 mm) Rainy and cold	7°C	14.6	1.4

FIG. 16–6. Correlation of incidence with environmental factors in ascariasis and trichuriasis. The map of Chile may be used as an ordinate for plotting incidence. In both diseases, incidence decreases with arid heat and extreme cold, but *Trichuris* is more sensitive than *Ascaris* to factors of humidity and temperature. The implications of these facts for programs aimed at control and/or eradication are quite clear. [After Neghme and Silva, 1964.]

They merit attention. Ascaris affects a vast number of people—an estimated 500,000,000—most of whom live on the edge of starvation. Although the number of victims who show symptoms is relatively small, perhaps between 1–10 percent, even that percentage is large in actual numbers. However, the "symptomless" cases are really not unaffected; for failure to grow properly, loss of weight, apathy, irritability, vague discomfort, and insomnia or restless sleep are some of the conditions that undoubtedly are caused in part by ascariasis. The interference with nutrition, a particularly serious but hidden aspect of this disease, has been amply demonstrated in both human and hog ascaris infection. The effect of such un-attributable symptoms, occurring in hundreds of millions of children and adults, must be very great indeed. It is only in contrast with such horrifying conditions as filariasis, schistosomiasis, epidemic malaria, etc., that ascariasis seems relatively unimportant. In the long run, it may have affected humanity more severely than the above-mentioned scourges. *Trichuris* affects fewer people than *Ascaris*, but it harms its victims more directly. It feeds upon its host. In children, trichuriasis is similar in many ways to hookworm infection and causes anemia with the many severe consequences of that condition. Taken by itself, it deserves attention. Con-sidered with *Ascaris*, *Trichuris* is capable of being attacked in much the same way, at the same time. With good prospect of eventual control and eradication, it presents to public health planning a challenge that should be met. Control of *Ascaris* and *Trichuris* will not be easy, for this involves deliberately inducing a change in cultural patterns. We know, however, in this century of accelerating invention and ubiquitous novelty, that human nature can and does change. Educational tech-niques, combined with medical skills, make it possible to alter age-old patterns in less than a generation. There is no reason for our children to wait long for deliverance from the diseases that are maintained in filth.

SUGGESTED READINGS

BEAVER, P. C., "Larva Migrans," *Parasit. Rev. Sec. Exptl. Parasit.*, 5:587–621, 1956.

BIAGI, F., and J. R. GOMEZ MEDINA, "Cuadro clinico de la tricocefalosis," *Bol. Med. Hosp. Infantil de Mexico*, 19:467–470, 1962.

HOEKENGA, M. T., "Treatment of Ascariasis in Children with Hetrazan Syrup," *Am. J Trop. Med. Hyg.*, 1:688–692, 1952.

JUNG, R. C., and P. C. BEAVER, "Clinical Observation on *Trichocephalus trichiurus* (Whipworm) Infestation in Children," *Pediatrics*, 8:548–557, 1952.

PIZZI, T., and H. SCHENONE, "Hallazgo de huevos de *Trichuris trichiura en* contenido intestinal de un cuerpo arqueologico incaico," *Bol. Chileno Parasitol.*, 9:73–75, 1954.

SPRENT, J. F. A., "The Life History and Development of *Toxocara cati* (Schrank 1788) in the Domestic Cat," *Parasitology*, 46:54–78, 1956.

SPRENT, J. F. A., "The Dentigerous Ridges of the Human and Pig Ascaris," *Trans. R. Soc. Trop. Med. and Hyg.*, 46:378, 1952.

SWARTZWELDER, J. C., "Clinical Ascariasis—an Analysis of 202 Cases in New Orleans," *Amer. J. Dis. Child.*, 72:172–180, 1946.

SWARTZWELDER, J. C., et al., "Therapy of Trichuriasis and Ascariasis with Dithiazanine," *Am. J. Trop. Med. Hyg.*, 7:329–333, 1958.

TAKATA, I., "Experimental Infection of Man with *Ascaris* of Man and Pig," *Kitasato Arch. Exp. Med.*, 23:49–59, 1951.

CHAPTER 17

TRICHINOSIS

INTRODUCTION

Trichinosis, a disease about which nearly one thousand published reports exist, is not a major problem in parasitology; yet it is an interesting problem, which after a hundred years of study still escapes solution, while it affects in some degree the health and income of perhaps one sixth of the people of the more prosperous regions of the world. Its history is one of early recognition, erroneous interpretation, and eventual understanding. Its significance is subject to debate. There are curious aspects of this disease—questions about its epidemiology, its subclinical effects on man, and the propriety of expensive control measures—which are worth considering for the light they may shed upon other, perhaps more important, problems.

Recognition of the "spiral threadworm," *Trichinella spiralis*, is credited to an English medical student, Paget, in 1835. Paget dissected some gritty particles from the muscle tissue of a cadaver and, wishing to examine them with a microscope, persuaded Robert Brown (the botanist who had discovered the nucleus) to assist him. Paget sketched the worms and took the sketches to the anatomist Owen, who published a description of the worm, naming it "Trichina spiralis," but failed to recognize it as one of the Nematoda. By 1844–45 several good descriptions of the encysted worm had been published which made it clear that "Trichina" was a nematode. The reports prompted von Siebold (one of the discoverers of the alternation of generations in animals) to state that the encysted worms were young forms of some already known species such as *Ascaris* or *Trichuris*. Because of the difficulty which this suggested life cycle presented and the fact that there is no ordinary way for such encysted parasites to be transmitted from one host to another, von Siebold decided that the larvae were "aberrant forms" which had lost their way. He was partly right, of course. In 1846, Leidy, an American physician, found encysted *Trichinella* larvae in pork and published a note stating the probable identity of human and porcine trichinae. This note did not influence continental scientific opinion; most workers believed that the worms in swine were a species peculiar to those animals. Leidy's observation was crucial but was not utilized in the contemporaneous attempts to understand the life cycle. Experimental work in Germany began by accident in 1850. Herbst threw some scraps of dissected cats and dogs to a caged badger, and when the badger died, Herbst found its muscles teeming with trichinae. He wisely fed portions of the badger to dogs, which then became heavily infected with larvae. This very clear demonstration was unnoticed by Herbst's contemporaries, probably because of the above-mentioned opinion that the trichinae of human muscles were unique to man. Shortly afterward, Leuckart found intestinal nematodes in mice which had been fed trichinous meat and recognized these as larvae which had been freed from their cysts. Unfortunately, the larvae had not developed to sexual maturity. Leuckart next fed trichinous meat to pigs. Upon finding, five weeks later, mature *Trichuris* (the whipworm, see Chapter 16) in the pig intestine, he erroneously concluded that the whipworm was the mature form of trichina, and he announced this "discovery." However, in 1859, Virchow fed some trichinous meat to a dog, and the dog died. In its intestinal mucus were many adult *Trichinella*, males and females, which Virchow recognized both as a new kind of nematode and as the mature forms of the spiral threadworm. In 1860, Zenker confirmed this fact clinically by finding at autopsy many adult *Trichinella* males and females in the intestinal mucosa of a girl who had died supposedly from typhoid fever. This girl's muscles were filled with *Trichinella* larvae. Thus after a mixture of good observation, misleading speculations, happy (and unhappy) accidents, and a fair amount of cooperation by scientists of several nations, the life cycle of *Trichinella spiralis* was successfully described. Clinical diagnostic methods were later developed; biopsy of muscle from suspected victims can be examined microscopically for the definitive larvae or cysts. Unfortunately, as will be seen below, neither diagnosis nor treatment can be said to be sufficiently effective at the present time.

INCIDENCE

Most information about the incidence and distribution of trichinosis today comes from autopsy data. The disease is widespread among the peoples of the northern hemisphere, particularly in the temperate and subarctic zones where meat consumption is relatively high. The tropics are almost free of trichinosis. In the United States, some estimates put the number of infected individuals at close to 60,000,000, or about 30%. However, clinical cases, at least those diagnosed as trichinosis, stand at only 600 per year in the U. S.; of these, only about 5% are fatal. The economic effect of trichinosis is unknown. Undoubtedly there are many undiagnosed clinical cases; it will be shown below that this disease has misleading symptoms. Until the economic effect is better known, it may be impossible to consider control measures realistically, for those measures usually advocated are very expensive.

The cost of preventing or controlling trichinosis is only one of several bits of unfinished business in the history of this disease. Is trichinosis important enough for the spending of a great deal of money? It will be seen later that the above public health dilemma depends for its solution on several other questions, such as (1) the possible value of widespread, low-grade infections in providing protective immunity (or, conversely, the risk in such infections of producing occasional hypersensitive states), (2) the potential usefulness of new and recently developed treatments (for treating 600 cases per year may, in the future, be more economical than prevention), (3) the eventual reduction of costs of prevention and control to "reasonable" size, and (4) the practicality of eradication as a substitute for control (a matter which depends on more knowledge of the reservoirs of trichinosis among wild animals). In short, more needs to be known about the nature, the extent, and the importance of trichinosis.

THE LIFE CYCLE

Trichinosis is caused by the invasion of the voluntary muscles of the body by larvae of *Trichinella spiralis*. The adults of this worm live partially embedded in the mucosa of the host small intestine. The males are small (about 1.5 mm in length) and short-lived and usually disappear from experimental hosts in less than a week. The females (3–4 mm) live for 2 or 3 months. Both sexes (Fig. 17–1) show a characteristic of the trichuroid nematodes (see Chapter 13); i.e., the slender esophagus runs through a single column of glandular cells. The anterior end of the worms is thinner than the posterior. The vulva of the female is near the anterior end. Eggs hatch within the uterus, and larvae pass out of the vulva into the mucous tissues of the host. About 1500 larvae are probably deposited by each female during her lifetime. These larvae enter the lymphatic vessels and blood stream through the mucosa of the host intestine. They are at this stage quite small (about 100 by 5 or 6 microns) and are easily carried through various capillary systems. They are to be found in any organ supplied with blood shortly after being deposited in the tissues, but they eventually are filtered out (or actively escape) in the voluntary muscles. There they grow rapidly to a length of 250–500 microns; they first move about among the muscle fibers and later become surrounded with a capsule laid down by the host. Cyst formation is complete seven or eight weeks after the larvae have entered the blood. In order to develop further, the encysted larvae must be eaten by a second suitable host.

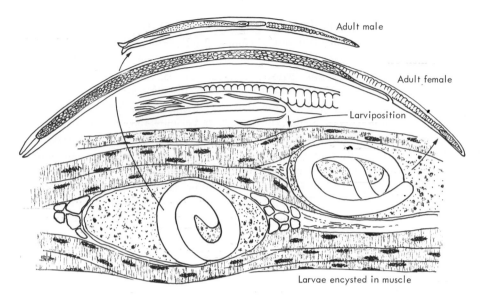

Adult male

Adult female

Larviposition

Larvae encysted in muscle

FIG. 17–1. Development of *Trichinella spiralis*. Male and female wórms mate in the host's intestine; then the female deposits larvae which eventually come to rest in skeletal muscle tissue, surrounded by host connective tissue and phagocytes.

The concept "host" is not easily applicable to animals harboring *Trichinella*, for, as can be seen above, the same animal that harbors mature forms (a criterion for employing the term "final" host in ordinary usage) also harbors infective larvae, and hence can be called (by conventional standards) an intermediate host. Thus in nature the worm must be transmitted by either predation or cannibalism, and such predation is distinctive in that the prey must be able to assume, at times, the role of predator, or at least scavenger, upon some host of *Trichinella*. Hosts in nature include such cannibalistic animals as rats. Thus a rat-to-rat cycle occurs. In the Arctic regions other rodents have been shown to be hosts. Under domestication, and under the special conditions by which hogs raised for market are fattened on slaughterhouse scraps (of hogs), an important cycle of *Trichinella* is maintained, i.e., the hog-to-hog cycle. Thus natural cannibalism, as in wild rodents, or forced cannibalism, as in domestic hogs, maintains trichinosis in nature and under animal husbandry. Probably minor cycles involve, for example, rats which have access to hog carcasses or slaughterhouse offal. In the far north, *Trichinella* larvae are found in the flesh of marine mammals, including the walrus and the white whale. Studies of these surprising sources have recently revealed a scavenger-marine mammal cycle. The carcasses of walrus or whales dying from natural causes are consumed by a variety of animals ranging from crabs to seagulls. Because the trichinella larvae, which may be in the flesh of the large mammals, are able to survive in the intestine of at least some of these scavengers, notably birds, and are dropped with bird feces over wide areas of sea and shore, the larvae can be ingested by fish, which then serve as transport hosts to infect fish-eating mammals. Fish and presumably crustacea may also acquire transmissible larvae from large mammal carcasses. Thus the complex food chain or food web of arctic waters maintains *Trichinella* as a common and readily acquired parasite of the Eskimo, whose food habits explain an incidence of about 25% in the Eskimos in Canada and Alaska.

In addition to the natural hosts mentioned above, *Trichinella* has many "unnatural" hosts. "Accidental" would be a more appropriate term than "unnatural" if infection of such hosts were very rare. But man, wolves, bears, and various other carnivores, including domesticated cats and dogs, acquire *Trichinella* by eating infected meat; and these carnivores, while usually failing to transmit the parasite (hence "unnatural hosts"), are, as we have seen, quite commonly infected. In all the above carnivores, it seems very unlikely that any evolutionary significance can be attached to the survival and encystment of *Trichinella* larvae; these larvae are at a dead end; they are not parts of a cycle but are victims of an excursion.

Of course, this excursion has consequences for man. The preference for raw or partly cooked meat—not perhaps an atavism going back to fireless prehistoric man, but a civilized interest in highly spiced, smoked, or salted meat products—exposes man to trichinosis wherever the above epidemiological conditions occur. Thus the Alaskan who eats walrus and polar bear, the sportsman who hunts bear and boar, the guest at a pork feast, and the purchaser of sausages of many kinds, as well as the consumer of partly cooked "hamburgers" (supposedly beef but often containing pork), all share a risk of acquiring *Trichinella* larvae. The protection of these people is complicated by strong cultural patterns, economic factors in the meat industry, and the lack of effective education aimed at warning people of the risk in eating uncooked meat of various kinds.

As we have seen, the risk is statistically large yet not clinically serious. Clinical trichinosis in the U. S. may be as low as 600 cases per year, although, of course, *Trichinella* occurs in a vastly greater number of people. However, the nature of trichinosis is such that no one should willingly expose himself to the danger.

THE DISEASE

Trichinosis is a painful, sometimes fatal, disease which may last for months, and is incurable. The first symptoms are diarrhea, pain, nausea, and other effects which mimic the symptoms of several other diseases, such as bacterial dysentery, typhoid fever, food poisoning, etc. These symptoms last from about the seventh day after ingestion of infective larvae to the second week. A second stage of symptoms, lasting from the second to the fifth week, is brought on by the migration of larvae to the voluntary muscles. Inflammatory reactions in the latter cause intense pain and interfere with the action of the affected muscles. Breathing, chewing, vision, etc., may be affected. Edema becomes pronounced, and there is constant sweating as well as fever. Eosinophile counts rise to over 50%. In this second stage, death frequently occurs. The third stage begins about 6 weeks after infection. The larvae have undergone encapsulation, and calcification of larval cysts begins; however, the symptoms of the second stage may return. The skin of the whole body may swell and itch. Anemia, skin eruptions, and pneumonia may occur. If the patient survives the seventh week, he usually recovers.

The reasons why some patients suffer so severely, while others may merely feel ill, are unknown. Laboratory experiments suggest that the number of larvae ingested determines the severity of the disease. The rat, for example, can safely ingest somewhat less than 30 larvae per gram of body weight. Assuming that one half of the infective larvae ingested are females, that each female produces 1500 larvae, and that all of these successfully invade their host, a rat's small body might survive up to 37 million larvae, or 22,500 per gram, in its muscles. It is estimated that a man would be killed by ingestion of 5 larvae per gram. Hogs have more tolerance than man but less than rats; they can survive the ingestion of no more than 10 larvae per gram of body weight. While the data on rats and hogs are convincing, the estimate for man is based on counts of larvae in fatal cases of trichinosis and in biopsies of other cases. There is so much variation in human symptoms that it is perhaps unwise to attribute the severity of clinical cases entirely to numbers of larvae ingested. Many of the symptoms of trichinosis suggest allergy, or hypersensitivity; inflammation, edema, etc., are quite common allergic responses. The very large number of mild infections recognizable only by autopsy after death from natural causes (a number estimated at 30% of the U. S. citizens, for example) makes it quite likely that a sizable proportion of the clinical cases had a history of earlier infection. Supportive therapy with cortisone, a noted suppressant of inflammation and allergic symptoms, is helpful in clinical trichinosis; this fact suggests the presence of allergic reactions due to preinfection.

Successful treatment of trichinosis is obstructed initially by the difficulty of making diagnosis. The disease masquerades as any of a dozen possibilities. The two most valuable diagnostic clues, an elevated eosinophile count (not conclusive but highly suggestive) and the recovery of larvae from muscle tissue, are not available until the first stage of the disease is quite far along and large numbers of larvae have entered the blood stream. Such delay is also seen in serological tests or skin tests; time must elapse before the formation of antibodies upon which such tests depend. Although anthelmintic treatment to remove adults is not very helpful at such a late period, it should be carried out. The recently developed drug, Thiabendazole, can be administered as soon as diagnosis has been made, for it will greatly reduce the severity of subsequent symptoms. In most cases, however, diagnosis will be made late, and the only effective treatment will be supportive, using cortisone or similar anti-inflammatory drugs to reduce swelling, stop much of the pain, and make the patient generally more comfortable. Problems of fluid balance, fever, pain, and muscle disturbance can be met with specific or general remedies. Time will eventually cure the patient, if he does not die.

PROSPECTS

Apparently trichinosis is a disease that cannot be easily diagnosed, treated, or controlled. What are the prospects for the future, or, more practically, what peculiar problems remain to be solved?

First, the epidemiology of trichinosis is not completely understood. In domestic situations it is clear that the enforced hog-to-hog life cycle, in which hogs are fed infective garbage (slaughterhouse scraps, scraps from hotel kitchens, etc.), is the major cause of human infection. For the most part, the rat-to-rat cycle is probably closed, since hogs do not feed extensively on rats, nor rats on hogs. Other cycles (the Arctic rodent complex, and the astonishing cycle by which seals, walruses, and whales acquire trichinosis, for example) have not been extensively investigated. Hope for eradication of trichinosis depends on better knowledge of its reservoirs in nature. Second, the development and effects of immunity in trichinosis need further study. As was suggested above, immunity may have both helpful and harmful results; practical artificial immunization must wait until its benefits can be intelligently weighed against its dangers. Third, better diagnostic techniques, especially for very early diagnosis, are needed, since the effectiveness of treatment depends on promptness. Fourth, the control of trichinosis is a problem of unique nature. It may yield to a many-factored analysis which takes into account such things as actual cost (damage, suffering) of the disease, probable cost (in money and effort) of a public health program of control, attitude of the public toward the above factors, means of altering that attitude if necessary, economic pressures (from the meat industry, for example) for or against governmental control, practicality of a program of regional or continental eradication, and perhaps other factors.

These potential factors are illustrated by an interesting accident which brought about a measure of control against trichinosis in the United States. Advice by public health officials in favor of laws to require the heating of slaughterhouse garbage before feeding it to hogs was for many years ignored by state and federal legislators in the U. S. Then in 1952, there occurred an extremely costly outbreak of vesicular exanthema of hogs; this serious virus disease began spreading rapidly via garbage from a few states within which the disease was formerly thought to have been localized. Under lobbying pressure from the meat industry, federal and state laws were soon enacted in the U. S. requiring that garbage for pigs be heated at the boiling point for 30 minutes; this process inactivates the vesicular exanthema virus. Similar laws had been enacted earlier in Britain and Canada. These laws protect the swine industry from losses due to epidemic vesicular exanthema. They also protect millions of people from a major portion of the risk of trichinosis.

CONCLUSION

This brief chapter has discussed a minor health problem which illustrates peculiarly well some of the historical idiosyncrasies of science. The discovery and early studies of trichinosis were marked by dark ignorance; the "gritty particles" in human muscle had been seen but not noticed by many teachers and students of anatomy, we may be sure, before Paget happened to look with true curiosity and enthusiasm (perhaps the best qualities of any scientist) upon these cysts. Unfortunately, the several hasty conclusions and false speculations that followed are also characteristic of science or at least of scientists, for even today the desire for priority causes some researchers to hurry into publication before their data are sufficiently complete. Still a third aspect of the scientific culture is illustrated by the problem of prevention and control raised by the discovery of the cause of trichinosis. Even a well-informed public could not be persuaded (except in a few countries) that inspection of pork would be worth the cost. Perhaps the public was right; the lives of a few hundred people dying in misery each year from a preventable disease might not be worth the expenditure of millions of dollars plus considerable inconvenience to the meat industry. Thus scientific wisdom came into conflict with public wisdom, and the latter prevailed. Such an event is far from unique; witness the failure of the American public, for example, to adopt safety devices proven to reduce substantially automobile-caused injuries. Witness the reluctance of smokers in England, the United States, and no doubt many other lands to alter a habit (or injure a vast industry) which demonstrably and substantially contributes to death from heart disease and cancer. In our time, there have been other examples of persistent stupidity in connection with sanitation and tropical amoebiasis, feces disposal and hookworm disease, and the failure of population control, which is a human disgrace of vast significance. In trichinosis, however, the story does not end with an impasse in which the forces of public health education are on the one side, and popular custom allied with a prosperous industry is on the other. When threatened with large losses due to epizootic vesicular exanthema, the prosperous swine industry in the U.S. readily accepted legislative control of the feeding of garbage to hogs. Thus the control of trichinosis was in great measure due to its accidental connection with a disease of hogs and to economic, not humanitarian, considerations. There is no reason, however, to disparage such a result. In a world where interests often seem to conflict, there is little point in claiming purity of motive or exclusive rationality of aim; the value in the example of trichinosis as a public health problem solved by irrelevancy is to suggest that unexpected allies may exist in other contests and that humanitarian scientists should offer cooperation with such unlikely helpers.

SUGGESTED READINGS

BRITOR, B. A., "The Role of Fish and Crustaceans in the Transmission of Trichinosis to Marine Mammals," *Zool. Zhur.*, 41:776–777, 1962. (English summary.)

FAY, F. H., "Carnivorous Walrus and some Arctic Zoonoses," *Arctic*, 13:111–122, 1960.

GOULD, S. E., *Trichinosis*. Thomas, Springfield, Ill., 1945.

KAGAN, I. G., "Trichinosis in the United States," *Publ. Health Reports*, 74:159–162, 1959.

LUKASHENKO, N. P., and W. W. BRZESKY, "Trichinellosis in Wild Animals in Siberia, Arctic, and Far East U.S.S.R." *Wiadomosci Parazytolog.*, 8:589–597, 1962. (English summary.)

REINHARD, E. G., "Landmarks of Parasitology. II. Demonstration of the Life Cycle and Pathogenicity of the Spiral Threadworm," *Exptl. Parasit.*, 7(1):108–123, 1958.

CHAPTER 18

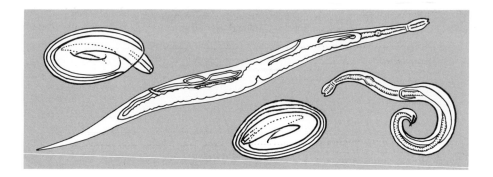

ENTEROBIASIS

INTRODUCTION

An estimate of world pinworm incidence made in 1947 was 208.8 millions. Numerous but not random diagnostic surveys have been made before and after the above estimate. Some regions (for instance, Holland, which had reported 100% incidence of pinworms in the late 1940's) show tremendously high frequencies of infected persons. Others, such as some tropical countries where crowded populations experience seasonal extremes of humidity, have fluctuating pinworm incidence. The United States, and, by extension, most of the western world, may have an incidence approaching 50% or even higher. Incidence in much of the tropics is not so high. Racial factors have been suggested to account for the generally low incidence of pinworms in tropical Africa. The estimate of over 200 million mentioned above is almost surely too low, for the conditions that permit pinworms to flourish in the Netherlands and in the United States are presumably found throughout most of the industrialized temperate zones, in which there are nearly 1 billion persons. The number of hosts of *Enterobius vermicularis* may actually exceed 500 million; this parasite may be as common as hookworm or *Ascaris*; it may be the most common worm infection of mankind.

At first glance, the prevalence of pinworms among the most materially advanced peoples of the earth seems surprising. The temperate zones have seen the disappearance of most of the serious diseases and parasitisms during the last 50 years, so that *Ascaris* and the body louse are strangers to industrial society, and malaria, typhoid fever, and plague are no longer the frequent threat to public health that they had been for ages. Yet pinworms are still found in every school, and in many homes, where sanitation is excellent, medical care is available, and parasites other than pinworm are indeed rare. It seems that pinworms, like stomach ulcers and mental illness, are a disease of civilization. Yet this paradox is readily explained when we consider two facts which will constitute the body of this chapter. The first of these facts is the life cycle and ecology of *Enterobius vermicularis*, a cycle ideally adapted to modern society with its comfortable year-round artificial climate, its friendly person-to-person contact, and its preference for clothing, draperies, and other pinworm-egg catchers and spreaders. This life cycle, moreover, is in little danger of interruption by medical efforts despite the existence of excellent diagnostic and therapeutic methods. The second fact is the unimportance of pinworms. Like all well-adapted parasites, they cause little obvious damage. The difficulty and cost of controlling them has seemed to most people, including physicians, a good reason to ignore the problem. After all, even in a prosperous midtwentieth-century world, there are many problems more worthy of attention than an annoying little worm living in the large intestine. But it will be our purpose to consider whether, in fact, the pinworm is both so effective and so unimportant that mankind will continue to harbor it in the midst of scientific progress and unprecedented prosperity.

LIFE CYCLE

The life cycle of *Enterobius* is a model of simplicity (Fig. 18–1). The adult worms live in the cecal region of the large intestine, where they presumably feed on tissue juices of the mucosa. After the males and females mate, eggs develop in the two uteri until about 11,000 eggs may be present. Then the female migrates to the anus of the host and crawls out, laying eggs freely among the perianal folds of skin. Whether she makes only one visit to the anus or perhaps returns to the colon after one or more egg-laying excursions is not yet known. Frequently the female is trapped by dryness or folds of skin in the perianal region, where she dies and usually bursts in a shower of eggs. The deposited eggs, as well as the wriggling worm, cause local irritation to the host. Often severe pruritus occurs. Of course, the resulting scratching transfers pinworm eggs to clothing, the hands, and eventually to food, door-handles, the hands of uninfected persons, etc. The eggs are infective very soon after being laid. When eggs are swallowed, they are believed to hatch in the duodenal region. The young worms hide in the crypts of the mucosa and move toward the colon as they grow. After two moults they reach the caecum, where they attach themselves to the mucosa and become mature. An entire cycle takes 15–43 days.

The above is believed to be the usual manner of development and transmission. Autoinfection, or the hatching of eggs within the host intestine, probably does not occur, but "retrofection," or hatching in the perianal region, with reentry of the larvae into the rectum, has been demonstrated. In some very long-lasting cases of adult enterobiasis, in which reinfection by ordinary means was carefully guarded against, retrofection seems to have been the only acceptable explanation. Since it takes about six hours for the larvae in the new-laid eggs to reach the infective stage, perianal washing at less than six-hour intervals can remove the threat of retrofection.

The horse pinworm, *Oxyuris equi*, while it is much larger than the human pinworm, has much the same cycle and effect. These worms deposit their eggs in the perianal skin and cause itching. Infected horses can be recognized by the abraded appearance of the tail and by scratches around the anus caused by the animal's rubbing against fences to relieve the itching. Oxyuriasis is spread by the dissemination of the nematode eggs through drying and scaling off of patches of eggs around the anus. These "flakes" of eggs fall to the ground to contaminate grass or fodder, from which horses obtain new infections. The chief damage due to pinworms is the host-inflicted lesions of the perianal region, which becomes susceptible to bacterial invasion. Other animals—rats, mice, rabbits, and opossums—have cecal worms which may presumably have similar life cycles to that above. Their chief interest is for investigators of human enterobiasis; rat and rabbit pinworms are useful in drug-screening programs. The horse pinworm is shown in Fig. 20–3D.

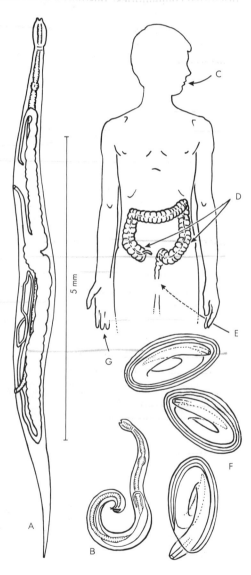

FIG. 18–1. Life cycle of *Enterobius vermicularis*. Female (A) and male (B) adults mate in the colon (D). The female migrates periodically to the anus (E) where she lays eggs (F) on the perianal skin. In a short time (6 to 8 hours) the eggs become infective. If carried (G) to the mouth and swallowed (C) they hatch, and the larvae grow to maturity in about two weeks.

CONTROL AS PRACTICED

The life cycle of the pinworm, while very simple and direct, presents real problems in the matter of control. Pinworm is a group infection; if one member of a family has pinworm, the other members either have the worm or will acquire it. In an institution such as a hospital, orphanage, or prison, enterobiasis is apt to be nearly universal. These facts mean that nothing short of eradication of the worm from the family or the institutionalized group can prevent widespread reinfection after treatment. The same principle applies to neighborhoods and other more loosely organized communities; dustborne eggs, as well as those spread by other means, are a threat so long as any resident harbors pinworms. We thus have a situation where knowledge and action are apparently not sufficient to eradicate pinworms, although medical techniques for diagnosing pinworms and for removing them are highly effective.

DIAGNOSIS

Pinworm eggs found in the perianal skin or elsewhere are diagnostic. Also, mothers frequently observe the little white worms creeping out of a child's rectum, and sometimes the worms can be seen in loose stools. Several techniques for finding the eggs have been invented. The Dutch workers recommend a "wet pestle," i.e., a glass rod applied with a rocking motion to the perianal skin and then rolled in a drop of saline on a slide. The U. S. National Institutes of Health have developed a cellophane swab which is used in a manner similar to the rod. However, Graham's "scotch-tape" method, as modified by several workers including Beaver, seems to be best, for by it a parent can conveniently obtain specimens at ideal times for examination, that is, when the child wakes up and before defecation or washing. A piece of adhesive transparent cellophane tape is applied, sticky side down, to the perianal folds, especially to the crevices between the folds. Then the tape is flattened on a slide in a drop of toluene, which allows air bubbles to be driven out and tends to make transparent most of the objects collected on the tape. The characteristic eggs can then be recognized under low magnification. Incidentally, the development of the above methods reflects a considerable interest in enterobiasis over a number of years. The public health agencies of several nations have supported research on enterobiasis and are continuing to do so.

TREATMENT

This research extends to methods of treatment. For many years gentian violet in enteric-coated capsules was used to remove pinworms. This remedy is available in the United States as a patent medicine, "Jayne's P-W Pills." It requires a series of doses over a one- or two- week period; however, it is not highly successful, although it is not difficult to administer and is reasonably well tolerated. Much better modern remedies are the piperazine citrate Antepar (mentioned above for ascariasis), and pyrvinium pamoate, Povan. Antepar, like gentian violet, must be given in a series of doses. Povan is effective in a single dose. Neither of these drugs produces side effects in the doses recommended (unless staining of clothing by the purple gentian violet or the red Povan be considered a harmful effect). Dithiazanine is another recommended drug. Other anthelmintics, such as tetra-chlorethylene and hexylresorcinol, are reasonably effective but may have un-desirable side effects. Thiabendazole, currently being tested as a wide-spectrum anthelmintic, may become an important enterobicidal drug.

It is clear that neither diagnosis nor treatment presents any real problem in the medical care of enterobiasis. The question then arises: if both diagnosis and treatment are easy, why should pinworms prevail throughout most of the developed world? The reason would seem to be that pinworms are not considered important, and the effort required to eradicate them does not seem justified. Whether this view is correct we shall now consider.

IMPORTANCE

The importance of any disease depends on such things as the number of its victims, its pathology and symptoms, and its harmful effects in terms of misery and economic loss. It also depends upon such intangibles as the relationship it bears to other improvable conditions, to available resources, and to attitudes of mind of the deci-sion-making groups involved. Misery and economic loss are not usually connected with pinworms.

While in a few remarkable cases such effects as convulsions, eczema, and in-tolerable pruritus leading to secondary infection have been observed, the great majority of pinworm sufferers do not suffer much. Restlessness, inattention, mild itching in the anal region, loss of appetite, dark circles around the eyes are fairly frequently observed in pinworm-infested children. These symptoms are incom-parably less severe than those due to other helminthiases. Mild ascaris infections alone may be equally harmless, but there is, in ascariasis, always the risk of severe, even fatal, symptoms from "wandering" worms. *Enterobius* never causes death. But physical symptoms are not the sole misery-inducing effect of pinworms. Mothers are often dismayed to find worms near the anus of a little child. The resulting consultation with a physician often leads to attempts at household sanitation which are time-consuming, expensive, impractical, and, because of the last, frustrating. The "pinworm neurosis" (so-called by a leading parasitologist) can be more irritating than the worms themselves. Therefore psychological trauma, the shame and anxiety occasioned by the presence of worms in the household, must be con-sidered as one of the harmful effects of *Enterobius*.

Economic loss appears to be slight until two facts are considered. The first is the undoubted fact that pinworms are very often diagnosed and treated; perhaps "repeatedly" is a better word than "often." Each episode can represent a doctor's fee and a purchase of prescribed medication. The money that public health agencies of several nations have spent on research on enterobiasis can be added to the cost.

The medical costs may not, in Keynesian economic terms, be counted as absolute loss, for the medical and pharmaceutical industries are important parts of the flow of goods and services, and they benefit from the prevalence of pinworms. Possibly, however, the money spent on pinworms could be better spent for other goods and services. The second fact, after medical costs, is the probable slow-down of education due to enterobiasis in the classroom and of labor in the factory. No study of the former has been made, but a study of enterobiasis in a German munitions factory during World War II is instructive. There, among workers who lived in underground shelters, pinworms created such distress that work was severely interrupted. Anthelmintic therapy quickly restored normal production. It is likely that the mild symptoms of pinworms, multiplied by the hundreds of millions of infected children and adults, cause considerable unmeasured losses of learning and working efficiency.

In addition to the above measures of economic loss and the misery caused in small measure to a large number of persons, there is an intangible effect which might be called affront or challenge. The pride of a civilized society suffers, or should suffer, when a tiny worm successfully challenges the doctors, the public health educators, and the research parasitologists, and flourishes, as does *Enterobius*, in the midst of modern plumbing and the world's highest standard of living.

It may now be apparent that upon at least three bases—prevalence, pathology, and economics—pinworms really are important. But importance, in disease as elsewhere, is a relative matter. Pinworms have not appeared to be a problem of general importance in public health, which is rightly concerned with such major diseases as syphilis and the common cold. Public opinion, which in many nations has some effect on the way in which the common wealth is spent, is not very likely to support a campaign to eradicate pinworms. Therefore, unless some accident, such as the relationship of the control of swine erysipelas to trichinosis (Chapter 17), should affect the transmission of pinworms, they can be expected to continue to flourish.

A PLAN FOR CONTROL

Nevertheless, I should like to discuss the control of pinworm in an optimistic spirit, in order to suggest that there may be ways to approach such a problem indirectly but effectively. The orthodox approach to the control of disease is based on two assumptions: (1) that the disease is controllable, and (2) that its control is necessary. The need for control is related to the disease's importance and publicity. The orthodox approach does not always work. As we have seen, several undoubtedly important diseases are controllable, but they remain, for various reasons, relatively frequent. In the case of pinworms, the orthodox approach is bound to fail, for the importance of pinworms cannot be demonstrated in a simple way understandable to the public. What I would suggest here is a commercial approach (see Fig. 18–2). In a capitalistic society the public can be sold many things, some of which to a rational person may seem undesirable or even harmful. If money can be made by

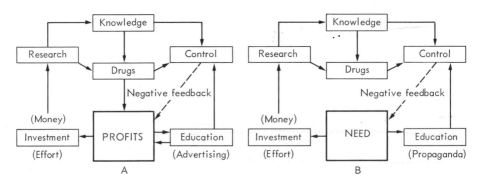

FIG. 18–2. (A) Capitalism. The central factor, profit from the production of pharmaceuticals, is the dominant factor in cyclic activities aimed at control, but success (control) tends to limit profits. (B) Socialism. The central factor, need for improvement in health, is dominant in generating the activities aimed at control, but success (control) tends to reduce need and stabilize the cycle short of completion.

selling a product, the advertising industry will find ways to publicize the product, to inform the potential consumer, to repeat endlessly the name of the product, and even, it is rumored, to assault subliminally the emotions and will of the reader, the listener, or the viewer of "commercials." Two products can be suggested for such expansive promotion aimed at commercializing the control of pinworms. One is an existing anthelmintic, harmless, palatable, readily available, e.g. Antepar or Povan, or a combination of both. The other is a remedy which probably exists only as an idea. That is an ointment, suitable for perianal application, which would either destroy pinworm eggs on contact or would cause hatching and immediately destroy the larvae. Such an ointment, used regularly as a preventive of transmission, would stop the spread of pinworm while the short life span—about 40 days—would permit the spontaneous cure of most infections. If a perianal ointment cannot be developed, then universal use of the one-treatment drug, Povan (pyrvinium pamoate), over a period of only a month or so would accomplish the same result. There would have to be a preliminary study, no doubt, of the probable reactions of a not very fastidious public to advertising about worms and perianal ointments. The public relations industry might have to assess, by a pilot study, the shock effect of such advertising, and its writers might have to invent euphemistic phrases in place of plain words about pinworms. But a potential market of hundreds of millions would spur businessmen toward solutions of these minor problems, and eventually pharmaceuticals for the control of enterobiasis might share the giant markets of toothpastes, cosmetics, depilatories, and deodorants. The only risk in this proposal is the risk of success, the danger that saturation of the population with pinworm remedies would actually eradicate pinworms and destroy the market so cleverly created. Of course, socialist countries could not hope to emulate the commercial method of control. Perhaps public health authorities in these nations could point with envy to the capitalist success and by substituting political zeal for the profit motive, achieve voluntary control of their own pinworms. Stranger things have happened.

It may appear that a tongue-in-cheek suggestion like the above unorthodox approach to control of helminthiasis has no place in a serious text. Yet there is no reason not to speculate freely in science. Where there appears to be as in enterobiasis an impasse—a condition both controllable and important yet not controlled and not recognized by many as important—the impasse may be broken by some bizarre approach. The suggested use of profit-seeking to realize improved public health has its parallels in the health programs of industries which own land and hire labor in underdeveloped countries; no one could object to the better health brought by these investors on the grounds that their motives were mercenary. Perhaps the defeat of misery and disease has no politics or economic theory of its own; to most of us the welfare of human beings comes first, and must be considered basic to the better world which everyone desires.

SUGGESTED READINGS

Beaver, P. C., "Methods of Pinworm Diagnosis," *Am. J. Trop. Med.*, 29:577–587, 1949.

Beck, J. W., D. Saavedra, G. D. Antell, and B. Tejeiro, "The Treatment of Pinworm Infections in Humans (Enterobiasis) with Pyrvinium Chloride and Pyrvinium Pamoate," *Am. J. Trop. Med. Hyg.*, 8:349–352, 1959.

Brady, F. J., and W. H. Wright, "Studies on Oxyuriasis. XVIII. The Symptomatology of Oxyuriasis as Based on Physical Examination and Case Histories of 200 Patients," *Am. J. Med. Sci.*, 198:367–372, 1939.

Graham, C. F., "A Device for the Diagnosis of *Enterobius* Infection," *Am. J. Trop. Med.*, 21:159–161, 1941.

Sawitz, W., V. Odom, and D. R. Lincicome, "The Diagnosis of Oxyuriasis. Comparative Efficiency of NIH Swab Examination and Stool Examination by Brine and Zinc Sulfate Flotation for *Enterobius vermicularis* Infection," *Public Health Reports*, 54:1148–1158, 1939.

Schuffner, W., and N. H. Swellengrebel, "Retrofection in Oxyuriasis. A Newly Discovered Mode of Infection with *Enterobius vermicularis*," *J. Parasit.*, 35:138–146, 1949.

CHAPTER 19

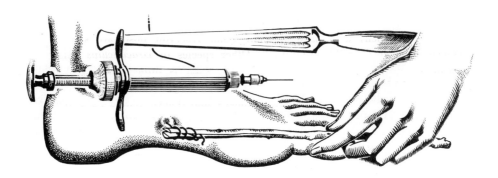

DRACUNCULIASIS

INTRODUCTION

The guinea worm, *Dracunculus medinensis*, lives in the subcutaneous tissues of more than 50 million inhabitants of warm and arid lands. Its history may be traceable to the chronicle in the Hebrew-Christian Bible known as Numbers 21, where it is stated that the Israelites, complaining about the inhospitable desert through which Moses was leading them, were punished. "Then the Lord sent fiery serpents among the people, and they bit the people, so that many people of Israel died." Then Moses was instructed by the Lord: "Make a fiery serpent, and set it on a pole; and every one who is bitten, when he sees it, shall live." While the actual meaning of this story must remain hidden, the "fiery serpents" of a desert region would likely be guinea worms, tiny "serpents," causing burning pain. And, as we shall see, the remedy to set a serpent "on a pole" is certainly the prehistoric treatment for dracunculiasis. In modern times, dracunculiasis has interest in the field of medicine as well as in evolutionary parasitology. In medicine, dracunculiasis is one of the more interesting diseases. Long before its manner of transmission was known, the disease was successfully treated by removal of the worm. Tribal doctors still catch the extended part of the worm in a cleft twig, on which it is then patiently rolled up, a little at a time, for 10–15 days. Patience is indeed essential, for if the worm is broken, its withdrawal into the flesh carries dangerous contaminants into the body and causes sometimes fatal infection. Modern methods are minor improvements on the ancient twig, or, perhaps, on Moses' magical bronze serpent. In evolutionary theory, *Dracunculus* provides some insights of value. *Dracunculus medinensis* is chiefly a parasite of man. Some reservoir hosts are known, for instance, dogs are possibly important. Other species of *Dracunculus* exist, the parasite of North American raccoons, *D. insignis*, is quite common in these mammals, which are known for their habit of washing their food before eating. Species of *Dracunculus* occur in various snakes, and related genera parasitize certain birds and mammals. Man is by far the best host for *Dracunculus medinensis*, with infection rates of 25% in many villages of the drier parts of India and similarly high incidence in Africa, Arabia, and other dry lands. Several writers have called attention to the apparent paradox of a water-transmitted parasite being most commonly found in desert areas. That this is neither a paradox nor a surprise will become clear when we consider the principle of contact-frequency, or parasite density, a principle of major significance in the success of parasites. A desert concentrates its inhabitants near water; a parasite both water-dependent and density-sensitive finds desert conditions ideal. *Dracunculus* provides an evolutionary model for understanding many other parasitisms, including Gambian sleeping sickness, coccidiosis in fowl, and several helminthiases of domestic animals.

LIFE CYCLE

Dracunculus medinensis is acquired by drinking water which contains small crustaceans, the intermediate hosts of the guinea worm. (See Fig. 19–1.) Larval worms escape from the crustaceans into the human intestine and spread from the intestine to the body tissues. They grow from about 600 microns in length to about 20 millimeters in several months. Males and females mate at this time. Then the male dies, but the female continues to grow until she reaches a length of a meter or more and the thickness of a piece of string (about 1 or 2 mm). This size is reached in about a year, during which time the female worm migrates in the soft connective tissues under the skin; it is noticeable, at times, as an undulating ridge, somewhat comparable to a varicose vein. Little or no trouble is given by the worm until it becomes fully grown. It is now merely a tube filled with embryos, for the intestine has atrophied and the uterus occupies the entire length of the organism. The worm moves at this time to a wrist or ankle, where it discharges histolytic enzymes under the skin, punctures the deep layers of the skin, and raises a blister. The blister breaks, and at the base of the area of exposed dermal tissue is a little pit or hole through which the head of the worm may project very slightly. When the blister is wet, the uterus of the exposed worm bursts, and a milky fluid is discharged from the worm into the water. This fluid contains thousands of active

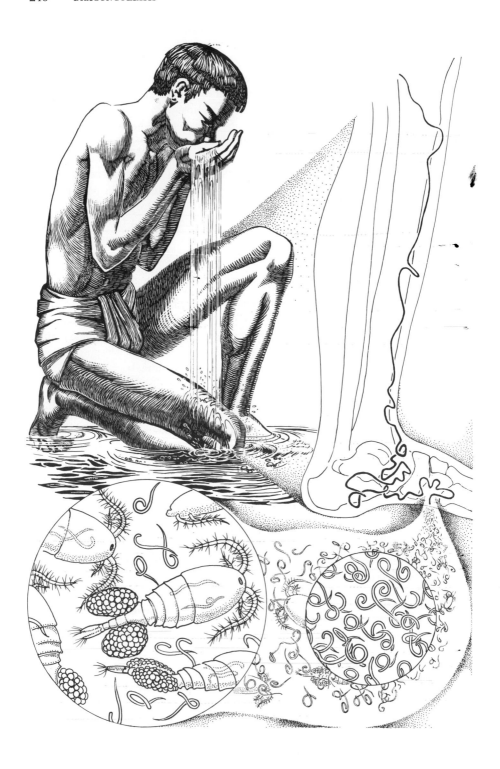

FIG. 19–1. Life cycle of *Dracunculus medinensis*. A laborer in the tropics is drinking water which contains infected copepods, the intermediate host of the guinea worm. A hypothetical x-ray enlargement of the man's leg shows an adult female worm in the intermuscular and subcutaneous tissues. The head of the worm emerges from the foot at an ulcer or broken blister, where the worm's uterus discharges a cloud of milky fluid into the water. To complete the cycle, the thousands of larval worms in this fluid can be ingested by copepods. [After Manson-Bahr, 1945.]

larvae, notable for their long and very slender tails. Much like the bristles on many kinds of plankton, the tails serve both as locomotor organs and as flotation devices. The larvae swim about for several days. If eaten by a species of *Cyclops* (minute copepod crustacea which inhabit most bodies of fresh water), the tiny worm bores into the *Cyclops'* body cavity, where it moults twice to become no longer but somewhat more robust than it was. *Cyclops* can tolerate only a few larval guinea worms and perhaps no more than one. After about three weeks in the *Cyclops*, the larva is infective for man. Ingested with water, the *Cyclops* with its larva reaches the duodenum, where digestion frees the larva and a new cycle of penetration, fertilization, growth, migration, and larviposition begins.

It is interesting to note that dracunculiasis, known for so long to the practitioners of medicine, waited until the end of the nineteenth century for the elucidation of the above life cycle, which, of course, was information absolutely essential for control. It is perhaps ironical that a good cure for dracunculiasis was discovered before the dawn of history but that even in the midtwentieth century millions are still ignorant of the simple ways to prevent this unpleasant affliction.

SYMPTOMS AND PATHOLOGY

The clinical course of a guinea worm infestation is as follows. No symptoms and no damage accompany the early stages. Not even when the female worm has reached its full growth can it be noticed, except (as noted above) visually and tangibly, as a threadlike subcuticular ridge. Then, quite suddenly, symptoms of great severity supervene. Just before the blister forms, allergylike reactions occur. Rash, nausea, diarrhea, itching, asthma, dizziness, and faintness are common, persisting sometimes for several days. When the blister breaks, the above symptoms subside. Probably the worm causes the symptoms by secreting whatever enzymes it produces to cause the blisters; such enzymes and metabolic substances may have been subtly sensitizing the body for the whole year preceding. Evidence that such may be the case is the observation that dracunculiasis ordinarily occurs no oftener than once a year in a person, although that person may harbor many worms. Also, after about four years of repeated infections, some people become refractory to reinfection. Both of the last two facts support the view that (in addition to the allergic reaction mentioned above) two kinds of immunity develop: a premunition, which protects against superinfection, and a more lasting immunity, which requires considerable time to be built up. After the blister breaks, a secretion of toxic substances may continue, but these substances are discharged to the surface of the blister area and can no longer circulate harmfully. The fluid which lies in the blister and later oozes from it is sterile. But after the dermal tissue is exposed to air and dust, contaminating bacteria often infect the lesion. Secondary bacterial infection can be a serious complication. Usually the later course of an attack of dracunculiasis is clinically uneventful. Each time the blister becomes wet, the worm exudes larvae; when the blister dries, the uterus of the worm dries also and seals itself until the next wetting. In about three weeks the worm has emptied itself of all its embryos, the blister heals, and the dead worm is resorbed by the tissues. Many cases, however, do not cure themselves so harmlessly. Inflammation occurs, sometimes at the blister, often along the course of the body of the worm. Such inflammation may be a local immune response or may be due to bacterial invasion. Its results may be quite serious, for joints may become fused (ankylosed), muscles and tendons may be injured, and permanent weaknesses or deformities may result. The guinea worm is dangerous enough so that medical treatment is definitely indicated.

TREATMENT

Treatment by the split-twig method is potentially very dangerous. If the worm is broken, the stretched remainder draws back into its hole. It brings with it surface bacteria from the blister and also discharges into the surrounding tissues a large amount of harmful antigenic material. Gangrene, amputation, and even death may result. The ancient method sometimes includes moistening the worm while winding it on the stick; the moistening probably induces discharge of larvae, making the worm less turgid and therefore less susceptible to breakage. In India suction is sometimes employed in preference to traction. A funnellike cone is placed over the blister, sucked vigorously, and then closed tightly to maintain a partial vacuum at the surface of the blister. Pressure of the body fluids and tissues around the worm then forces it out of its tunnel without any danger of breaking the worm during extraction. Modern medicine adds antibiotics and disinfectants to the old methods to make them safer. Modern anthelmintics, either applied hypodermically (emulsion of phenothiazine) or given by mouth (Hetrazan in large amounts), can kill the worm *in situ*, making it easier to remove by traction or suction. Surgical removal of worms visible throughout most of their length has been recommended. It can be said that in this disease, at least, the wisdom of ancient and primitive physicians is still in use and has merely been augmented by the techniques of modern medicine.

CONTROL

No such ancient wisdom was ever applied to the control of dracunculiasis. Like the present methods of managing mental illness, for example, the solution of dracunculiasis by cure is an imperfect solution. Continuous study of such problems is required even after partial solutions have been discovered, and in dracunculiasis such study has finally revealed the possibility of a complete solution, i.e. control. However, success in control may be as slow in coming as was the knowledge upon which control must be based. The life cycle of *Dracunculus* could hardly have been worked out without the aid of the microscope. Imagination and ingenuity sufficient to suspect and test the water-borne transmission of a parasite which appears no oftener than once a year are probably the possession of few geniuses in history and certainly were not the attributes of ancient physicians. Invention and technology are needed to make certain ideas possible. Now we know that in the fluid exuding into water from a dracunculiasis blister is a crowd of tiny worms, and to educated persons who have heard of malaria and filariasis, this idea does not seem strange, nor does the subsequent *Cyclops* transmission seem remarkable. Informed persons would refuse to drink the water from step-wells and "tanks" in India, from draw-wells improperly rimmed to prevent back-drainage, or from community ponds and water holes in Africa. The sight of people using such water sources to bathe their arms and legs which bear obvious blisters of the guinea worm must be shocking to the informed observer. It is indeed shocking to realize that year after year in many villages attacks of dracunculiasis affect from one-fourth to one-third of the people, and that year after year local physicians prescribe their simple but dangerous remedies for removing several feet of worm from the body. But until the number of "informed observers" increases, this disease, like many others, will follow its immemorial pattern. The facts that *Dracunculus medinensis* can be eradicated by stopping contact between infected persons and public sources of drinking water, that *Cyclops* in wells can be controlled by certain small fish, and that the individual may protect himself by heating his drinking water or, more simply, by straining out the copepods with cloth are not simple enough for the people who must practice control and self-protection. Still credulous in terms of witchcraft and talisman and still capable of taking literally the story of the serpent on a pole which cured the Israelites more than three thousand years ago, most people of the desert regions and elsewhere can not easily be persuaded to believe in invisible larvae encased in the bodies of almost invisible *Cyclops*.

While the above paradox of the persistence of ignorance in the presence of knowledge is so characteristic of mankind as to merit little attention here, another paradox—the success of a water-dependent parasite in arid regions—has some evolutionary interest. Let us recall Whitehead's dictum that a contradiction (paradox) in formal logic is the signal of defeat, while in the history of science it shows the way toward knowledge. It may be that apparently contradictory facts (such as the data on incidence of dracunculiasis in the light of knowledge of its life cycle) can lead to better understanding of the evolution and control of parasitism.

The dracunculiasis situation seems to be a reversal of a well-known rule—Mac-Donald's principle of critical density. (See Chapters 2 and 14, for examples.)

Applied to mosquitoes and the transmission of malaria, this principle means that unless the number of contacts between anopheline mosquitoes and malarious hosts in a given region is maintained at a certain level (the critical density) malaria will diminish until it becomes extinct. The factors from which a definite figure (the critical density number) can be derived include (1) the biting habits of the particular species of mosquito involved, (2) its rate of reproduction and the effects of predation, senility, etc., upon it, (3) the level of gametocytes in the blood of malaria patients, (4) the average distance (or transmission range) between people (a function of population density), and (5) the average time of development of the parasites in the mosquito (a function, of course, of the environmental temperature). Critical densities have been worked out for some malarious regions; the disappearance of endemic malaria from the United States was the result of reducing the mosquito and parasite populations over a number of years until the critical density was reached. Malaria eradication throughout the world depends on the same principle being deliberately applied.

In other parasitisms critical density is certainly important but has not been studied as thoroughly as in malaria.

Examples of the practical use of the above principle are found, however, in screwworm control by release of sterile males (see Chapter 25) and in the control of schistosomiasis and Gambian sleeping sickness by the trapping of vectors. In Chapter 25, we shall see that screwworm flies fluctuate seasonally in numbers and that matings occur only once for each female fly. By irradiating vast numbers of laboratory-reared male flies and releasing these sterile flies into the environment, entomologists can cause the majority of matings to be without offspring. This results in populations so low that they cannot maintain themselves against the rigors of cold or the constant attack of predators, and extinction follows. Trapping snails (see Chapter 9) in submerged brush at fords or other places where people can be penetrated by schistosome cercariae and removing the trapped snails from which the local cercariae escape is an effective way to lower the density of schistosomiasis in a region where this practice is suitable. In time, such a control measure by itself would greatly reduce the incidence of blood-flukes; with mass treatment available to prevent infection of new generations of snails, the density of schistosomes in certain areas might be reduced below the critical or extinction level. Gambian sleeping sickness (see Chapter 4) is being controlled by a similar method. This disease has long been concentrated along rivers and trails and at the wooded borders of cleared areas—any place where tsetse flies breed and where people travel or work. Despite the fact that they are the sole vectors, tsetse flies are not particularly "good" carriers, because only a small proportion of them are capable of harboring developmental forms of *Trypanosoma gambiense*. Therefore this parasite depends on rather high concentrations of flies and vertebrate hosts; such a condition is satisfied by trails and riverbanks. The high critical density of Gambian sleeping sickness makes it possible to control this disease without actually eradicating tsetse flies or parasites.

Examples of the converse of control by the use of critical density are seen in such diseases as coccidiosis in fowl (see Chapter 3) and human dracunculiasis. In these diseases, human factors have increased the frequency of parasite and host far above what is probably a very low natural critical density. Coccidiosis attacks flocks of chickens and causes extremely high mortality, for fowl under domestication are concentrated in small areas which become so heavily contaminated with infective coccidial oocysts that overwhelming infections occur. In nature, it is very likely that the jungle fowl ancestors of domesticated chickens would acquire coccidial oocysts at the rate of one or two at a time—a rate afforded by the sparse scattering of infective feces over large areas. Immunity, established by light initial infections, would protect the wild fowl from further attacks. (The principle of coccidiostasis, the use of drugs to reduce the numbers of developing coccidia in fowl in order to allow time for immunity to be acquired, is applied widely in the poultry industry. In this way the natural conditions are simulated and coccidiosis is controlled.) It appears that the coccidia are adapted to survival under extremely difficult conditions for parasitism, i.e., wide dispersal of relatively small numbers of infective stages. When conditions change, as in domestication, the parasite changes from mild to dangerous merely through increased frequency and extent of inoculation. In terms of density-dependence, dracunculiasis is a similar disease. While it may be impossible to eradicate *Dracunculus* because of reservoir hosts, the present prevalence of guinea worms is unreasonably high. This parasite has been concentrated by human agencies—by trade, by population pressures which cause men to invade the inhospitable desert, by the over-use (for bathing, laundry, and drinking) of limited and concentrated water supplies, and by failure to recognize the importance of these factors. This concentration of people and parasites has resulted in epidemiclike frequencies of a parasite which would cause little harm in a hunting or food-gathering society.

CONCLUSION

In conclusion, dracunculiasis remains a puzzling, an interesting, and a challenging disease. Its history shows both progressive and conservative aspects of medicine. The prehistoric discovery of a simple cure has achieved nothing of value in public health, because the cure is a treatment of symptoms only and has no effect in prevention or control. Modern medicine for nearly a century has known how to prevent and control dracunculiasis, yet the prevalence of this disease is probably greater than ever. One hundred years may not seem long in the history of the human species, but in that time perhaps 500 million *Dracunculus* infections have brought pain, deformity, or death. The modern staff of Aesculapius seems less effective than the legendary rod of Moses. For the speculative parasitologist, dracunculiasis shows an interesting evolutionary pattern, which has been mentioned briefly in earlier chapters. It shows how the civilizing activities of man may bring misery through the concentration of parasites and the multiplication of the risks of infection. Moreover, such parasites as *Dracunculus* and *Trypanosoma gambiense* evolved successfully before man learned to crowd his planet, and hence they have a low critical density. Since these parasites are almost impossible to eradicate, they must therefore be accepted but kept under control. Control will be accomplished through education, which should be able to make simple preventive measures acceptable to the millions who must adopt them.

SUGGESTED READINGS

LINDBERG, K., "La dracontiasis en Asie, particulierement en Moyen-Orient, avec liste des Cyclopides recueillis dans des regions epidemiques," *Rev. de Paludisme*, 8:87–111, 1950.

MOORTHY, V. N., "A Redescription of *Dracunculus medinensis* Velsch," *J. Parasit.*, 23:220–224, 1938.

ONABAMIRO, S. D., "The Transmission of *Dracunculus medinensis* by *Thermocyclops nigerianus*, as Observed in a Village in Southwest Nigeria," *Ann. Trop. Med. Parasit.*, 45:1–10, 1951.

ONABAMIRO, S. D., "The Early Stages of the Development of *Dracunculus medinensis* (Linnaeus) in the Mammalian Host," *Ann. Trop. Med. Parasit.*, 50:157–166, 1956.

ROUSSET, P., "Essai de prophylaxie et de traitement de la dracunculose par la notezine en Adrar," *Bul. Med. Afr. Occid. Franc.*, 9:351–568, 1952.

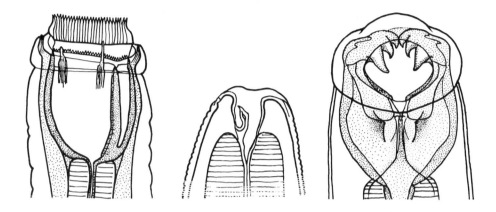

CHAPTER 20

OTHER NEMATODES

The final chapter on nematodes will include several noteworthy parasites which are perhaps not significant enough to merit individual chapters. The first of these is *Strongyloides stercoralis*, a dangerous parasite of man and dogs. This genus is the only one of medical or veterinary importance in the superfamily Rhabditoidea, which contains mostly free-living nematodes. Various species of *Strongyloides* inhabit domestic animals and fowl. The second is a group of parasites of agricultural animals—the strongyles, trichostrongyles, and metastrongyles. These bursate nematodes are members of the superfamily Strongyloidea, the same group to which the hookworms (Chapter 15) belong. The third is the eyeworm (*Thelazia* spp.) and various stomach- and esophagus-worms—all members of the superfamily Spiruroidea. Last, *Dioctophyma renale*, the giant kidney worm, will be mentioned.

STRONGYLOIDES STERCORALIS

This is one of the most versatile parasites known. It can live indefinitely in warm, moist, fecally polluted soil. In such an environment it can continue from generation to generation. Some of its larvae, however, can penetrate the skin of man, dogs, foxes, cats, or chimpanzees in the same manner as hookworm larvae. It is then carried by the blood to the lungs, where further development occurs. Sometimes females mature in the lungs and lay eggs either after fertilization by males or by parthenogenesis. Even in the intestinal mucosa, where the majority of the parasitic adults are found, many females are believed to be parthenogenetic. Once they reach maturity, the males cannot survive in the tissues, but they reach the intestinal lumen and are evacuated. Eggs laid in the tissues of lung or intestine promptly hatch. To complete a generation in a single host, the larvae may then moult, become infective, and reinvade the tissues. (Such autoinfection makes possible very large worm burdens and correspondingly severe illness.) Many larvae which hatch in the intestine, however, reach the soil with feces; in the soil they reestablish a free-living cycle with the potentiality of invading new hosts via the skin. The above life cycle, surely one of the most complex in terms of "options," is illustrated in Fig. 26–3.

The distribution of S. stercoralis follows the hookworm pattern wherever moderate temperatures, moist loose soils, and fecal contamination occur. Unlike human hookworms, however, this parasite does not depend on continuous seeding of the soil with eggs and larvae; its independent cycle of growth and reproduction frees it from this reliance on insanitation. Also, because dogs are reservoir hosts, the absence of human disease in an area does not mean that there is no danger of acquiring Strongyloides. Enough isolated cases of human strongyloidiasis have shown up to implicate dog-feces pollution of soil as a source of human infection. Strongyloides stercoralis frequently occurs in dogs in regions where human sanitary practice precludes human sources.

The effects of strongyloidiasis are quite varied. Typical skin irritation, similar to the itch and rash caused by hookworm larvae, is a symptom of larval invasion. Pulmonary symptoms (cough, pain, fever) accompany involvement of the lungs, and severe diarrhea results from intestinal invasion. As in most parasite-caused diseases, some hosts suffer relatively little injury, while others may be severely affected. The disease, with acute febrile and dysenteric symptoms, sometimes runs a rapid course. Eosinophilia is pronounced. Many of the varied effects of strongyloidiasis are believed to have resulted from allergic response to the invading larvae after a state of immunity has developed. Without treatment, recovery may be possible, but a chronic state of ill health may accompany for years the continued presence of the worms. Death may occur.

Treatment of strongyloidiasis has only recently been developed. Dithiazanine iodide, recognized now as a broad spectrum anthelmintic, is effective in treating strongyloidiasis as well as hookworm disease and, notably, trichuriasis.

S. stercoralis of men and dogs is not the only species of importance. S. papillosus occurs in sheep, goats, cattle, and wild ruminants. S. westeri parasitizes horses, pigs, and zebras. S. cati in cats, S. ransomi in pigs, and S. avium in various domestic and wild fowl have been described. So far as is known, the life cycles of the above species resemble that of S. stercoralis. Pathogenicity is also similar. Although dithiazanine is probably effective, treatment of veterinary animals has not been well worked out. As in the case of man, prevention of veterinary strongy-loidiasis depends on sanitation. Dry, clean stables, corrals, or cages will not permit the survival of *Strongyloides* outside an animal's body.

SUPERFAMILY STRONGYLOIDEA

Cattle, sheep, and horses, as well as other valuable animals, are afflicted by a large number of species of worms belonging to the superfamily Strongyloidea. (See Figs. 20–1 and 20–2.) (The reader should take warning that *Strongyloides*, the genus just discussed, is not a member of the Strongyloidea and does not resemble these worms.) The hookworms, strongyles, trichostrongyles, and metastrongyles are various groups of bursate nematodes. The copulatory bursa of the males not only serves to distinguish the superfamily from others but can be used in identifying species among the many forms which are found in farm animals.

STRONGYLES

The family Strongylidae contains a number of subfamilies and genera. All are characterized by cuplike mouth cavities surmounted by "leaf-crowns" of spikelike papillae. The males, like all Strongyloidea, are bursate. Collectively, these worms are called strongyles.

Over 40 species of strongyles parasitize stock; it is convenient to refer to the condition in general and then to consider a few of the more important worms. The disease-complex, strongyliasis, is caused by the chiefly mechanical trauma due to the invasive larvae and voracious adults, and is accompanied by secondary infection. The larvae of strongyles are acquired from pasture grass, upon which the tiny worms have climbed from feces-contaminated soil. Within the digestive tract these larvae move to the walls, where they bore into the tissues; in some species they progress to mesenteries and the walls of arteries, and in other species they remain in swellings or nodules of the intestine. The pathological effects include aneurysms (weak and swollen places in the arteries) and abscesses (caused by the introduction of bacteria into the intestinal nodules or lesions). The adult worms return to the digestive tract and feed actively upon the mucosa; they usually suck blood and cause innumerable small hemorrhages. Each species of strongyle invades a particular group of tissues and inhabits as an adult a special region of the intestine; but because more than one species is usually involved, strongyliasis is a general disease of intestinal and related tissues. Specific effects of certain strongyles will be mentioned when these species are discussed below. Since strongyliasis is serious and widespread, many studies have been undertaken to improve methods of prevention and cure. Prevention depends on protection of young animals from infection. This is a very difficult problem, because the larvae of strongyles survive desiccation and ordinary temperature fluctuations and can be found in puddles of water in low pastures, in hay cut and stored for winter fodder, and, of course, in the grass of open grazing land. The adult worms can be removed

FIG. 20–1. Hookworms, strongyles, and trichostrongyles. (A) *Ancylostoma caninum* and (B) *A. braziliense*, dog hookworms causing creeping eruption in man. The heads show teeth characteristic of the genus. (C) Head and tail of *Bunostomum* sp. (D) *Strongylus vulgaris* and (E) *S. equinus;* two common strongyles of equines, showing large ornate buccal capsule characteristic of the family Strongylidae. (F) Typical copulatory bursa of a strongylid worm, showing bilobed appearance. (G) Head and tail of a member of the family Trichostrongylidae, *Trichostrongylus colubriformis*, a parasite of various herbivores and occasionally man. (H) *Haemonchus contortus*, head and tail of male, an important "wireworm" of livestock. Note the typically trichostrongylid mouth [cf. hookworms (A, C) and strongyles (D, E)]. [After LaPage, 1962.]

by phenothiazine, carbon tetrachloride, or other anthelmintics. Mass treatment of stock, carried out routinely, will greatly reduce the numbers of larvae in pasture and corral. Prophylactic use of phenothiazine has been recommended. But unless continuous and well-planned sanitary measures are followed, strongyliasis will remain a threat especially to young animals, which may acquire fatal infection from symptomless carriers.

In horses, strongyliasis causes various symptoms, which depend on the parasite or parasites involved. The most serious damage results from larval invasion by *Strongylus vulgaris*. These worms, entering the lumen of intestinal arteries wherein the larvae live, cause thrombi which adhere to the artery walls. As an artery becomes progressively blocked with such fibrous clots, its walls weaken and stretch. These weak places may rupture and cause fatal internal hemorrhage. The arterial obstruction blocks circulation to the intestine and causes the intestine to deteriorate. Nervous tissue, such as mesenteric ganglia, may also be affected. Often an acute and destructive form of colic results. Treatment of the above conditions is palliative or supportive. Other strongyles affect primarily the mesenteries (*S. edentatus*), the pancreas and liver (*S. equinus*), or the intestinal wall itself (*Trichonema* spp.). Since more than one species of strongyle may be present in a single host, any combination of symptoms may occur. Other horse strongyles, sometimes called the "small strongyles," are species of *Triodontophorus* (affecting the right colon and causing ulcers), *Craterostomum* (in the colon), and *Oesophagodontus* (rather rare, in the colon).

Ostriches suffer from parasitism with strongyles of the genus *Codiostomum*. *C. struthionis* lives in the large intestine of the bird. Little is known about the life cycle or the effect upon the host, but presumably this is a dangerous parasite. As in other strongyles, the life cycle is probably direct. Therefore heavy infections might build up under conditions of domestication.

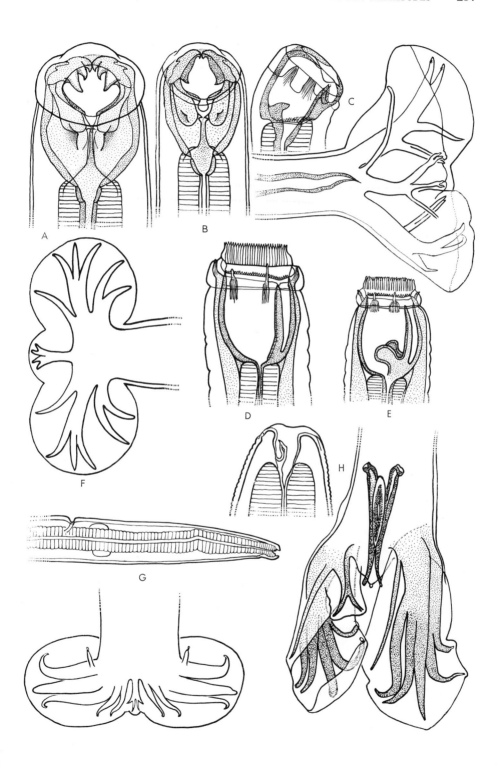

Sheep and goats harbor the "nodular worm," *Oesophagostomum columbianum.* This species lives in the colon, where the larvae penetrate the mucosa and cause nodules to form. Within the nodules, the larvae develop. Adults leave the nodules and inhabit the lumen. Symptoms occur only if a large number of nodules are present; then peristalsis and absorption are interrupted, and persistent diarrhea follows. The adults irritate the mucosa of the colon. The disease progresses from diarrhea and waste to extreme emaciation, prostration, and death. Phenothiazine in a fine-particle suspension is effective in removing the worms and curing the disease. Since the life cycle is typical of strongyles, treatment at times when the pasture has been cleared of larvae by extreme dryness or cold will help prevent reinfections of the grazing areas. Other nodular worms of the same genus occur in sheep, goats, cattle, camels, and other ruminants. Pigs also harbor a nodular worm.

Chabertia is a second genus of strongyles affecting the colon of ruminants. These worms feed actively on the mucosa; they do not suck blood but chew and irritate the lining of the colon. Severe infections cause sheep to lose condition and die.

A strongyle of hogs infects the kidneys. This "kidney worm" is *Stephanurus dentatus;* it lives in the fat around the kidneys, the pelvis of the kidney, and the walls of the ureters. Occasionally kidney worms are found in other organs. Swine acquire the infective larvae either orally or through the skin. The larvae are relatively sensitive to dryness and cold; *S. dentatus* is a parasite of the tropics and subtropics. While kidney worms seldom cause severe symptoms, the wanderings of the larvae and adults cause local inflammatory reactions which make the kidneys and sometimes the whole carcass unfit for market. Thus economic loss ensues.

TRICHOSTRONGYLES

The family Trichostrongylidae includes strongyloid worms without the well-developed cuplike buccal capsule or mouth cavity of the members of the Strongylidae. The bursa of the males is very broad. These small worms are parasitic in the digestive tracts of sheep, cattle, and horses. They appear also in other vertebrates, including man. The worms suck blood from the wall of the stomach or intestine to cause anemia and also toxic reactions. Death results more frequently from toxic reaction than from anemia. Like the stronglyes, however, the trichostrongyles often contribute to the effects of a worm burden consisting of several species. The known life cycles of trichostrongyles are direct. Either eggs containing sheathed third-stage larvae or the larvae themselves are eaten by grazing animals. In the latter, the larval worms bore into the lining of the digestive tract, and further growth occurs often in lesions or blood clots caused by the worms. However, the mature worms generally live free in the lumen. They feed by piercing the mucosa with their small mouths, injecting anticoagulants, and voraciously sucking blood. Often occurring by the thousands, they cause the collapse of some animals by toxemia and the chronic wasting away of others by loss of blood. A discussion of the more important trichostrongyles follows.

Trichostrongylus contains eight or nine species. *T. colubriformis* occurs in the anterior small intestine and abomasum (posterior stomach) of sheep, goats, cattle, camels, and antelopes. Several instances of human infection with this worm have been reported, and dogs and rabbits can harbor it. *T. axei* of horses is an important species. The effects of *Trichostrongylus* result from the presence in the intestine of a large number of worms which cause desquamation (loss of intestinal epithelium) with consequent diarrhea and weakness. Heavy infections cause death rather quickly. Acute symptoms are due to toxic effects; the animals weaken and collapse. Chronic cases show emaciation, dry skin, occasional diarrhea, and mild anemia.

Ostertagia, a genus of trichostrongyles, contains the "brown stomach worms," so-called because of their color when alive. These worms, appearing in the abomasum of sheep, goats, cattle, etc., live in nodules which are produced by invasion of the mucosa by infective larvae; the adults often remain in the nodules. Adults are also found free in the lumen, where they live on the blood of the mucosa. Their chief effect is anemia as a result of blood-sucking.

Cooperia, another important genus, contains small reddish worms with a transversely ridged cuticle; they occupy the anterior part of the small intestine of ruminants. Five or more species are known. The symptoms are similar to those caused by *Trichostrongylus;* however, anemia is more evident in *Cooperia* infections.

Nematodirus is a genus of rather long worms (longer than most trichostrongyles) which lay relatively large eggs. These are intestinal worms, which cause symptoms similar to those of *Trichostrongylus*. It is noteworthy that *Nematodirus* is relatively resistant to the widely used anthelmintic, phenothiazine, which, by eliminating "competing" parasites, sometimes permits *Nematodirus* to thrive. *N. spathiger,* the commonest species, infects many ruminants, including sheep and cattle. *N. battus* and *N. helvetianus* occur in sheep and cattle in Great Britain.

Haemonchus contortus, living in the abomasum of ruminants all over the world, is known as the "stomach worm" or "wireworm." The red male and the red- and white-striped female (the white ovaries coil around the red intestine) are distinctive. The cuticle is both transversely striated and longitudinally ridged. These worms are 10–20 mm (males) and 18–30 mm (females) in length. Like other strongyles and trichostrongyles, the species of *Haemonchus* possess a larval stage which involves the tissues (mucosa of the stomach) and an adult stage which feeds on blood. Blood loss due to wireworm infection may be very severe. In chronic anemias caused by *Haemonchus*, edema occurs and produces the watery swelling of the jaws called "bottle-jaw" and other swellings of the abdomen. In haemonchiasis, emaciation is masked by replacement of fat with gelatinous, watery tissue. Weakness leads to prostration and death. Fortunately, this severe trichostrongyle infection can be treated with a number of remedies, including phenothiazine as well as several recently developed drugs such as organophosphorous compounds. Since sheep and cattle acquire protective immunity from haemonchiasis, a vaccine consisting of irradiated larvae (cf. *Dictyocaulus* vaccination below) has been developed and has proven useful. Prevention and control depend on pasture rotation, control of grazing, and climatic factors such as humidity.

Mecistocirrus digitatus infects sheep, cattle, zebu, buffalo, pigs, and rarely man in tropical regions. It is similar in effect to *Haemonchus*.

In conclusion, the family Trichostrongylidae, like the Strongylidae, contains a number of genera of somewhat similar worms affecting the tissues of the intestinal tract of ruminants and other animals. These worms are all directly transmitted. Therefore very heavy infections can be built up in animals restricted to limited pastures, feed lots, or pens. Such infections, like the coccidial plagues which affect domestic fowl, are "unnatural" in the sense that they are produced by man-made conditions. They can, of course, be controlled or prevented, but only by careful sanitation, by methodical treatment to reduce the number of worms and larvae in the environment, and by special attention, via nutritional supplement or artificial vaccination, for the purpose of building or maintaining resistance to these parasites.

METASTRONGYLES

The family Metastrongylidae (Fig. 20–2) includes parasites of the air passages of vertebrates. Like the trichostrongyles, the metastrongyles have an inconspicuous mouth and buccal capsule. In addition, the bursa is smaller and simpler than that of the other families of the Strongyloidea.

These parasites of the lungs, bronchial tubes, and alveoli of domestic animals include the metastrongyles of swine; *Dictyocaulus* of cattle, horses, sheep, and goats; *Muellerius* and *Protostrongylus* of sheep and goats; *Crenosoma* of foxes, dogs, etc.; *Aelurostrongylus* of cats; and *Angiostrongylus* of dogs, rats, and man. Because the lungs and air passages are essential to life and are also unusually susceptible to damage by secondary invaders, lungworms are important pathogens. Of the above parasites, only *Dictyocaulus* can directly infect its host; the others require an intermediate host, usually snails or earthworms, for a part of their developmental cycle. Treatment is more difficult for lungworm disease than for most helminthiases, because it involves intratracheal injections in some types of disease, or systemic drugs like emetine. Because of the essential intermediate host, control involves destruction or avoidance of that host; but although in one sense intermediate hosts make control easier, such hosts also are effective distributors of these parasites and bring about reinvasion of cleared pens or pastures.

FIG. 20–2. Lungworms. (A) Lung of sheep afflicted with "hoose," a congestive condition caused by *Dictyocaulus filaria*. (B) Larvae of (1) *Dictyocaulus*, (2) *Metastrongylus*, (3) *Protostrongylus*, (4) *Muellerius*. (C) Male and female specimens of *Metastrongylus pudendotectus*, a lungworm of pigs in most parts of the world. (D) *Crenosoma vulpis*, a lungworm of the fox. (E) *Syngamus trachea*, the fowl gapeworm. These worms, in permanent copulatory union, obstruct the trachea of poultry. [After LaPage, 1962.]

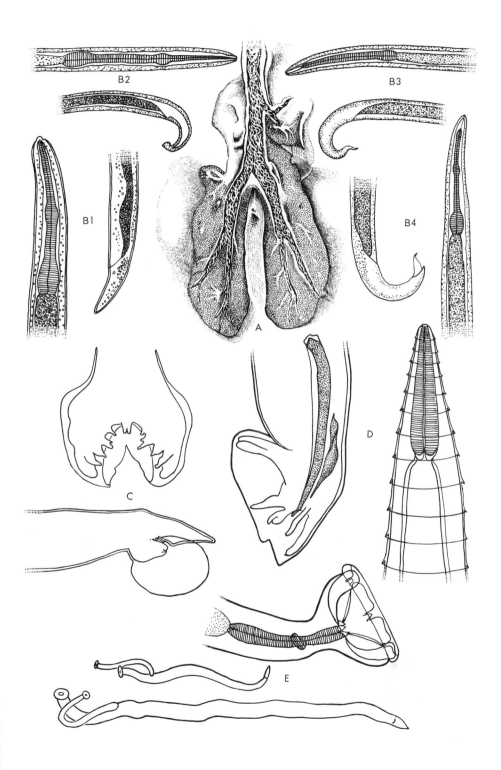

DICTYOCAULIASIS

Dictyocauliasis is acquired by horses, burros, sheep, or cattle which graze in pastures where larvae of this lungworm have been deposited with the feces of infected animals. The eggs hatch either before the feces are deposited or very soon thereafter; and after two moults the larvae, ensheathed in two exuviae, may be found widely scattered on grass, on the soil, and in low-lying spots where water collects. Although there is some evidence that infective larvae can survive mild to severe winter weather, it appears that the chief source of infection of stock in the spring is not such survivors but larvae being continually deposited in the pasture in the feces of infected carriers. After being ingested with grass or water, the infective larvae penetrate the wall of the intestine and make their way to the lungs via lymphatics and blood stream in a way similar to the larvae of other nematodes (*Ascaris*, hookworm). Various symptoms, including fever due to introduced secondary infections, occur during the above migration. After reaching the lungs, the parasites enter the alveoli, bronchioles, and bronchi, where they mature; in some species the females reach a length of over 50 mm, and the males up to 40 mm. About three to four weeks elapse between ingestion of larvae by the host and egg-laying by the female parasites. Depending on the condition of the host, the adult worms inhabit their host for a period ranging from a month or two to a year; well-nourished animals undergo self-cure.

Treatment of dictyocauliasis by intratracheal injection of iodine solutions has been recommended. Since this procedure is rather dangerous and since self-cure usually occurs in well-nourished animals, some workers suggest improved feeding as the only treatment needed.

Control of *Dictyocaulus* spp. depends on identifying carriers or chronically affected animals (usually adults) and removing them from the pastures where young animals graze. Such identification is by immediate examination of feces for the larvae. Since old feces may contain the larvae of other worms (e.g. hookworms, strongyles), freshly passed feces or even feces extracted from the rectum should be used. Pastures may be rotated, since infective larvae decrease in numbers unless constantly augmented by larvae from carriers. Both American and Russian experts recommend keeping calves or foals away from infected animals or pastures; as the young animals mature, they become less susceptible to clinical disease and, if infected, generally throw off the infection in a few months. Prophylactic use of phenothiazine in feed or with salt has been recommended as a way of reducing the intensity of infection and allowing recovery with immunity to occur. Ingenious experiments by Jarrett and others have led to a method of artificial immunization against the cattle lungworm, *Dictyocaulus viviparus*. Collected larvae are irradiated with gamma rays from Cobalt 60, a radioactive isotope widely used in experimental and clinical radiobiology. Radiation prevents the larvae from reaching full development but permits their growth when fed to susceptible animals. As a result, these animals develop protective immunity. As we shall see in Chapter 27, residence of a living helminth for a considerable time within its host is usually necessary in order that immune responses become effective. In other conditions (for example, haemonchiasis), larvae "attenuated" by radiation have been used effectively. Whether it is better to attempt to eradicate dictyocauliasis from a farm or to control it by sanitary, prophylactic, and quarantine measures depends on economic factors which must be determined for each particular case. Unfortunately, such economic problems are not easily solved, and more research in economic parasitology is badly needed.

OTHER LUNGWORM DISEASES

Metastrongyliasis, lungworm disease of swine, is similar in effect to dictyocauliasis (above) of other hosts, but its causative organism has a relatively complex life cycle, and occasionally severe outbreaks occur among young animals. The mortality rate of those infected may reach 30%. The female worms in the lungs lay eggs which are swallowed and passed by the host. When ingested by earthworms, these eggs hatch, and the larvae undergo development to the infective stage in the body of these intermediate hosts. When the latter are ingested by pigs, the larvae escape, enter the intestinal wall of their host, and reach the lungs in the manner of *Dictyocaulus* larvae. The complete cycle requires 35–55 days. During larval migration, pathological changes occur in the intestinal wall, lymphatics, and lungs. Lung damage is rather general and includes cellular infiltration, nodule formation, and tissue proliferation in alveoli. Hemorrhages occur in the lungs. Symptoms, in addition to evidence of poor condition, are chiefly a cough; sometimes emphysema is present. Treatment by intratracheal injection of iodine solution is effective in removing *Metastrongylus* from the lungs. This important disease can be controlled by worming infected animals, collecting and storing the feces to allow biothermic destruction (at 55°C) of embryonated eggs, and using temporary, frequently ploughed pastures (to reduce earthworm frequency). Earthworms with viable lungworm larvae have been found in contaminated pastures as long as four years after the last use of such pastures by swine.

Muelleriasis of sheep and goats is caused by a minute lungworm (Fig. 20–2B4). In the United States it is not considered an important disease, but in Europe and the USSR considerable attention is given to it. Larvae hatch from eggs laid in the lungs; they are carried to the pharynx and are swallowed; then they pass out with feces. Such larvae moult twice after they penetrate the skin of various slugs and snails. They either emerge from the snail as infective larvae or remain indefinitely within. Ingested with grass by sheep or goats, the larvae characteristically migrate to the lungs via intestinal wall, lymphatics, and blood. In the lungs they mature. Several months are required for the complete cycle. In lambs, heavy infections cause severe but nonspecific symptoms, which are chiefly the result of damage to the lungs. Treatment with intramuscular injections of emetine (cf. amoebiasis) is said to be effective. Protection of young animals by restricted grazing is perhaps more practical than attempts at eradication, since the larvae of *Muellerius* are as durable as those of some strongyles mentioned earlier. (Protostrongyliasis of sheep and goats is similar to muelleriasis; the parasite's infective larva retains its exuviae as a hard cystlike shell, which protects the worm.)

Other lungworms include those of carnivores and fowl. Crenosomiasis is a disease of foxes, dogs, and other carnivores caused by *Crenosoma*, a small nematode bearing annular ridges on its cuticle. Females deposit larvae in the bronchial tubes of the host, and these larvae behave like those of *Protostrongylus* and *Muellerius* and complete their development in snails. Instead of escaping from the snails, however, the worms remain indefinitely in these intermediate hosts. Ingested by foxes, etc., they migrate to the lungs. Young foxes suffer from the disease more than the old. In fox farms, sanitary precautions, including the use of snail barriers or repellents, reduce the risk of outbreaks or heavy infections. Treatment by intratracheal injection has been suggested. Another lungworm of foxes is *Eucoleus aerophila* (syn. *Capillaria aerophila*), a trichuroid, which has direct transmission. This parasite is important in North America. Carriers serve to maintain the infection, which causes losses among young animals. Poor sanitation is the chief element in the development of outbreaks in fox farms.

Aelurostrongyliasis, caused by *Aelurostrongylus abstrusus*, is a pulmonary disease of cats. The adult worms live in the smaller pulmonary arteries. The eggs reach the capillaries of the alveoli, where they hatch. Larvae enter the air passages, ascend to the pharynx, and are swallowed. They pass out with feces. They develop to an infective stage in many kinds of snails and slugs. Transport hosts are important in transmission, for cats do not often feed on snails. Various rodents, lizards, and birds acquire the infective larvae of *Aelurostrongylus*, and the larvae encyst in the tissues of these hosts. When ingested by cats, the larvae penetrate the mucosa of the upper digestive tract, are carried by the blood to the lungs, where they complete development. In heavily infected animals, the large numbers of eggs and the lesions caused by escaping larvae produce a severe cough. Diarrhea and emaciation are present in heavy infections, which sometimes result in death. Similar worms from the leopard and tiger, as well as domestic cats, belong to the genus *Bronchostrongylus*. In dogs and foxes, *Angiostrongylus vasorum* lives in the pulmonary artery, where it lays eggs which lodge and hatch in the alveolar capillaries to cause breakdown of the smaller arteries in the lungs, obliteration of interalveolar septa with resulting emphysema, and long-range effects such as liver congestion, ascites, and heart failure.

Another species of *Angiostrongylus*—*A. cantonensis*—has been implicated in several cases of human meningitis and encephalitis. Larvae have been found in brain lesions of such victims. *A. cantonensis* is a lung parasite of rodents. Its third-stage larvae are the result of a necessary development in various species of slugs. Transport hosts may also occur. When ingested by its final host, the infective larva penetrates to the blood system and may be carried to any part of the body. Heavy infections (50 or more larvae) are fatal to laboratory rats. All organs of the body, including the brain and spinal cord, may be involved. The lesions seem to be granulomatous; that is, they involve local cellular immune responses causing local damage. Because of the dangerous nature of the larval migration, more study is needed on the life cycle of *Angiostrongylus cantonensis* especially on the possible ways in which man can be infected.

A relatively rare strongyle, *Filaroides osleri*, which inhabits nodules in the trachea and bronchi of dogs, causes coughing and produces a chronic infection with loss of appetite and some emaciation.

Syngamus trachea, the gapeworm of fowl, inhabits the larger air passages of chickens and other poultry. Unlike most other lungworms, this worm is not a metastrongyle; it belongs to the family Strongylidae with the true strongyles, or perhaps it belongs to a family of its own. *Syngamus* feeds on the mucosa and blood. Large enough to be visible when the bird's neck is held next to a source of light, the worms show distinctly within the translucent trachea. They appear to be Y- or T-shaped, because the males are attached in permanent copulation to the larger females (Fig. 20–2E). Young chicks may suffocate from tracheal blockage by the gapeworm. The cycle involves no intermediate host, but the eggs, passed out with feces, require about a week to become infective. At this stage the eggs may hatch and the larvae may be ingested. Or the eggs may hatch within the intestine, and the larvae may migrate to the lungs and bronchi. The cycle takes about thirty days to complete. Adult worms do not live more than 45–50 days in the final host. Infected chicks may be treated with intratracheal introduction of iodine. Since older birds do not suffer much and will lose the parasites spontaneously after a month or two, treatment is not necessary. Protection and control depend on sanitation; birds raised on wire netting are not apt to acquire tracheal worms in large numbers. Since various insects, snails and earthworms may ingest the eggs or larvae of *Syngamus* and transport them, and since numerous reservoir hosts, including starlings, harbor gapeworms, eradication of the parasite from poultry farms is probably impossible.

Species of *Syngamus* also infect Asiatic cattle and buffaloes, as well as sheep, cattle, goats, and deer in various tropical and subtropical regions. A related form, *Cyathostoma bronchialis*, lives in the trachea and bronchi of ducks, geese, and swans, in which losses result especially among the young.

AMIDOSTOMUM

The small family Amidostomidae is somewhat similar to both the Strongylidae and the Trichostrongylidae. Its members have a broad, shallow buccal capsule, which lacks the ornate features of a strongylid mouth but is a more prominent structure than that of the trichostrongyles. One species of importance is *Amidostomum anseris*, the gizzard worm of ducks and geese. The horny layer of the gizzard is eroded by migrating worms and then the host becomes sluggish and emaciated. Geese, especially the young, die from amidostomiasis. Larvae develop within eggs which are found in the feces. Two moults occur; then the infective larva hatches from the egg and swims actively in the water of a pond or puddle or in the moisture of the soil. It is able to crawl about on the surface of grass. Larvae usually live 15–25 days, and continuous infection depends on new eggs being hatched. Amidostomiasis is maintained by carriers—the adult birds. Wild birds serve as reservoirs and as agents of dissemination. Treatment of infected birds (including carriers) and rotation of pasturage and ponds can effectively control gizzard worms.

SPIRUROIDEA

The superfamily Spiruroidea (Fig. 20–3) is primarily a collection of parasites of fishes. Two lateral lips, simple or subdivided, are a basic feature; there is a partly glandular and partly muscular oesophagus, and the tail of the male is usually spiraled.

Habronema and *Drascheia* are spiruroid worms which inhabit the stomach of horses and donkeys to cause gastrointestinal symptoms. The former nematode also causes "summer sores," a widespread disorder in which third-stage larvae of *Habronema* inhabit skin wounds and cause aggravation of the sores. The larvae of both parasites occur in flies, the maggots of which feed on feces containing active larvae. Fly control is a means of controlling these diseases.

Hartertia gallinarum affects bustards and domestic fowl in Africa. Living in the small intestine, these worms, if numerous, cause diarrhea, weakness, and death. Termites are the necessary intermediate host.

Spirocerca lupi lives in the walls of the esophagus, stomach, and aorta of dogs, wolves, foxes, and jackals. Larval migration causes lesions which heal; the result is stenosis (narrowing) of the blood vessels. Adult worms live in tumors in the walls of the affected organs. The worms are transmitted by dung beetles, in which the larvae develop to the infective stage, and by various transport hosts such as rodents, lizards, birds, etc., in which larvae freed from ingested beetles reencyst.

FIG. 20–3. Some veterinary ascaroids and spiruroids. (A1) *Ascaris lumbricoides*, var. *suis, en face*. The lips with their "dentigerous ridges" are slightly different from the lips of the human variety of *Ascaris*. (A2) Tail of male. (B) Head of *Toxocara canis*, the dog ascarid. (For life cycle, see Fig. 16–4.) (C1, 2) Tail and head of *Heterakis gallinae*, a caecal worm of poultry, probably a transmitter of histomoniasis (see Chapters 5 and 6). The preanal sucker found in members of the family Heterakidae is seen anterior to the cloacal aperture, through which the unequal spicules protrude. The pattern of perianal papillae is, as usual, diagnostic of the species. (C3) Tail of *Ascaridia galli*, a poultry worm. (D1) Adult female of the horse pinworm, *Oxyuris equi*, and (D2) tail of male. (E) A spiruroid, *Habronema megastoma*, which occurs in tumors of the stomach wall of equines. The bilateral symmetry of the mouth, with its paired, lateral lips, is characteristic of the superfamily Spiruroidea. (F) *Tetrameres*, distended with eggs. This worm feeds on blood in the walls of the proventriculus (gizzard) of poultry. The small, spinose male worms usually are found free in the lumen; the females cause most of the injury. (G) *Acuaria* (*Dispharynx*) *spiralis*, a spiruroid of the walls of the proventriculus and crop of poultry. Sinuous ridges called "cordons" are characteristic of some spiruroids. (H) *Thelazia*, the eye worm. These spiruroids affect the eyes of cattle, occasionally man. [After LaPage, 1962.]

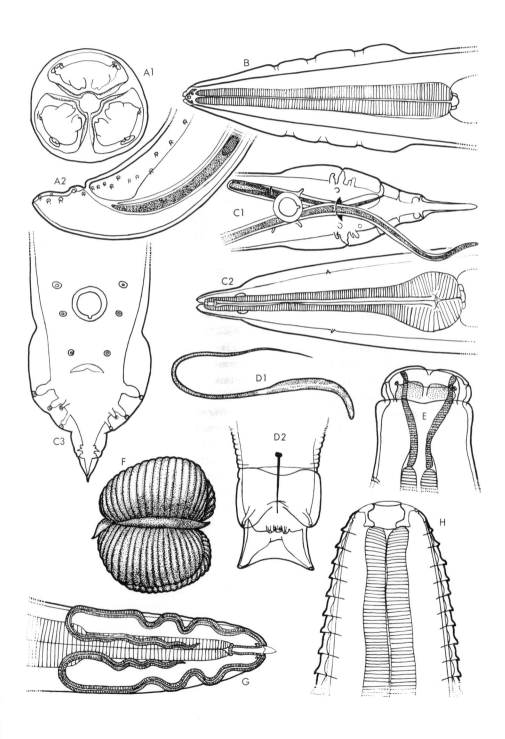

Included among spiruroids affecting stock are the stomach worms of pigs. These worms, members of the genera *Ascarops* and *Physocephalus,* irritate the gastric mucosa but cause little damage unless they become very numerous. Dung beetles serve as intermediate hosts.

Gongylonema, characterized by warty thickenings in the anterior region, lives in the esophagus of cattle, pigs, sheep, various other grazing animals, and occasionally man. As it weaves itself in and out of the mucosa, it causes little damage. One species, *G. neoplasticum* in rats, stimulates the development of cancers in these hosts.

Acuaria, which bears "cordons" (looped or sinuous ridges near the mouth), contains species in the gizzard and esophagus of fowl. (See Fig. 20–3G.) There the worms have somewhat the same effect as *Amidostomum* (see above) but are not so pathogenic. By softening the musculature, they weaken the gizzard and interfere with the birds' nutrition. The esophagus worm, *Acuaria spiralis,* affects the crop and esophagus of turkeys, pigeons, chickens, etc. It uses isopod crustacea (pill-bugs) as an intermediate host. *A. uncinata,* infecting water fowl, utilizes *Daphnia* species (water-fleas) as intermediate host.

The curious worms belonging to the genus *Tetrameres* are remarkable for the greatly swollen condition of the gravid female. (See Fig. 20–3F.) While such swelling is seen in a number of plant-parasitic nematodes, *Tetrameres* is the only common nematode of animals showing this development. *T. americana* inhabits the proventricular gland of birds. The female lives in the gland. while the spiny, slender male lives free in the lumen of the crop. The young worms migrate in the tissues of the crop and cause injury which may kill young birds. Various orthopterans (grasshoppers, cockroaches, etc.) serve as intermediate hosts.

Physaloptera praeputialis is a stomach worm of cats in the tropics. The cuticle forms a sheath which extends beyond the worm's anterior end. The worms live attached to the mucosa, which may become eroded and inflamed. Various species of *Physaloptera* infect canines, badgers, and other animals. A number of insects transmit the parasites. *Physaloptera* is important in zoos, where confinement and the presence of scavenging insects facilitate heavy infections.

Gnathostoma is a stomach worm of cats, dogs, and wild carnivores. It occasionally infects the skin of man as a cutaneous *larva migrans* (see under Hookworms). This parasite is highly pathogenic; its larvae migrate throughout the liver of its normal hosts, and the heavily spined adults excavate cavities in the stomach lining. The life cycle is complex. The first-stage larvae is ingested by copepods, in which some growth occurs. The larvae develop further in freshwater fish, various frogs, or reptiles—true second intermediate hosts in which they encyst. The encysted forms emerge in the carnivore host to migrate before taking up residence in the stomach.

The spiruroid family Thelaziidae contains two genera of eye worms. *Thelazia* (about seven species) affects mammals, and *Oxyspirura* (several species) affects fowl.

Eye worms cause irritation to the eyes of sheep and other animals as well as man. Conjunctivitis in cattle, sheep, horses, pigs, and camels is caused by the presence of adult *Thelazia* in the lacrimal ducts and conjunctival sacs of these hosts. It is not certain that the worms alone cause that condition for many infected animals are without symptoms. The worm is transmitted by species of flies of the genus *Musca* which commonly feed on the eye secretions of the final host by lapping up lacrimal fluid containing the larvae of the viviparous parasite. The larvae require a month to moult, develop, and become infective; they migrate to the proboscis of the fly, and creep out, presumably as do filariids, through thin parts of the fly's exoskeleton when the fly is feeding near the eye of the final host. Deposited larvae enter the lacrimal duct and conjunctiva and mature within several weeks. In some parts of the world summer infection rates become very high, so that nearly all the animals of a herd may be infected. If untreated, thelaziasis causes scarring of the cornea, hypersensitivity to light, and sometimes permanent loss of sight. In the USSR, economic loss due to required slaughter of blinded animals is said to be very high. Diagnosis may be made not only by recognizing symptoms, which resemble those of other kinds of conjunctivitis, but also by examining the lacrimal fluid for larvae. Eye worms migrate frequently but rapidly over the surface of the eyeball. They can be mechanically removed during this migration. Anesthetization of the eye aids in the search for eyeworms. Washing the eye with a strong jet of boric acid solution is useful in removing worms which may be concealed beneath the nictitating membrane.

Eye worms occasionally affect man and can be diagnosed and removed by the methods developed for veterinary thelaziasis.

Eye worms of the genus *Oxyspirura* live under the nictitating membrane of chickens, turkeys, and peafowl all over the world. Symptoms are similar to those caused by eye worms of mammals, but the life cycle is different. *Oxyspirura* adults lay eggs, which pass down the lacrimal ducts to the digestive tract and are voided with feces. Cockroaches and probably other insects serve as intermediate hosts. Presumably the larvae leave the ingested insect and migrate via the esophagus, the pharynx, and the lacrimal duct to the eye.

DIOCTOPHYMA RENALE

The superfamily Dioctophymoidea has unusual characteristics already mentioned (Chapter 13). *Dioctophyma renale*, the kidney worm of carnivores, is the largest known nematode. The male may be 35 cm by 3–4 mm, and the female may be as large as 103 cm by 5–12 mm. The worms are blood red. Like those of *Trichuris*, the eggs are barrel-shaped. They are covered with pits. The adults are found in a kidney (usually the right), which may be totally destroyed except for its distended capsule. Worms frequently escape from the kidney into the abdomen, where they cause adhesions, peritonitis, and various mechanical injuries. Yet despite the extensive lesions, infected animals sometimes show few symptoms, because a single kidney suffices for renal function. The life cycle of *Dioctophyma renale* involves several hosts. At present it is believed that the cycle starts with ingestion of the eggs by a branchiobdellid worm, an annelid ectoparasite of the gill chamber of crayfish. After hatching in the annelid, the larva escapes and encysts in the crayfish. When the crayfish is eaten by a fish, the larva encysts in the viscera of the fish. Mink, cats, or other carnivores, as well as man, can acquire the parasite by eating infected fish.

CONCLUSIONS: NEMATODES AND ANIMAL HEALTH

This final chapter on nematodes has been chiefly concerned with some veterinary parasites. Other worms affecting stock have been considered briefly in earlier chapters more or less as footnotes to human disease. Such are the agents of the animal filariases, the hookworms of dogs and other animals, *Ascaris* and *Trichuris* as causes of animal unthriftiness, and *Trichinella* as a zoonotic problem. The following remarks on veterinary disease caused by nematodes are applicable to the above as well as to the many parasites mentioned in this chapter.

In a review of veterinary disease caused by nematodes, a few principles stand out. First, the wide range of diseases (from nutritional deficiencies caused by mild ascariases to fatal arterial lesions caused by *Strongylus vulgaris*) and the great variety of animals affected (practically all domestic animals, including fowl) require that each nematode be studied separately; there is no general treatment for veterinary nematodiases. Continuing research on nearly all such diseases is desirable.

Second, methods of chemotherapy are neither fully understood nor completely developed. The valuable drugs phenothiazine and piperazine are effective in prophylactic as well as curative roles, in ascariasis, and against some strongyles, lungworms and filariae; strongyliasis in its more serious forms is resistant to therapy. Diet remains an important element in treatment, or rather, in the suppression of symptoms, just as it is in human hookworm disease.

Third, control of nematodes is dependent on particular conditions. Basic to control are accurate diagnosis, knowledge of life cycles, and understanding of the kinds of engineering and husbandry practices needed; but such understanding depends on education, which is inadequate in many agricultural areas of the world. Agricultural extension service should emphasize parasite control, and its experts should be able to teach individual farmers how to control parasites under the conditions of their particular stock-raising operation.

Fourth, the relationship between veterinary nematodes and human health is explicit in trichinosis and in such diseases as creeping eruption caused by larval hookworms, visceral larva migrans of ascarids, or angiostrongyliasis. Moreover, the above relationship is implicit in many more nematodiases of stock, such as the filariases and intestinal ascariases (which have their counterparts in human disease), the study of which, in veterinary medicine or in medicine proper, has its value in both areas.

Finally, it should be emphasized that in order to reduce the losses due to veterinary nematodes, research ought to be undertaken or continued in the fields of economics (to justify expensive measures against nematodes), therapy (to perfect practical methods of treatment), and vector control (to prevent spread as well as local outbreaks).

SUGGESTED READINGS

ALICATA, J. E., "*Angiostrongylus cantonensis* (Nematoda: Metastrongylidae) as a Causative Agent of Eosinophilic Meningoencephalitis of Man in Hawaii and Tahiti," *Canad. J. Zool.*, 40:5–8, 1962.

BEAVER, P. C., "Larva migrans," *Exptl. Parasitology*, 6:587–621, 1956.

ERSHOV, V. S. (ed.), "Parasitology and Parasitic Diseases of Livestock," State Publ. House for Agr. Lit., Moscow, 1956. (English Translation, Israel Program for Scientific Translations.)

FAUST, E. C., and A. DEGROAT, "Internal Autoinfection in Strongyloidiasis," *Am. J. Trop. Med.*, 20:350–375, 1940.

FAUST, E. C., and P. F. RUSSELL, *Craig and Faust's Clinical Parasitology*, 7th ed. Lea and Febiger, Philadelphia, 1964, pp. 354–365.

JARRETT, W. F. H., F. W. JENNINGS, W. I. M. McINTYRE, W. MULLIGAN, N. C. C. SHARP, and G. M. URQUHART, "Immunological Studies on *Dictyocaulus viviparus* Infection in Calves. Double Vaccination with Irradiated Larvae," *Am. J. Vet. Res.*, 20:522–526, 1959.

KRASTIN, N. I., A Study of the Developmental Cycle of the Nematode *Thelazia gulosa* (Railliet and Henry, 1910), a Parasite of the Eyes of Cattle," *Vohl. Akad. Nauk. SSSR*, n.s., 70:549–551, 1950. (Russian text: English abstract in *Helminth. Abstr.*, 19:90–91, 1951.)

LAPAGE, G., *Monnig's Veterinary Helminthology and Entomology*, 5th ed. Williams and Wilkins, Baltimore, 1962.

McNEIL, C. W., "Pathological Changes in the Kidney of Mink Due to Infection with *Dioctophyma renale* (Goeze, 1782), the Giant Kidney Worm of Mammals," *Trans. Am. Micr. Soc.*, 67:257–261, 1948.

MIYAZAKI, I., "On the Genus *Gnathostoma* and Human Gnathostomiasis, with Special Reference to Java," *Exp. Parasit.*, 9:338–370, 1960.

MORGAN, B. B., and P. A. HAWKINS, *Veterinary Helminthology*. Burgess, Minneapolis, Minn., 1949.

SWARTZWELDER, J. C., *et al.*, "Dithiazanine, an Effective Broad Spectrum Anthelmintic," *J. Am. Med. Assn.*, 165:2063–1067, 1957.

THOMAS, L. J., "*Gongylonema pulchrum*, a Spirurid Nematode Infecting Man in Illinois, U.S.A.," *Proc. Helm. Soc. Washington*, 19:124–126, 1952.

WEINSTEIN, P. P., L. ROSEN, G. L. LaQUEUR, and T. K. SAWYER, "*Angiostrongylus cantonensis* in Rats and Rhesus Monkeys and Observations on the Survival of the Parasite *in vitro*," *Am. J. Trop. Med. and Hyg.*, 12:358–377, 1963.

YOELI, M., H. MOST, H. H. BERMAN, and G. P. SCHEINESSON, "The Clinical Picture and Pathology of Massive *Strongyloides* Infection in a Child," *Trans. R. Soc. Trop. Med. and Hyg.*, 57:346–362, 1963.

CHAPTER 21

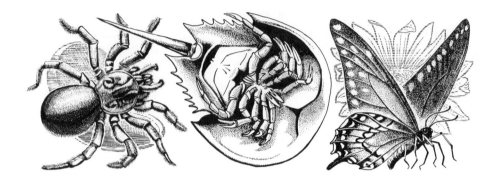

ARTHROPODS

INTRODUCTION

Among approximately one million species of arthropods, there are many truly parasitic forms. Some of these, such as various ticks and mites, live their entire lives upon or within other animals. Other arthropods utilize a host during only a part of their life cycle; often they are parasitic as larvae, but sometimes, as in the case of fleas, they become parasites as adults. Parasitic arthropods may be damaging in themselves, but their importance is often due to the fact that they transmit pathogens (viruses, bacteria, protozoa, or even helminths) from one host to another. Nonparasitic arthropods may also be very important transmitters of disease either by sucking blood or merely by acting as effective and dangerous contacts between the contagiously infected and the healthy. The latter is typical of the common housefly and cockroach. Some arthropods are venomous.

STRUCTURE OF ARTHROPODS

Arthropods (meaning jointed legs) are organisms with a relatively hard outer covering which serves not only as protection for internal organs but also as attachment for muscles. Because the arthropods have flexible, soft joints in their bodies,

272

the appendages, antennae, wings, etc., can move. One disadvantage in the arthropod pattern of structure is the inability of these animals to grow continuously. The exoskeleton is permanently formed and hardened, and its possessor may increase in size only by discarding the hard covering after first secreting a new, soft, expansible exoskeleton to replace the old. Thus arthropods are periodically deprived of the support and protection of their armor. The risks and inconveniences of growth by moulting have surely limited the size of arthropods; only a few aquatic forms (such as the larger decapod crustacea) ever become as large as the ordinary vertebrate.

The cockroach will serve as an illustration of arthropod structure. Variations from this pattern in the five or six groups of living arthropods can be pointed out where each group is discussed.

Cockroaches (Fig. 21–1) are medium-sized arthropods. Like all insects, they have a body consisting of three parts: head, thorax, and abdomen. The head bears a pair of long, slender tactile antennae, a pair of compound eyes, three simple eyes or ocelli, and the chewing mouthparts characteristic of this group of insects (the Orthoptera). The head is the result of evolutionary fusion of six body segments. The thorax consists of three body segments, to which the legs and wings are attached. Each leg has five parts: the thick, basal coxa; a short piece, the trochanter; a rather long femur; a slender tibia; and the five-jointed tarsus, which terminates in a double-hooked claw. The front wings are tough, flattened wing-covers, which protect the pleated or folded membranous hind wings. The latter expand and vibrate during flight. The abdomen bears appendages only on the tenth and eleventh segments; these appendages are the jointed cerci and the external reproductive organs (claspers in the male, ovipositors in the female). A tiny twelfth segment, the telson, bears no appendages. The cockroach, like all arthropods, has an exoskeleton of chitin, which is relatively impervious to moisture. Attached to the chitin are the body muscles. In addition to the muscular system, internal structures include the nervous system, the digestive tract, two circulatory systems (air and blood), and the reproductive system. The nervous system is a chain of ventral ganglia (one for each body segment) and a dorsal brain (consisting of the fused ganglia of the head segments) just above the mouth and esophagus. The double nerve cord is fused, so that the primitive ladderlike nervous system of arthropods is not apparent. The digestive tract consists of mouth, esophagus, crop, gizzard, midintestine, hind intestine, rectum, and anus. Around the mouth are the mouth appendages. Anteriorly and posteriorly the mouth is closed by two flaps of tissue, the labrum and labium, respectively. Two pairs of lateral structures—the rather delicate maxillae and the more powerful mandibles—serve as jaws; the mandibles are capable of cutting, scraping, and biting very tough materials, such as dried food, meat, and tendon. Jointed labial and maxillary palps help locate, taste, and manipulate the objects to be eaten. The maxillae and labium (the latter actually a fused pair of mouth parts) are typically primitive appendages such as are found in the crustacea; each of these mouth parts has a lateral leglike portion (the palp) and a medial portion specialized for feeding. Among arthropods the mouth parts

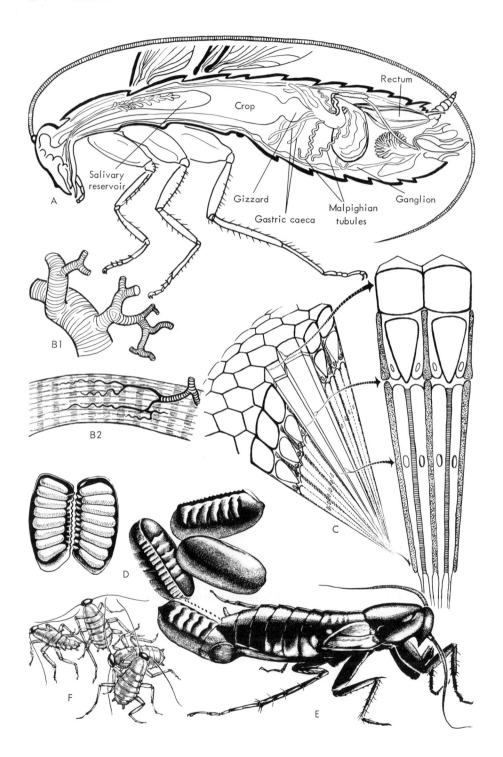

Rectum

Crop

Salivary
reservoir

A

Gizzard

Gastric caeca

Malpighian
tubules

Ganglion

B1

B2

C

D

F

E

FIG. 21–1. The cockroach, a typical insect. (A) Lateral dissection of adult cockroach (male *Blatta orientalis*). (B1) Tracheae, showing annular thickenings of chitin, and (B2) a muscle fiber with a branch of a trachea ending in several tracheoles. (C) Portion of compound eye, showing hexagonal facets and ommatidia. The latter consist of sensory elements (retinal cells) surrounded by pigment which separates each unit (ommatidium) from its neighbors. (D) Egg cases, open and closed. (E) The female roach carries the egg. (F) The eggs hatch into tiny nymphs. [(A) After Metcalf, Flint, and Metcalf, 1962; (D) and (E) after Buchsbaum, 1938.]

show great diversity of form and function. (See Fig. 21–2.) In the walls of the esophagus salivary glands open. Saliva lubricates and softens the food while it is being chewed for swallowing. The crop is a large cavity in which digestion actually takes place. It is lined with toothlike spines projecting from the chitinous wall, and it receives digestive juices secreted by cells lining the midintestine. Food is thus chewed and mixed with enzymes in the crop, and only finely divided liquid substances pass back to the absorptive regions. Between the midintestine and the hind gut the excretory malpighian tubules are attached; they secrete semisolid uric acid into the intestine. Some absorption of food occurs in the hind gut, and the fecal mass slowly passes into the rectum. The body cavity of the cockroach is filled with blood, which is pumped forward by a dorsal heart. Blood carries sugar, amino acids, etc., to the tissues and removes nitrogenous wastes, which are absorbed and excreted by the malpighian tubules. Insect blood has no respiratory function. Oxygen reaches the tissues by the tracheae and tracheoles, a system of tubes which extend throughout the body. Oxygen diffuses into these tubes from lateral spiracles or air openings in each body segment. Breathing motions force exhalation of exhausted air and remove carbon dioxide. The tracheal system, like both the anterior and posterior chambers of the intestine, is lined with chitinous material which is discarded when the insect moults. Growth of the cockroach, incidentally, is gradual. The eggs, layed in capsules containing 16 to 40, hatch into tiny cockroaches which are wingless and without sex organs. After a number of moults, the young reach the size and appearance of adults with functional wings and mature sex organs. Males have paired testes, an ejaculatory duct, copulatory organs consisting of claspers for holding the female, and a penis for injecting sperm. The females have paired ovaries and also ovipositors, which are capable of holding the hardened egg capsule until the female finds an appropriate hiding place for the eggs.

The above sketch of the anatomy of one kind of insect cannot, of course, serve as a description for all insects, let alone all arthropods. Nevertheless, the basic principles of arthropod structure—a jointed, hardened body covering, appendages specially modified in various regions for different functions, and separate sexes—are as well illustrated by the cockroach, a generalized and highly successful insect, as perhaps by any other arthropod.

FIG. 21–2. Some mouthparts of arthropods. (A) Lateral view of the head of a blood-sucking fly, the tsetse fly, genus *Glossina*. Hairy maxillary palps are spread to reveal the lancelike labrum, the extremely slender, needlelike hypopharynx, and the hemicylindrical labium. The labium contains a narrow trough through which blood flows upward into the esophagus. (B) Head of *Culex*, frontal view. At center is the hairy labium. At right are the hypopharynx and labrum-epipharynx, which together can form a tube for the flow of blood. On either side are a lancelike mandible and a sawlike maxilla. When a mosquito bites, the labium bends and allows the piercing structures to penetrate the skin and lacerate the capillary bed; the blood flows through the troughlike labrum. (C) Lateral view of head of flea. Sheltered by comblike

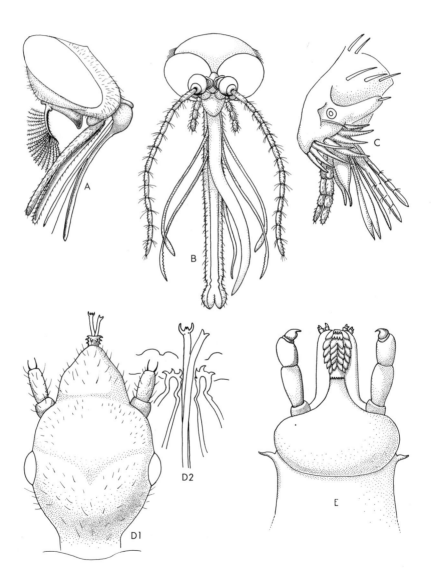

ctenidia at the sides, the mouth bears the short mandibles, the long piercing maxillae, a single lancelike epipharynx, an anterior pair of jointed maxillary palps, and a posterior pair of labial palps. (D1) Dorsal view, head of a sucking louse (Anoplura), with (D2) enlargement of the lacerating and rasping structures. The dorsal and ventral stylets bear cutting or tearing points. The sucking tube bears retractable teeth at its end, by which the louse anchors itself while feeding. (E) Ventral view of mouth parts of an ixodid (hard) tick. Two pointed pedipalps arise from the base of the capitulum, which also supports the rasplike hypostome and the closely associated cheliceral sheaths, from which protrude the terminal teeth of these protractile structures. [(A) After Horsfall, 1962; (B), (C), (E) after Herms, 1950.]

THE KINDS OF ARTHROPODS

The above example of a generalized insect differs in various ways from other kinds of arthropods. The once very numerous trilobites are an ancient and extinct class of the phylum Arthropoda. They had the head fused with the thorax and a rather large number of gradually tapering abdominal segments. The modern animal most similar to a trilobite is *Limulus,* the horseshoe crab, the larvae of which certainly resemble trilobites, with the adults also showing structural resemblances. *Limulus,* however, is considered a member of the class Arachnida, which includes spiders, scorpions, ticks, and mites. These terrestrial and often parasitic animals are descended from trilobitelike ancestors.

AQUATIC ARTHROPODS, CLASS CRUSTACEA

Nearly all aquatic arthropods in existence today are members of the class Crustacea. This class includes crabs, shrimp, and lobsters, as well as minute ostracods and copepods (mentioned in several chapters of this text as hosts for tapeworms, guinea worms, and acanthocephala), and curious animals such as the terrestrial isopods or pill bugs. Crustacea have various numbers of appendages. The smaller crustaceans obtain oxygen through the body surfaces, and the larger crustaceans, by feathery gills supplied with blood rich in the dissolved respiratory pigment hemocyanin. Crustacea possess two pairs of antennae and may have movable, compound eyes. They inhabit many ecologic niches and comprise a major part of the rich plankton of the sea. They include in their numbers predators, scavengers, and parasites.

TERRESTRIAL FORMS

Terrestrial arthropods are chiefly insects, but there are also the spiders, scorpions, ticks, and mites, as well as the still less numerous centipedes and millipedes. Other forms include such curious kinds of animals as tardigrads (water-bears) and pentastomids (parasites perhaps related to the mites).

CLASS ARACHNIDA

The arachnids (spiders, etc.) have bodies variously compressed; all have fusion of head and thorax, and some (ticks and mites) have fusion of cephalothorax with abdomen. They lack antennae or compound eyes. The more minute forms (aquatic and parasitic mites, for example) "breathe" through the body surface; spiders and scorpions utilize "book lungs" (partially enclosed arrays of moist plates, of which the separated surfaces are exposed to air). Some arachnids have tracheal systems composed of air tubes similar to those of insects. Spiders and scorpions have some importance in medicine because of their venom. They have no importance as vectors of disease. Ticks and many of the mites are true parasites. They can be dangerous pests, for they can suck blood in large quantities, invade the skin with resulting irritating lesions and secondary infections, or transmit various viral, rickettsial, bacterial, and protozoan agents of disease. Other noninsect arthropods are of little importance in medicine or parasitology.

CLASS CHILOPODA

The Chilopoda, or centipedes, are flattened, many-segmented, predatory animals of moderate size (up to about 12 cm). They have antennae, tracheal systems, simple eyes, and a single pair of legs per segment. They possess poison glands in the bases of their first pair of legs and can cause painful "bites" with sharp, fang-like claws.

CLASS DIPLOPODA

The harmless Diplopoda, or millipedes, are similar to Chilopoda in many ways, but they are nearly cylindrical in form and have two pairs of legs per visible segment. They are scavengers; other than their membership in various animal communities and their part in food chains, their only effect upon man is to annoy him occasionally with secretions from the repugnatorial glands which some millipedes possess.

CLASS PENTASTOMIDA

The remaining arthropods to be mentioned here are the Pentastomida, a group of wormlike parasites living in the nasal passages of carnivores, in a few birds, or in the lungs and tracheae of various reptiles. When they are larvae, the pentastomes have two pairs of jointed legs. The four clawlike hooks by which they attach themselves when adults may be homologous with arthropod appendages.

The great arthropod phylum therefore contains several successful variations of a basic structural pattern, illustrated above by the cockroach. Insects, with a three-part body, antennae, compound eyes, and, above all, wings, are the most numerous representatives of the phylum. The rather diverse arachnids range from the scorpions (with long flexible abdomen, claws, etc.) and the spiders (two-part body, with highly developed instinctual behavior patterns) to the single-bodied ticks and mites and their interesting life cycles. Centipedes and millipedes have a wormlike body structure involving repetition and reduplication. The aquatic Crustacea are varied in structure; they include the largest of arthropods. From these diverse and numerous animals, the arthropoda of medical and veterinary importance will be discussed in the following chapters.

SUGGESTED READINGS

BARNES, R. D., *Invertebrate Zoology*, Chapter 13. Saunders, Philadelphia, 1963.

HORSFALL, W. R., *Medical Entomology*, Chapter 2. Ronald Press, New York, 1962.

CHAPTER 22

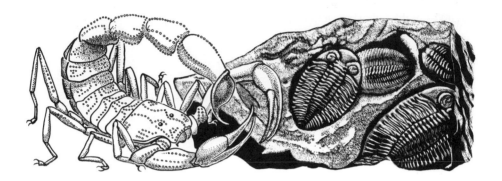

PRIMITIVE
ARTHROPODS

INTRODUCTION

A discussion of arthropods of medical and veterinary importance does not fall into any easy system of organization. Some writers of textbooks on medical entomology arrange these animals under the ways in which they cause injury. Thus section headings might be "venomous and urticating arthropods," "vectors of disease," "invasive and parasitic insects," etc. Other writers prefer to discuss each taxonomic group in turn by listing the important organisms and describing the effects they have on human and animal welfare. In this chapter and the next two chapters a general taxonomic outline will be followed, but within that broad outline particular conditions, injuries, or diseases will occasionally be discussed rather fully. Thus after brief mention of the primitive arthropods such as crustaceans, spiders, scorpions, etc., which have only slight medical importance, this chapter deals with the significant acarines (ticks and mites) and emphasizes their role in transmission of virus and rickettsial disease. Then succeeding chapters will discuss (1) insects other than the Diptera, with emphasis on lice, fleas, and epidemic disease, and (2) the Diptera in relation to their many important roles as parasites and vectors. Finally, medical and veterinary entomology will be reviewed in terms of the control of arthropods.

CRUSTACEA

Crustacea are valuable to man directly as food, and indirectly as very important parts of the food chains of various other animals. The limitless plankton of the sea are harvested by man through fisheries. All of the fishes which do not depend directly upon microscopic animals and plants still subsist on these primary foods through the intermediate predators and herbivores which they eat. Crustaceans of various sizes are a large part of both the primary plankton and the intermediate consumers. But several crustaceans are harmful to man and his stock.

Members of the subclass Copepoda which transmit the fish tapeworm, *Dibothriocephalus* and the guinea worm, *Dracunculus*, are minute crustaceans found in all waters rich enough in other smaller organisms to support these animals. Their heads bear two pairs of antennae, the first of which is very long and is used for swimming and in keeping the organisms suspended in their planktonic habitat. They have as mouthparts one pair of mandibles, two pairs of maxillae, and one pair of maxillipeds. There are usually six pairs of swimming legs on the cephalothorax, and the jointed abdomen bears at the end two bristlelike structures which, like the antennae and legs, aid in swimming and "floating." There are no gills, for these animals are small enough to respire through the delicate exoskeleton. Copepods go through a series of moults from egg to adult. Their early larvae are quite unlike the mature form but resemble the early larvae of other crustaceans. Some copepods are highly specialized parasites of other Crustacea or fishes, and these forms are variously modified as to both structure and life cycle.

In addition to the above copepod Crustacea, certain freshwater crabs and crayfish are important in medical parasitology. Decapod crustaceans are perhaps too well known to justify any description here. They have, as their name indicates, ten walking legs. They also possess two pairs of antennae and a pair of compound eyes on movable stalks. The mouthparts include one pair of mandibles, two pairs of maxillae, and three pairs of maxillipeds. They respire by gills sheltered by the prominent carapace or dorsal shield of the cephalothorax. The chief difference between crabs and crayfishes is in the form of the abdomen; in the former the rather insignificant abdomen is folded tightly under the ventral surface of the cephalothorax, which is usually laterally expanded, while in the latter the abdomen is an important part of the body, is highly muscular, and bears paired appendages (swimmerets) on each of its segments. The abdomen is used by the crayfish in rapid swimming, for escape from predators. Crabs and crayfishes are esteemed as food in many parts of the world. The lung fluke of man and carnivores, *Paragonimus westermani* (see Chapter 10), a common and dangerous parasite of man in the Far East, is transmitted by decapod crustaceans.

CLASS MYRIAPODA

Millipedes and centipedes, as was stated above, are not particularly important; the former are only mild and occasional pests, and the latter, while capable of injuring man, are not as effectively venomous as scorpions or spiders.

Millipedes (subclass Diplopoda) are heavily armored and are usually cylindrical, slow-moving creatures. They are common in the damp litter of forest floors. Each of their many externally single but actually double segments bears two pairs of legs. They breathe by spiracles and tracheal systems. They have a single pair of antennae and short, blunt mandibles with which they feed on vegetable mold. They are moderately large; some representatives in the tropics reach 25 or 30 cm in length. Members of one group, the juliform millipedes, have repugnatorial glands which open by dorsal pores. When the animals are disturbed, the repugnatorial secretions are ejected, sometimes as jets, from these pores. Often quite volatile, the secretions cause odors similar to those of garlic, chlorine, bitter almonds, or urine. The odor may be literally sickening. Millipede populations sometimes grow so immense that mass migrations take place and result in millipedes fouling roads and open wells or entering houses. But aside from the bad odor of molested millipedes and the inconvenience caused by their migrating hordes, millipedes cause no damage and doubtless form a significant part of the forest community.

Centipedes (subclass Chilopoda) are definitely harmful, for they are venomous predators capable of injuring man. Superficially like the millipedes, the centipedes are composed of many segments, breathe by tracheae, and have a single pair of antennae. Yet the centipedes differ in several respects from their harmless relatives. They have only one pair of legs per segment (there may be 20 or 25 or more), and their first pair of legs is modified into enlarged, sharp, and poisonous claws, with which they seize and kill their prey. On the ventral surface of the body are the openings of glands, the secretions of which are very irritating to human skin. Centipedes are not sluggish as are the millipedes; while hunting, they run about actively. Dorsoventrally, centipedes are usually rather flat; they live in moist crevices in forest litter, under fallen logs, or in piles of stone and rubble. The bite of a centipede usually shows twin puncture marks (as do the bites of spiders, and, of course, venomous snakes). The injury is painful, but symptoms are usually local and brief.

CLASS PENTASTOMIDA

Although often considered a phylum, the linguatulids (tongue worms) or Pentastomida are undoubtedly arthropods. They are considered by some to be related to the mites (small acarines); by others, they are said to be related to the unique microscopic tardigrads (water bears). In larval stages the pentastomids have four jointed claws or legs; in the adult, these claws are represented by four hooklike structures by which the animals anchor themselves in pharyngeal mucosa or in the lungs. The pentastomids feed on blood. Their simple body is superficially segmented; it is sometimes flattened (hence "tongue worms") and sometimes ringed with ridges. Judging from the few cycles known, we find that the life cycle of pentastomids involves (1) ingestion of eggs by some herbivore or omnivore, (2) development of larvae within the tissues or body cavity of this intermediate host, then (3) ingestion of the larvae along with the latter host by some predator (snake, carnivore, etc.). The ingested larvae cling to the pharynx, or in some cases the lung cavities, of the final host. Eggs layed in this host reach the exterior by being spit out with sputum or passed out with feces. Pentastomes of the genus *Porocephalus* live in the lungs of rattlesnakes (*Crotalus* spp.) and other pit vipers. The genus *Armillifer* is found in pythons; its intermediate hosts are monkeys. *Linguatula* spp. are found in carnivorous mammals. Man may presumably harbor the larval stages of various pentastomes. *Linguatula* has been reported as both a larva and an adult from human viscera and nasopharynx, respectively.

CLASS ARACHNIDA, ORDER SCORPIONIDA

The class Arachnida contains the relatively unimportant scorpions and spiders, which will be discussed next, as well as the extremely important acarines, the ticks and mites, which will occupy the bulk of this chapter.

Where scorpions appear they are much feared, and rightly so. Their stings caused over 1600 deaths in a 36-year period in the Mexican state of Durango. On the basis of 20 years of data, scorpions in the state of Arizona kill more than twice as many people as are killed by all other venomous animals in the region. There are about 300 species of scorpions; it has been estimated that perhaps 20 of these are quite dangerous. The scorpion is an elongated arachnid which is recognizable by its four pairs of walking legs, its prominent claws or pincers, and its jointed abdominal "tail" with a bulbous sting on the end. Adult scorpions range in length from 5 to 10 or 12 cm. They prey upon other arthropods, such as spiders and large insects; they seize their prey, sting it, and then ingest the juices of the victim after crushing and tearing it with the pincers. Scorpions usually have small simple eyes but sometimes have none. When disturbed, they strike forward repeatedly over their cephalothorax with the sting, which injects a neurotoxin, sometimes combined with a hemolysin, into the disturbing factor—often man. First aid for scorpion sting involves chilling the injured part of the body to slow down absorption of the toxin. Symptoms of scorpion sting are the single, not double, puncture wound with its pain, neuromuscular abnormalities of the face and throat, disordered movement of the limbs, distortion of vision and touch sensations, and convulsions (usually repeated many times during a period of one or two hours). Death may ensue, but if the victim survives more than three hours, symptoms usually subside and the patient recovers.

ORDER ARANEAE

Venomous spiders are found in the tropical and subtropical regions of the world. Although the great majority of spiders are harmless and are, in fact, beneficial as controllers of insect populations, nevertheless the estimated 19 genera containing species dangerous to man justify a brief discussion here.

FIG. 22–1. Spiders, ticks, and mites. (A) Lateroventral view of cephalothorax of spider, showing fanglike chelicerae with which spiders poison their prey. Above the chelicerae can be seen six simple eyes. [After Barnes, 1963.] (B) Generalized diagram of a spider, showing four walking legs of one side and the divisions of the body into two parts, cephalothorax and abdomen. [After Herms, 1950.] (C) A chigger (larval trombiculid mite) at a feeding tube (stylostome) in the skin of its host. [After Horsfall, 1962.] (D) The "apartment mite," *Allodermanyssus sanguineus*. [After Horsfall, 1962.] (E) A soft (argasid) tick, *Ornithodoros* sp., exuding a drop of coxal fluid. [After Horsfall, 1962.] (F) A hard (ixodid) tick, *Hyalomma marginata*. [After Horsfall, 1962.] (G) Larva, eggs, and nymph of an ixodid tick. Note that larva has six legs, and the nymph has eight.

Spiders have two-part bodies. An anterior "head-leg region" bears the four pairs of legs, the simple eyes, the chelicera or poison-fangs, the pedipalpi, and the mouth. The "hind body" bears the spinnerets, 4 or 8 openings of the silk glands. Also in the hind body are the "book lungs," moist respiratory surfaces protected by pouches and connected with a tracheal system which distributes air to the body organs. Some spiders have spiracles, i.e., tracheal openings to the surface; such openings may be in addition to book lungs or may substitute for those organs. The behavior of spiders varies too much for any brief discussion; some are wandering hunters, some fashion trap-door lairs in which they wait, and many construct elaborate webs to catch flying or jumping insects. All feed by first injecting venom into their prey through the chelicerae (Fig. 22–1) and then sucking the juices of the quieted victim. Many spiders inject not only a lethal venom (usually a neurotoxin) but also potent digestive enzymes, which lyse the tissues of the prey until its cells and muscles are converted into fluids which can be sucked out.

It is this killing-feeding behavior that makes certain spiders dangerous to man. When molested, or perhaps "by mistake," spiders sometimes strike quickly at a hand or other part of the body and stab with their poison-filled fangs. The neurotoxin may cause very severe symptoms. Hemolysins or histolytic enzymes may add local or extensive necrosis to the damage to nervous tissue. Nervous symptoms are confusion, paralysis, or convulsions. Some toxins act to stimulate secretory glands and cause excessive salivation, sweating, or nausea. Hemolytic or histolytic toxins cause sloughing of tissues at the site of the bite, both local and general hemorrhage, and sometimes extensive necrotic lesions. Severe illness, frequently fatal, results from the bites of the more venomous spiders.

Dangerous spiders include the "widow" spiders, *Latrodectus* spp. ("domestic" spiders common in houses, sheds, trash heaps, and natural underbrush), of North and South America and the worldwide tropics. Corresponding in habitat and behavior to *Latrodectus*, *Loxosceles* species are particularly important in South America and are present in North America also.

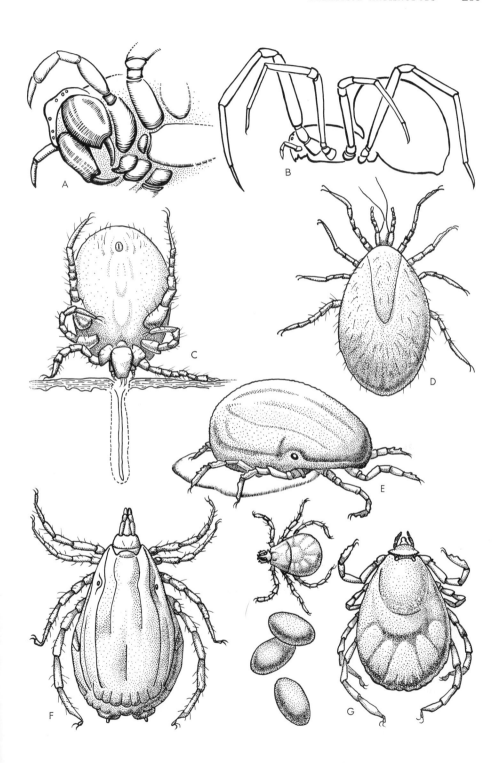

The effects of bites by these two kinds of spiders are quite different. The widows cause systemic poisoning, which results in a variety of alarming and serious symptoms including pain at the bite (a double-puncture wound) followed by dizziness, weakness and tremor of the legs, and abdominal cramps. Profuse cold sweat, nausea, severe headache, reduced heart beat, lowered blood pressure, acute nephritis, shock, delirium, and prostration have been observed in victims. The venom affects particularly the nerve endings, which explains the many neuromuscular symptoms noted. The *Loxosceles* species (brownish, velvety spiders with a "violin" marking on the cephalothorax) cause necrosis of the tissues near the bite but do not poison the whole body. However, necrosis may be severe and prolonged, and it may be quite disfiguring as a deeply eroded area is repaired with scar tissue upon healing. If much venom is injected, systemic reactions can occur from *Loxosceles* bites. Treatment for either kind of spider bite can be supportive only, for there is as yet no specific antidote to these venoms. In Australia a very large spider, *Atrax* species, lives by day in silk-lined tunnels under ground, and at night emerges to hunt beetles and other large insects. These pugnacious spiders attack people.

ORDER ACARINA: MITES AND TICKS

The most important arachnids are mites and ticks. They include many harmful pests which annoy and injure man and his domestic animals by sucking blood and by biting, chewing, and inhabiting the skin and tissues of these hosts. Being known transmitters of protozoa, bacteria, and viruses, many of these pests are important vectors of disease, and some of them also serve as intermediate hosts of worm parasites. "Mites" and "ticks" are the common names for the small and large members, respectively, of the order Acarina. Specialists prefer to divide this order into several groups on the basis of rather fundamental differences and similarities. In such a classification "ticks" would refer to the members of two families only (the Ixodidae and Argasidae), while "mites" would include all other acarines. Acarines (Fig. 22–1) have certain features in common. Their bodies appear to be compact, are without constrictions, and are not divided into separate regions (unlike spiders, various crustacea, and insects). Actually two body regions are usually distinguished: a head region (gnathosoma or capitulum), to which mouth parts are attached, and a posterior region. This latter portion, which is divided into a leg-bearing part and an abdominal part, is called the idiosoma, or body proper. The mouth parts of acarines are essentially two chelicerae, variously modified for cutting, tearing, seizing, or penetrating, and two lateral pedipalps, which are also appropriately varied. Acarines have eight legs as adults and six as the hexapod larvae. Some acarines are armored, while others are very soft and delicate. Some are quite active, even cursorial in habit, while others move very slowly. Some have simple eyes; others have none. They reproduce sexually (although in certain forms males are unknown); the females lay each egg by a rather elaborate process. All acarines have somewhat the same series of developmental forms. From the eggs a six-legged larva hatches. This, with or without several intervening moults, changes into an eight-legged nymph, which, again after one or more moults, becomes an adult. Parasitic forms, with which we are most concerned in this text, show interesting host-related adaptations in their life cycles; some live their entire lives, from egg to adult, upon or within the body of a single host, while others may remain on a host very briefly, may require a series of different kinds of hosts to correspond with a series of particular developmental stages, or may require a host at only one of many stages in the life cycle. Upon variations in the above general characteristics, several main groups of Acarina have been distinguished.

MITES

The small acarines, or "mites," are grouped in four suborders, each of which will now be discussed.

The suborder **Onychopalpida,** a group of primitive acarines with claws on the pedipalpi and two or more pairs of spiracles, is distributed in Australasia. They are of little significance in either medicine or agriculture.

The suborder **Mesostigmata** is a large group of rather active mites, including many forms which are pests and a few which are parasites or vectors of disease. Lung mites (*Pneumonyssus*) are found in the lungs of mammals, where they cause nodules which resemble tubercular lesions. Various rodent nest-parasites (including members of the genus *Laelaps*) feed on the blood of their hosts, but spend most of their time in the litter and debris of the host's habitat. Dermanyssid mites, the common mites of poultry, rats, and mice, are quite important because they affect man incidentally but frequently. Thus these blood-sucking, agile acarines may overflow from henhouses; or their populations in the holes of rodents around human houses or apartments may expand, and human hosts may be fed upon successfully. The chicken mite *Dermanyssus gallinae* (Fig. 22–2E), the northern fowl mite *Ornithonyssus sylviarum,* and the tropical fowl mite *O. buarsa* suck the blood from their hosts. The first species sucks blood intermittently, hides by day, and stays on the host only long enough for engorgement. The last two species are more or less permanent residents under the plumage of fowl. These mites, when numerous, cause anemia in addition to irritation and restlessness. The result is loss of weight, reduced egg production, or even death. They are known to transmit virus diseases of man and other animals as well as an American encephalitis in chickens. The "apartment mite," *Allodermanyssus sanguineus,* lives in house-mouse nests and often attacks man. This mite is a vector of the microorganism *Rickettsia akari,* which causes a disease similar to chicken pox (called rickettsialpox, vesicular rickettsiosis, or Kew Garden fever). Outbreaks of this disease occur where fairly large populations of mice or rats coexist with the "apartment mite" and moderately crowded human habitations. Usually a rodent extermination campaign brings about the epidemic, since destruction of their rodent hosts causes rickettsia-bearing mites to wander widely in search of new hosts. Under the above conditions, man is the nearest available substitute for the lost hosts.

The relationship of rickettsia organisms to blood-sucking or tissue-feeding acarines is significant and unusual. All acarines harbor characteristic symbionts (sometimes bacteria but usually rickettsiae) in their tissues. These symbionts

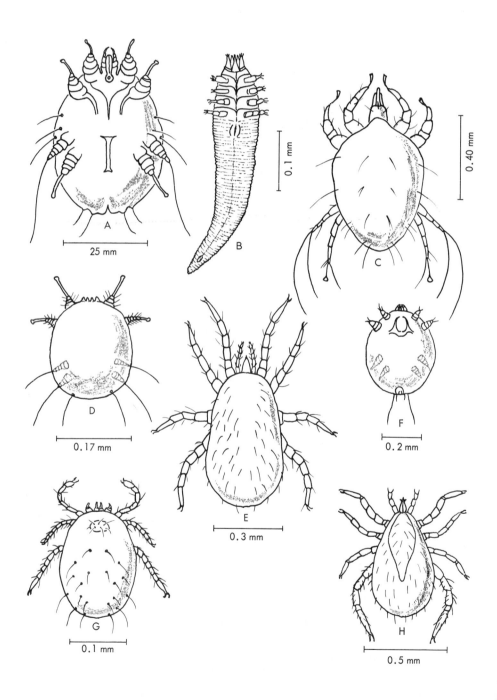

FIG. 22–2. Mites of medical and veterinary importance. [After LaPage, 1962; Chandler and Read, 1961; Horsfall, 1962.] (A) *Sarcoptes scabiei*, the itch mite. (B) *Demodex folliculorum*, the follicle mite. (C) A mange mite, *Psoroptes* sp., of the type causing psoroptic mange or "scab." (D) *Notoedres cati*, a mange mite of the domestic cats. (E) *Dermanyssus gallinae*, a fowl mite. (F) *Cnemidocoptes* sp., the scaly leg mite of chickens. (G) "Chigger," larva of a trombiculid mite. (H) *Ornithonyssus bacoti*, the tropical rat mite, a vector of filariasis in rats, and of various diseases of man.

are transmitted by egg from one generation of acarine to the next. It is believed that rickettsiae which are pathogenic to man may be harmless or valuable inhabitants in the body of the mite or tick. Thus acarines play a unique role as reservoirs of disease, for disease is an accidental and harmful consequence of the mutualism between mite or tick and its microorganisms.

The suborder **Trombidiformes** is a third group of mites, many of which feed on plants. This group includes also predators and true parasites. The plant-eating mites occasionally overrun greenhouses or gardens, or thrive on house plants. They may irritate and annoy by crawling upon the skin; sometimes such mites are inhaled by greenhouse workers. Other trombidiform mites are parasites of insects. If given the opportunity, some of these are capable of piercing human skin. Persons working with stored grain or hay where the insect hosts of mites may be extremely common are attacked by the acarine parasites of the latter. An unusual relationship, hyperparasitism, is shown by certain trombidiform mites, *Cheyletiella* and *Cheyletus* spp., which parasitize other species of mites. The former genus lives on mites which infest rodents; the latter is found on mites living in stored grain. These mite-inhabiting mites can attack man.

The genus *Demodex* (Fig. 22–2B) includes very minute mites which inhabit the oily skin glands of man and other mammals. Unless demodectic invasions are accompanied by bacterial infection, they are usually mild or symptomless. Demodectic mange in dogs is an example of bacterial complications following invasion by mites.

The family Trombiculidae includes the well-known "red bugs" or chiggers. These mites have a life cycle which has only one stage affecting man, i.e., the larval stage, a six-legged form with sharp chelicerae. Upon finding (or being brushed by) a mammalian or other vertebrate host, the larva crawls rapidly until it reaches the surface of the skin, where it inserts its chelicerae. Salivary juices are forced into the wound and create a sort of tunnel or tube of liquefied host cells. Next to this tube, which extends only part way toward the basement membrane of the epidermis, the host cells react by becoming somewhat hardened and thus preventing the digested cells from leaking away. A chigger (larval trombiculid) remains in this position and feeds for days or weeks on the digested cell products within the protective tube (Fig. 22–1C). The larva then leaves its host and moults to become an inactive nymph, which in turn moults to become an active nymph. From this stage on, the organism feeds entirely on arthropod eggs. Thus only the larval stage is parasitic. Yet because of the intense itching (some people are insensitive to chiggers and some chiggers do not irritate) resulting from the presence of the feeding larva and its food tube in the epidermis, this single parasitic stage is quite important. Also these parasites can transmit the rickettsial pathogens of the group of diseases called "scrub typhus," "tsutsugamushi disease," etc.—diseases which are contracted in wild territory and are caused by organisms found (perhaps normally symbiotic) in chiggers.

Sarcoptiform mites (suborder **Sarcoptifomes**) have chewing mouthparts. There are two groups of medical importance: the armored mites (oribatid) and the food and mange "acarid" mites (cheese, coconut, mushroom mites and the itch or mange mites of animals).

The oribatid mites abound in fields and pastures. They are protected by heavily sclerotized plates or by a cuticle which is tough and leathery. The oribatids are omnivorous, for they feed on vegetable detritus and scraps of animal origin, including the eggs of the tapeworms of the genera *Moniezia, Thysanosoma,* and other anoplocephalines (see Chapter 11).

The acarid mites occupy many habitats, i.e., in stored food products and under the scales or in the skin of various host animals. Cheese mites, etc., are of little medical importance, but they may cause allergic symptoms in people exposed to them. Warehouse personnel, laborers who handle grain, copra, or dried fruits, and even mushroom growers may suffer from contact with these mites. For example, "straw itch" occurs in grain or hay handlers, livestock attendants, etc., and is quite commonly recognized as an occupational hazard. A generalized eruption on forearms, neck, and trunk results from multiple punctures by the mite *Pyemotes ventricosus.* Similar diseases are due to other mites not inhabiting man but occurring in large numbers where men work. As in the case of the above mentioned Trombidiformes, the hosts on which such mites or *Pyemotes* develop are often insects infesting some material or commodity.

Other acarid mites are more directly harmful. The families Sarcoptidae and Psoroptidae consist of parasites which cannot survive away from integumentary structures or tissues of vertebrates. As these mites infest poultry and various domesticated mammals, they cause a variety of conditions.

The genera *Psoroptes, Otodectes, Notoedres, Chorioptes,* and *Sarcoptes* (Fig. 22–2) have various species which cause "itch," "scab," and mange in sheep, cattle, horses, dogs and cats, and other animals, including man. The life cycles of these mites are similar. The mites remain throughout life on their hosts and are true parasites. They mate, lay eggs, and develop to maturity in the lesions which they cause. The pathology depends on the nature and extent of these lesions. Psoroptic mange or "scab" is caused by mites (of the genera *Psoroptes* and *Chorioptes*) which feed on and destroy the surface of the skin. These mites remove portions of the epidermis as they live and spread upon the raw dermal layer under the scabs which form. Scratching by the host aggravates the injury; and as the condition spreads, large areas of skin may become hairless, thickened, scabby, and abraded. The abrasions permit secondary infections to occur. Sarcoptic mange or "itch" is caused by the burrowing of mites of the species *Sarcoptes scabiei* (Fig. 22–2A) throughout the epidermis. (The itch mites of cattle, sheep, hogs, dogs, man, etc., are thought to be varieties of this species.)

In man, the itch mite is a well-known parasite which tunnels within the epidermis. (See Heading, Chapter 28.) The female lays eggs as she digs, and the mites of both sexes and all sizes defecate as well as cause mechanical irritation in the excavated skin. These mites are spread from host to host by personal contact and by clothing; they become a problem during periods of crowding and when sanitary facilities are insufficient. There is a notable delay in host response to the mites which cause scabies (itch). About one month after invasion of the skin the mites may be numerous (20–50), but only mild itching may occur. The skin, however, becomes sensitized after 1½ months, so that itching may be intense and continuous. The peak mite population (100–500) declines after 2 to 4 months, until only a few mites (10 or less) are present, but the itch, now an allergic syndrome, persists. After treatment or sufficient time, the mites die out and the itch disappears. Reinfection of former victims cause a quick and intense response (usually within a day), although mite population growth is strongly inhibited, and no such severe and general itch develops as in the original infections. It is not known whether similar reactions occur in cattle, etc., but one would expect this to be true. The itching, which occurs in all hosts, results in violent scratching. The scratching causes pustules to break, lesions to form in the dermis, and secondary infections to develop. Similar, but less severe and quite temporary, infestation of man may come from mange or itch mites of domestic animals. These mites cannot establish populations in the human skin, although they can irritate and sensitize. Mites of the genus *Notoedres* (Fig. 22–2D) afflict cats with a severe mange or itch of the type caused in other animals by *Sarcoptes scabiei*. Follicle mites of the genus *Demodex* are truly systemic parasites. They are thin wormlike animals, which invade the skin by way of the hair follicles and live in these as well as in cystlike pustules or abscesses beneath the skin; as in other mite infections, condition of the host greatly affects the severity of the disease. In man, *Demodex folliculorum* is extremely common but very rarely causes any symptoms. It may produce severe symptoms, however, in poorly fed dogs. Follicle mites of various species afflict cattle, swine, sheep, goats, horses, and other animals. In cattle they cause substantial damage to the hides. This fact is reflected in significant economic losses.

Parasitic mites are not among the most important vectors of disease, although the lesions they cause are pathways by which other pathogenic organisms can enter the host. Certain poultry mites mentioned above, however, transmit virus encephalitis in chickens and can transmit encephalitis of equine and human strains. Mites of the genus *Trombicula* have larvae which bite man and other animals, and transmit virus and rickettsial diseases of man. *Ornithonyssus bacoti,* the tropical rat mite, transmits rat filariasis, and has been shown to transmit endemic typhus, rickettsial diseases, tularemia, and other diseases of man; its importance in transmission of animal disease is unknown. Oribatid mites, free-living in pastures, are the intermediate hosts of the anoplocephalid tapeworms found in rabbits, sheep and goats, horses, and other grazing animals.

Control of the animal losses caused by mites depends upon both the characteristics of the mites themselves and the nature of the host. Temporary chicken mites (*Dermanyssus* spp.) can be controlled by treating the chicken houses, roosts, and litter with penetrating insecticides. Other kinds of fowl mites, which remain on the host, may be killed by dusting the birds with malathion, DDT, or other chemicals in a suitable powder. But the scaly-leg mite, like other mange or itch mites which are protected by scabs or skin, must be specially treated by loosening the scales with warm water and then smearing with an insecticidal ointment. Mange of cattle and horses can be cured or treated with liquid sprays or dips. A special shed or chamber can be constructed for vapor treatment of an entire animal by using heat to produce sulfur dioxide, which is introduced into the chamber around the animal; only the animal's head is allowed to protrude through an opening. That this somewhat cumbersome and dangerous method, (the sulfur dioxide may condense, if the animal's skin is moist, to form sulfuric acid) is used at all is evidence, of course, of the importance and difficulty of controlling mange. Since the above and other methods are constantly being improved, state or national departments of agriculture should be consulted when mange problems arise. Being systemic or at least extending well below the epidermal layer of the skin, demodectic or follicular mange is very difficult to treat and cure. Systemic insecticides have shown promise. In general, the prevention of mange, itch, scab, and other mite-caused disease depends on keeping stock clean and avoiding the crowded conditions which encourage transmission of mites, their eggs, or their larvae. Where large or valuable animals are concerned, isolation of infected and suspected animals is effective in preventing transmission, especially if tools, harness, or even the clothes of workmen are not permitted to be taken from a mange-infested area to a clean one. The eggs, adults, and larvae of mites are not particularly resistant to extremes of temperature or drying; thus quarantine is relatively practical. In the treatment or control of mange, good food for the animals is essential. Accumulated evidence shows that condition, especially of the animals' skin, is highly significant in the occurrence, spread, and severity of mite-caused disease.

In summary, the parasitic and harmful mites are small, varied, and versatile. Although some are as large as small ticks (*Trombicula* spp.), most are barely visible. The acarologist distinguishes several major groups, some of which are medically important. In transmitting rickettsial organisms, in maintaining reservoirs of rickettsial disease among wild rodents, and in directly causing injury by their bites, their invasive abilities, and their tendency to incite allergic symptoms, the mites are significant contributors to human misery and animal disease. Little has been said here about control or prevention of attacks by mites, but the sources are so plain (contact with stored foods, exposure to chiggers in brush or field, contact with scabies, etc.) that avoidance of mites is not difficult. Control problems are specific; they depend on details of life cycles, behavior of the host, and advances in technical knowledge. Modifications of cures used for mite-infested animals can be used effectively for human victims.

TICKS

The Ixodides, a fifth suborder of the Acarina, contains animals which are definitely larger than members of the other suborders (the mites just discussed) and are commonly known as ticks. Ticks are distinguished from mites by several characteristics more significant than size. The structure of the mouth region (gnathosoma) is more complex than in the mites; the tarsus of the first leg bears a curious sensory pit (Haller's organ); only two spiracles are present; and there is an egg-laying structure (Gené's organ) in female ticks. All members of the suborder feed on liquids from vertebrate hosts, but some (the Ixodidae) are more clearly parasitic than the others (the Argasidae). Ticks are notoriously tough and long-lived. Details of their habits, life cycles, etc. will be discussed under each of the two families: the "soft ticks" (Argasidae) and the "hard ticks" (Ixodidae).

SOFT TICKS

The Argasidae are relatively quiet animals; they hide in crevices or debris found in nests or burrows of their (usually) avian or reptilian hosts and attack their hosts only briefly yet long enough to feed. In this pattern of behavior, they resemble the majority of mites, whose frequent movements onto and off of hosts, as we have seen, make them effective transmitters of disease as well as rather frequent "accidental" pests of man.

Argasid ticks (Fig. 22–1E) have a rather soft, but tough, bulbous idiosoma (body) which obscures the mouth parts (capitulum, gnathosoma) from above and extends laterally to partly obscure the legs. There is no dorsal plate (as will be described for the ixodids), and the sexes are similar in size and appearance. The females lay their eggs in small batches at widely spread intervals. Eggs from the genital opening pass against Gené's organ, a fold of secretory tissue which coats each egg with waxy materials and waterproofs the egg. The intermittent egglaying and the long life span (up to 20 years in captivity) are good insurance of survival for essentially a desert animal which may have only occasional opportunities to procure food. These ticks are analogous in the latter respect with some desert plants, which must live many months without water but can store water from rare episodes of rain in order to grow and reproduce successfully. Economy of effort and longevity complement each other.

The "soft ticks" are apt to attack any vertebrate which invades the cave, nest, or burrow where they live and hence are potentially dangerous to man. An important transmitter of disease, the argasid tick *Ornithodoros turicata,* is found in dry caves in the western United States, where it is a danger to explorers. Poultry ticks include *Argas persicus,* the fowl tick or blue bug. This tick remains attached to fowl from 3 to 10 days during its first (six-legged) larval stage and then drops off. Thereafter it attacks only at night and hides during the day in the nest straw, under dry detritus, and in cracks in the walls. It crawls out upon the perch when the birds are asleep. When large numbers of *Argas* are present in a poultry house, the chickens, repeatedly bleeding at night, become weak, unthrifty, and unprofitable. Moreover, a fatal disease, fowl spirochaetosis, is transmitted by this tick. Other argasids of importance are *Otobius* spp., the spinose ear tick of hoofed mammals; various species of *Argas; Ornithodoros erraticus* of Northern African toads, rodents, and foxes; and *O. moubata,* another African species, noted for its ability to transmit the spirochete (*Borrelia*) of relapsing fever. There are perhaps 18 or 20 strains or species of *Borrelia* which are transmitted by "soft ticks" in many parts of the world and produce the various diseases known as relapsing fevers. Some are mild, and some are quite severe; yet all are rather confusing symptomatology. These fevers will be discussed later in conjunction with the medical importance of ticks. The bites of these ticks are often painful and cause site hemorrhage and necrosis. Sometimes localized or general allergic reactions occur. Certain ticks (notably various *Ornithodoros* species) bite painlessly and make it difficult for the physician to obtain a history of tick bite in certain cases of relapsing fever. Although argasids occupy various ecological and climatic niches, many of them are remarkably well adapted to dryness, for they are able to tolerate desert humidities for months at a time. This survival factor, as well as the general longevity of these ticks, makes the argasid an unusually effective agent for maintaining a disease such as relapsing fever. Moreover, the ability of the spirochetes to pass via eggs from one generation of ticks to the next practically assures continuity and survival of these tick-borne pathogens.

HARD TICKS

Ixodid or hard ticks (Fig. 22–1F, G) are also important vectors of disease. They transmit viruses, rickettsial organisms, and bacteria, as well as the babesiid protozoa causing the cattle fevers. Some ixodids, moreover, cause "tick paralysis" if they bite the back of the victim's neck and are allowed to remain attached long enough for their toxic salivary secretions to diffuse deeply into the host's tissue. Ixodid ticks are distinguishable from argasids by their dorsal plate (scutum) as well as by the prominent gnathosoma or capitulum which projects in front of the flattened idiosoma. Female ixodids enlarge greatly when they feed; they increase in size several diameters during one feeding. These ticks are sexually dimorphic; the females are larger than the males. Ixodids live widely dispersed over the area in which their hosts range; in this as well as in the other above-mentioned respects they are different from the nest- and lair-dwelling argasids.

The life cycle of ixodids usually involves two or more kinds of hosts. *Dermacentor variabilis*, the American dog tick, has three serial hosts: (1) voles and shrews, (2) larger rodents, and (3) dogs and other large mammals (including man). The tiny, six-legged tick larva moves about in the passageways under leaves and grass which are frequented by semisubterranean voles (field mice) and shrews. The larvae attach themselves to these small mammals, insert their mouth parts into the skin, and feed upon tissue juices and blood for several days until engorged. Then they drop off and, after a time, moult in a curious fashion. Their outer cuticle appears to swell, while the internal mass shrivels away from the skin. Then the skin splits, and the flat nymph, with now eight legs instead of the larval six, backs out. The nymphs are larger than the larvae and feed on larger hosts than voles or shrews. They lie in wait along rabbit trails, usually clinging to blades of grass or to twigs, where the slightest motion in the vicinity, such as that made by a footfall or an animal feeding, causes the nymphs to extent their legs widely. If brushed by a host, they seize the hair of the animal and soon begin to feed. When engorged a second time, the young tick (nymph) drops to the ground and moults to become a sexually mature adult. The adults wander widely (sometimes crawling distances of over 100 meters) and finally come to rest on the grass or bushes beside a runway used by dogs and other large animals. Along such forest trails or pasture pathways the ticks wait—sometimes for many weeks. If brushed by the third host, they quickly climb upon it and usually hide within the ears, where the host has difficulty removing the ticks. The males and females copulate; the male drops off and dies, while the female engorges fantastically with blood. After ova have developed, she may have become twenty times her original size; this visible increase represents an even greater amount of ingested blood, because these ticks, like others, are able to excrete water while feeding and thus concentrate the blood into a highly nutritious material. The engorged and gravid female drops off, and lays a large number (thousands) of eggs. From these eggs six-legged larvae hatch to complete the cycle. The life cycle of *Dermacentor* spp., as of most ixodid ticks, may take one or more years. Unfed nymphs or adults can survive the winter.

The above example is of a three-host life cycle. Some ixodids do not leave the host to moult, and thus may complete the cycle with only one host; in such cases the dispersal phase is the egg-larval stages.

A one-host tick, *Boophilus annulatus*, completes its cycle on cattle, deer, etc. It has become so numerous in herds of grazing animals that it is known as the "shingle tick," because the ticks become crowded and even overlap on the necks of the hosts. Although this tick is an extremely successful parasite in nature, it has been eradicated in North America (except in certain refuge areas where deer herds maintain it) by the simple process of dipping all cattle in acaricidal solutions. This achievement is of historical significance, for *Boophilus* was the transmitter of cattle fever (babesiasis) in North America, and the eradication of the vector stamped out the disease.

Other ixodids utilize two or more serial hosts. *Haemaphysalis* spp., common in deserts and steppes in Asia and the Americas, feed upon birds of many kinds, upon reptiles during larval and nymphal stages, and later, as adults, upon rabbits or larger grazing animals. (See Fig. 22–3.) *Rhipicephalus turanicus*, of southern U.S.S.R., reportedly completes its cycle successfully on cattle, sheep, goats, dogs, cats, swine, and man by utilizing various other animals (five species of birds, two of reptiles, various rodents, and hedge hogs) during its larval and nymphal stages. Obviously, it is not so easy to control such a versatile tick as it is to control or eradicate a one-host tick, such as *Boophilus annulatus*.

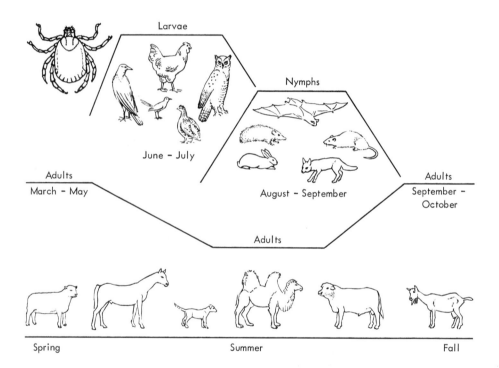

FIG. 22–3. Seasonal changes in populations of three stages in the life cycle of *Haemaphysalis punctata*, a "three-host" ixodid tick of central Asia. The number of adult ticks, which utilize large mammals as hosts, declines during the summer months when adults leave their hosts for egg-laying and die. In early summer there is a corresponding increase in larval ticks on their avian hosts; then in late summer there is a rise in the number of nymphs, which utilize small mammals. Each population maximum is separated from the others by decline during periods of crisis when the ticks must move from one kind of host to another. The large number of larvae suggests that this stage is subject to greater loss through predation or accident than any other stage. (Cf. Fig. 29–2, in which populations of trematodes are similarly graphed, to show the theoretical effects of genetic drift in such fluctuating species.) The implications of such fluctuation for control of a harmful species are obvious.

MEDICAL IMPORTANCE OF TICKS

The medical importance of ticks lies chiefly in their role as vectors of disease. Because the species of ticks are much more numerous than the diseases which they transmit, the following discussion is organized around the latter under viruses, rickettsias, spirochetes, bacteria, and protozoa. But first, tick paralysis, a direct injury in addition to bloodsucking, will be mentioned.

TICK PARALYSIS

Tick paralysis, a sometimes fatal injury to the motor nerves, appears to be an occasional but serious reaction of hosts to the injection of tick saliva. It occurs most frequently when the ticks have been feeding for some days near the base of the skull, and usually ceases when the tick is removed. This condition has attracted attention because it affects man directly, but it is also a factor in livestock damage from ticks.

VIRUS TRANSMISSION

Viruses are minute organisms which are submicroscopic in size and varied in structure and effect. They require intracellular habitats for growth and reproduction. Many of man's most important diseases (the common cold, influenza, yellow fever, small pox) are viral in origin. Tick-borne viruses include the pathogens of certain kinds of encephalitis, Colorado tick fever, and Nairobi sheep disease. These latter diseases are dangerous to man because they are occasional and unpredictable in their zoonotic attack. They exist in reservoirs of wild and domestic mammals and their ticks yet cause few or no symptoms in these carrier populations.

Certain serious virus diseases of livestock in which tick transmission occurs or is suspected include encephalomyelitis of various animals. For example, sheep in Scotland and northern Europe suffer from "louping ill," and sheep in Africa are victims of "Nairobi sheep disease." Many other mammals, including man, can harbor viruses of this group; "spring-summer encephalitis" in Russia and Siberia is of tick origin. This encephalitis occurs as an acute disease all across central Eurasia and extends to parts of India and southeast Asia. Swampy forests, savannas, or wet steppes harbor the reservoirs of this virus. Ticks, mites, small mammals, and birds, as well as large grazing animals, may all carry the virus in sylvatic areas; domestic animals and their arthropod parasites also maintain the virus. Ixodid ticks of at least five genera are important vectors; bloodsucking mites feeding on the young of rodents and birds are also significant. Man usually acquires the virus from the bites of adult ticks. Central European tick encephalitis ("louping ill" of northern Europe) in man is sometimes mild yet sometimes severely damaging to the central nervous system. Workers in farms or forests are most often attacked. Immunological evidence shows that about 18% of people in endemic areas have had this virus, that almost 50% of the sheep, goats, and cattle harbor or have harbored it. However, wild reservoir hosts show no symptoms, and no attempt has been made to use serological methods to determine incidence in that reservoir. Various strains of equine encephalitis occur in the United States and elsewhere, and can be transmitted by ticks. Fowl encephalitis causes widespread disease in poultry, pigeons, and wild birds; ticks of the genus *Argas* are its chief transmitters. Ticks should be suspected of transmission of other viruses, since the above viruses can so easily live and multiply in tick tissues.

A reservoir of dangerous virus disease exists in the Kyasanur Forest of India, where a bird tick (*Haemaphysalis spinigera*) has been found naturally infected, and where antibodies exist in numerous wild birds, monkeys, and small forest animals. The disease, which is similar to the central Asian "Omsk hemorrhagic fever," is neurotropic (affects the nervous system) with mucosal hemorrhage. Visitors to the forest become sick a week after exposure. Outbreaks of the fever among exposed populations may be sudden and severe. The disease is said to be relatively new, since there is no record of it before the midtwentieth century. It therefore illustrates the mutability of parasites—even of those of supposedly the simplest kind, i.e., viruses.

RICKETTSIAS

Rickettsias are a special group of bacteria which have certain similarities to viruses. They are very small, may be spheroidal or rodlike in shape, and possess, just as bacteria do, both RNA and DNA. But like the viruses, they seem to require for growth the environment of living cells. Also, they are characteristically found in arthropod tissue, and their occurrence elsewhere may be considered as accidental. Yet their common occurrence in bloodsucking arthropods makes it possible for rickettsiae to become adapted to two kinds of host: arthropod and vertebrate. The variation in the effects of rickettsial infection of vertebrates, from rather mild symptoms to fatal morbidity, suggests that many of these organisms are presently in the process of adaptation to new hosts.

Several groups of rickettsial diseases have been recognized. Two such groups, the agents of epidemic typhus and trench fever, are almost exclusively louse-borne, and a third, tsutsugamushi fever, is transmitted only by mites. Four rickettsial diseases normally transmitted by ticks, as well as the possible tick reservoir factor in epidemic typhus, will now be discussed.

Coxiella burnetii causes "Q-fever," a cosmopolitan disease with sometimes severe respiratory symptoms. Although Q-fever is transmissible by a great variety of infectious agents (contact with diseased animals or people, dust, contaminated food products, etc.), it is probably maintained in nature in a number of reservoir hosts and is transmitted from such hosts to others by ticks. About 20 species of ticks have been found naturally infected with *Coxiella*. Species of *Coxiella* are important agents of animal disease; sheep and goats are significantly affected. Thus Q-fever, an epidemic disease of man, has domestic reservoirs (farm birds and animals) as well as wild reservoirs, in which ticks play an important part.

Rickettsia prowazeki causes epidemic typhus, one of the few really devastating diseases of history. This disease will be discussed later in relation to its usual vector, the louse (an insect). There is evidence that man is not the only reservoir of infection, and it is known that certain ixodid ticks can harbor the pathogen.

Rickettsia mooseri causes murine typhus, a warm-climate form of typhus. The disease is endemic in populations of wild rodents, but it only sporadically afflicts man. Fleas as well as mites are ordinary vectors, but at least one ixodid tick, *Boophilus australis*, is part of a reservoir of murine typhus; however, other tick hosts probably exist.

Rickettsia rickettsi and a number of very similar species cause the American spotted fever. This disease has a high mortality (about 20%) and is characterized by a rash (sometimes with necrosis) and by a high temperature. Similar diseases occur in Eurasia and Africa, where somewhat different rickettsias (*R. conari* and similar species) are the pathogens. "Boutonneuse" fever, Abyssinian typhus, South African tick-bite fever, Indian tick typhus, and urban exanthematic fever of Southeast Asia are some of the names for these kinds of tick-borne rickettsiosis. An Australian disease similar to the above (with marsupial reservoir hosts) is known. The ticks responsible in the Americas are the ixodids (*Dermacentor*, *Amblyomma*, and *Rhipicephalus*) and the argasids (*Ornithodoros* and *Otobius*). In Europe, the vectors of *R. conori* are various ticks, including *Rhipicephalus*, *Amblyomma*, *Haemaphysalis*, and *Ixodes*. *Dermacentor* spp. in Russia and India have been found infected. In both Old and New Worlds, the tick-bite fevers have important reservoirs, many of which are as yet unknown. Various rodents, rabbits, and other small mammals probably serve as wild reservoirs; the ticks themselves can pass the rickettsias through several generations by egg transfer.

In Russia, *Rickettsia pavlovskyi* occasionally causes a kidney disease in man. Its vectors are ticks and mites, and its reservoir mammalian hosts are various rodents.

Other genera of rickettsias are *Cowdria*, which causes "heartwater fever" of South African cattle and Yugoslavian goats, and *Ehrlichia*, which affects dogs and cattle.

SPIROCHETES

Spirochetes are bacterialike single-celled organisms with a spiral form and without a cell wall. Well-known examples are the organisms causing syphilis and yaws. The soft ticks, argasids, are natural hosts for a number of spirochetes which cause "relapsing fevers." These pathogens are all members of the genus *Borrelia*, of which about 20 "species" have been distinguished. Since these species are morphologically identical, they must be defined by their hosts, their geographical distributions, and the symptoms they produce.

Relapsing fevers, in general, imitate many other diseases. Chills, fever, sweating, pain, nausea, and headache commonly occur. Symptoms often persist for several days, and then disappear for a week or more. They return in a series of such relapses over several months in untreated cases. Relapse is believed to be due to mutation of the pathogen after host antibodies form against the original strain, or due to selection, by antibody formation, of resistant strains (relapse strains) already present among a population of introduced pathogens. This situation seems similar to that observed in other relapsing diseases, such as certain forms of malaria and trypanosomiasis. Relapsing fever can be treated by the same group of antibiotics (e.g. penicillin) and arsenicals found effective against other spirochetal diseases.

There are about a dozen important *Borrelia*-caused relapsing fevers which are geographically distinct. "Tick relapsing fever" of the eastern Mediterranean lands and northern and central Africa is rather rare in man. Its reservoir is in various desert rodents, and its vector is uniquely *Ornithodoros erraticus sonrai*, although lice are capable of harboring the spirochete. A similar fever, found in Spain, in northeast Africa, and on adjacent Atlantic coasts, is transmitted by *O. erraticus erraticus*, a tick closely related to the above. The reservoir is a group of wild mammals, including burrowing rodents, wild rats, and jackal. Relapsing fever of tropical Africa and southern Arabia is primarily a disease of man. The spirochete involved is *Borrelia duttoni*, transmitted by ticks which inhabit the nests and roosts of fowl; these ticks (*Ornithodoros moubata*) transmit the pathogens to man and fowl indiscriminately. Lice may also carry the spirochete but cannot infect by biting; lice must be crushed to release the spirochetes. Asiatic relapsing fever occurs in central Asia in semiarid mountainous regions. This fever is not severe. Its reservoir is a group of burrowing vertebrates (lizards, tortoises, rodents) and the argasid ticks which feed upon them. Another Asiatic relapsing fever afflicts inhabitants or visitors to rocky central Asian regions where caves, ruins, etc., serve both as habitations for man and as shelter for argasid ticks and small rodents. Urban forms of this disease are known. *Borrelia persica*, therefore, has two reservoirs: (1) a complex of cave- and rock-dwelling semisylvatic vertebrates with associated ticks, and (2) a human reservoir in villages and towns. Since the tick host is very long-lived and may pass the pathogen to its offspring, the persistence of central Asian spirochetosis is assured. In the United States, occasional relapsing fevers occur in the western mountainous region, where *Borrelia hermsi* exists in a sylvatic reservoir of ground squirrels, other small mammals, and *Ornithodoros hermsi*. Other strains of *Borrelia* in the central and southwestern United States, Mexico, Venezuela, and Colombia are known to cause relapsing fevers. A cosmopolitan relapsing fever is due to *Borrelia recurrentis*, but this is a louse-borne disease. It may become epidemic in population where crowding and poor sanitation make louse infestation common and louse-transfer frequent. Probably this disease, like all the other spirochetal relapsing fevers, has a tick-mammal reservoir in nature. However, some strains of *B. recurrentis* (Ethiopian and Chinese) cannot be transmitted by ticks at all.

BACTERIA

Two important bacterial diseases, tularemia and brucellosis, are spread by ticks as well as by certain bloodsucking insects.

Tularemia is caused by *Pasteurella tularensis*, a plague-related bacillus. Symptoms are nausea, fever, and glandular involvement. Severe cases may be fatal. A common source of human infection in the United States is contamination of broken skin by blood or viscera of infected rabbits (hence a common name "rabbit fever").

Transmission occurs also by ingestion of contaminated food or water or by inhalation of dust. Thus tularemia may break out among susceptible vertebrates under a variety of conditions. The epidemiological chains of transmission already known are quite different in various parts of the world; they involve, for instance, voles, striped mice, and house mice in the Caucasus, a water rat (*Arvicola*) in connection with voles and mice in Russia along the river Don, and rabbits and wild ground-living birds in much of the United States. In each of these situations arthropods are essential vectors during times of low incidence when transmission by contact or contamination is unlikely. In addition to ticks (both ixodid and argasid), lice and fleas can harbor *Pasteurella tularensis*, but only the ticks are capable of maintaining the pathogen indefinitely and of passing it on to their progeny.

Another bacterial disease, brucellosis, caused by *Brucella abortus*, has epidemiological features somewhat similar to the above. This disease of cattle and man spreads throughout herds by contact and spreads to man usually through contaminated dairy products. Its persistence in nature and its recurrence in cattle herds long after infected animals have been destroyed are due to the arthropod-rodent reservoir. Burrowing rodents such as ground squirrels (*Citellus* spp) and many species of ticks maintain brucellosis. As in tularemia, the pathogens survive and multiply within tick tissues and are passed from generation to generation.

ANAPLASMA AND BARTONELLA

Other diseases transmitted by ticks to man or his domestic animals are anaplasmosis, bartonellosis, and piroplasmosis, the last of which was discussed in Chapter 3. *Anaplasma marginale* and one or two other species are very small organisms of unknown relationships and may be seen in the peripheral region of many red blood cells of infected animals. Anaplasmosis is a variable disease; it sometimes is subclinical, but often causes marked lack of condition, anemia, and even death. While it is easily transmitted by biting insects, surgical instruments, and other mechanical means, the organism is probably a normal parasite or symbiont of ticks; in such organisms the disease can pass through several generations and remain infective. *Haemobartonella* and related forms are another group of small organisms of uncertain place. They multiply in the red blood cells of mammals, including man, and are transmitted by arthropods, including ticks. (See also Chapter 24.) Their significance in disease is not well understood; some of them cause animal disease only in splenectomized hosts.

PIROPLASMS

Since the important blood protozoa, *Babesia* and *Theileria*, were discussed in Chapter 3 and their transmission by ticks was stressed there, the reader is merely reminded that ticks are vectors of these very harmful causes of sickness and death of valuable animals.

CONTROL

All of the above diseases for which ticks are vectors are peculiarly difficult to control because of the persistence of the pathogens in ticks (often as permanent symbionts) and because of the varied habitats and biting habits of the ticks themselves. Tick-borne diseases of stock are sporadic in outbreak, may occur on a new range just as easily as on an old, and, as discussed above under piroplasmosis, require thorough knowledge of the biology of ticks for tick-eradication programs to become successful. The fact that few ticks are one-host parasites, as in the case of *Boophilus annulatus,* which was eradicated in the United States simply by a program of dipping cattle, means that few of the tick-transmitted diseases can be so easily controlled as was babesiasis in the U.S.A. Fortunately, other approaches to controlling some of these diseases are practical; against *Rickettsia*-like pathogens, as well as some viruses (polio, yellow fever), vaccines have been developed for the protection of man and can probably be developed for the protection of farm animals also.

Since eradication of ticks seems generally impossible and since the vaccines just mentioned are not yet in existence, measures to control ticks and tick-borne diseases should be studied, improved, and utilized. Human disease transmitted by ticks can be avoided simply by avoiding the habitats of ticks, wearing protective clothing, or using tick-repellent chemicals. Animal disease due to ticks cannot be completely avoided. Its incidence can be regulated or controlled, however, by well established methods. Large concentrations of ticks, so harmful both directly and indirectly to stock, can be prevented. Buildings, in which argasid ticks take shelter, can be treated with residual insecticides. The host animals themselves can be periodically dipped or dusted to remove ixodid ticks and to prevent the latter from invading pasture land. Introduction of ticks to clean flocks or herds can be prevented by quarantine and treatment of infested animals before mixing the latter with clean stock. The various wild hosts which serve as reservoirs for both ticks and tick-borne diseases are difficult to control. A great deal of knowledge needs to be acquired before absolute protection from such sources of infection can be provided. The general methods outlined above, however, should keep tick damage to a minimum and should reduce the risks of general outbreaks of animal diseases carried by ticks.

SUGGESTED READINGS

ARTHUR, D. R., *Ticks: a Monograph of the Ixodoidea. Part V. On The Genera Dermacentor, Anocentor, Cosmiomma, Boophilus, and Margaropus.* Cambridge University Press, Cambridge, 1960.

BAKER, E. W., T. M. EVANS, D. J. GOULD, W. B. HULL, and H. L. KEGAN, *A Manual of Parasitic Mites of Medical or Economic Importance.* National Pest Control Association, Inc., New York, 1956.

CANNON, D. A., "Linguatulid Infestation of Man," *Ann. Trop. Med. Parasit.*, 36:60–167, 1942.

ERSHOV, V. S. (ed.), *Parasitology and Parasitic Diseases of Livestock.* State Publishing House for Agricultural Literature, Moscow (in Russian) and Israel Program for Scientific Translations, Jerusalem (in English), 1956.

HERMS, W. B., and M. T. JAMES, *Medical Entomology*, 5th ed. Macmillan, New York, 1961.

HORSFALL., W. R., *Medical Entomology: Arthropods and Human Disease.* Ronald Press, New York, 1962.

LAPAGE, G., *Mönnig's Veterinary Helminthology and Entomology*, 5th ed. Williams and Wilkins, Baltimore, 1962.

METCALF, C. L., W. P. FLINT, and R. L. METCALF, *Destructive and Useful Insects, their Habits and Control*, 4th ed. McGraw-Hill, New York, 1962.

PHILIP, C. B., "Ticks as Purveyors of Animal Ailments: a Review of Pertinent Data and of Recent Contributions," *Advances in Acarology* (J. A. Naegele, ed.). Cornell University Press, Ithaca, New York, 1963, Vol. 1, pp. 285–325.

RIVERS, T. M., and F. L. HORSFALL, JR. (eds.), *Viral and Rickettsial Infection of Man*, 3rd ed. Lippincott, Philadelphia, 1959.

CHAPTER 23

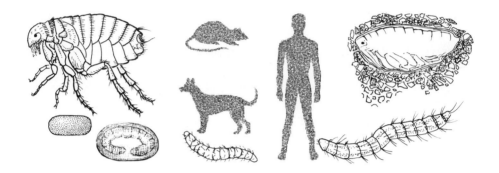

NONDIPTEROUS INSECTS

INTRODUCTION

The previous discussion of primitive arthropods was organized around the relatively few organisms affecting man and his domestic animals directly as pests, predators, parasites, or vectors. Ticks and mites make up a major portion of injurious non-insects, and they are, in most cases, clearly parasitic. It will be difficult to treat insects in the same simple way because of the extraordinary size of the group and because of the fact that insects are much more varied in form, life cycle, and relationship to disease than are the parasitic mites and ticks.

Therefore this and the next chapter will deal with insects of medical and veterinary importance. Omitted will be the whole range of insects "parasitic" on plants, the fascinating groups of insect parasites of insects, or any discussion of social parasitisms found among the Hymenoptera. Included here will be the chief insects which parasitize man and his stock and also the insect vectors of disease. A small group of wholly parasitic flies—the louse flies of mammals and birds—will be mentioned because they are interesting. In order to make this rather arbitrarily restricted treatment somewhat orderly, the insects will be arranged for discussion in a taxonomic pattern.

In the class Insecta, the following orders have medical importance: Orthoptera, Coleoptera, Lepidoptera, Hymenoptera, Mallophaga, Anoplura, Hemiptera, Siphonaptera, and Diptera. The Diptera are probably equal in importance to all the rest together. These groups are better known by their most common or well-known members and by one or two distinguishing characteristics. Thus Orthoptera (straight-winged) include cockroaches, mantids, grasshoppers and crickets. Coleoptera (sheath-winged) are the beetles, including dung beetles sacred in ancient Egypt, stored-grain beetles, the boll-weevil pest of cotton, and the large, showy predatory tiger beetles. Lepidoptera (scaly-winged) are the butterflies and moths. Hymenoptera (membrane-winged) are the ants, bees and wasps. Mallophaga and Anoplura are the biting and sucking lice, respectively. Hemiptera (half-winged) are the true bugs—stink bugs, squash bugs, assassin bugs, harlequin bugs, etc.—which suck the juices of plants, animals, or other insects. Siphonaptera (sucking and wingless) are the fleas, the bloodsucking parasites whose wings have been lost. The members of the great order Diptera (two-winged) are distinguished by having only one pair of wings; the posterior pair is modified into balancing organs called halteres. The Diptera include the flies, gnats, midges, and mosquitoes; some of these are man's worst enemies (besides himself). There are many other orders of insects in addition to the above ten.

The remainder of this chapter will be devoted to insects other than the Diptera, the subject of the succeeding chapter.

ORDER ORTHOPTERA

At the beginning of Chapter 21 a cockroach (Fig. 21–1) was used to illustrate the anatomy and development of a generalized arthropod. The cockroaches are members of a group called the Blattaria, which sometimes is considered an order in itself and sometimes is included in the order Orthoptera. The Orthoptera are an ancient group which has changed little in over 50 million years. As might be expected from their long history as a group, these insects are very adaptable and are found in a wide range of habitats all over the world. Many of them, especially the "domestic" cockroaches, share rather intimately and successfully the life of man. These insects develop gradually and hatch from the egg as tiny replicas of the adult. They differ from the adult only in their lack of wings and sex organs. Essentially omnivorous, they are extremely common in garbage or refuse dumps, cellars, kitchens, sewers, and other places where food or food scraps are found. From these habits it is clear that cockroaches, like houseflies, must be effective transmitters of all sorts of enteric pathogens. It has been shown that such pathogens pass through the cockroach gut in a viable state and are deposited widely with the soft feces of the insects as they wander about over utensils, plates, and dishes during their nightly invasions of urban dwellings. Cockroaches enter houses nightly through drains and return by day to their subterranean man-made home. Since they often live in vast populations in sewers, they must ensure the wide dissemination of most sewage organisms into residential kitchens and dining rooms.

ORDER COLEOPTERA (BEETLES)

Beetles have curved and usually hard wing covers as modifications of the anterior pair of wings, and they have folded, membranous hind wings under the covers. Of the 250,000 species of beetles, very few have medical importance. Blister beetles (Meloidae) when crushed irritate the skin. Some beetles (such as dermestids, "museum beetles," which may become quite numerous and may contribute substantially to "dust") are allergens which produce asthmalike reactions when fragments or hairs are inhaled. A few beetles are intermediate hosts of helminths. The cestodes *Hymenolepis nana* and *H. diminuta*, parasites of murine rodents or of man, can develop to an infective stage in flour beetles (*Tribolium* spp.). The giant thorny-headed worm of swine (*Macracanthorhynchus hirudinaceus*) utilizes various scarabeids (large beetles with subterranean grublike larvae) as intermediate hosts. The accidental ingestion of such hosts can produce infestation of man with these helminths.

ORDER LEPIDOPTERA (BUTTERFLIES AND MOTHS)

Butterflies, moths, and their larvae affect man medically only by shedding scales which may cause allergies or, in the case of larvae, by stinging man with poisonous or nettling hairs.

ORDER HYMENOPTERA

Hymenoptera (ants, bees, wasps, and many less-well-known insects) affect man adversely chiefly by their venom. (Of course the well-known values of these insects in the balance of nature—pollinating flowers, controlling other insects by parasitism and predation, etc.—should be acknowledged, for the Hymenoptera are surely an important ally of man.) The stings of Hymenoptera may cause only local irritation, but some, especially the larger hornets and wasps and certain ants, can injure larger animals severely. Also, some people become sensitized to the protein in certain bee or wasp toxins, so that subsequent stings may produce even fatal anaphylactic shock. Ants, as we have seen in Chapter 10, may serve as intermediate hosts for an important sheep trematode, *Dicrocoelium*.

ORDER PHTHIRAPTERA (LICE)

The Mallophaga and Anoplura should probably be considered members of one order, the Phthiraptera, although many texts refer to these two groups as completely separate kinds of lice. In all lice there are no wings, the body and head are flattened dorsoventrally, the antennae have 3–5 segments, the spiracles are dorso-lateral, the tarsi have 1 or 2 segments, and there are no simple eyes (ocelli). De-velopment is gradual. They are so modified, especially as to the legs and claws, that they can press closely against hair or feather. All stages of lice live permanently on the surfaces of their host. Particular species of lice show preferences for certain species of host; however, host specificity is variable. Site specificity is also charac-teristic of lice, as shown in the head lice, pubic lice, and body (clothing) lice of man. Lice are sensitive to environmental factors such as differences in temperature in different hosts or in different parts of the same host, and condition of the host. Louse populations may fluctuate sharply with seasons, changes in host hair or feathers, and other factors.

SUBORDER ANOPLURA

Human lice belong to the suborder Anoplura, members of which parasitize placental mammals only. The legs of these lice are modified for clasping hair by having the tarsal claw fold anteriorly against a projection of the next (tibial) segment. The space between folded claw and tibial plate is approximately the right size to accommodate a hair of the appropriate region of the body. Thus the pubic louse, *Phthirus pubis*, has relatively large tibio-tarsal apertures, while the head-body louse, *Pediculus humanus*, has somewhat smaller hair-grasping claws. The head of the human louse (Fig. 21–2) has piercing, sucking mouth parts consisting of the labium (a ventral, tubular stylet), the hypopharynx (two stylets, one salivary and one sucking) and the prostomium, which has hooks to anchor the feeding louse. The life cycles of these lice are typical of the order; eggs are cemented to hairs, the eggs hatch, and gradual development ensues. Lice never leave the host except under conditions of close bodily contact, or when trapped in clothing which is removed. An interesting feature of the human-body louse is its occurrence as two habitat-specific strains. One strain lives only on the head. The other strain lives in clothing and cements its eggs to cloth fibers instead of hairs. These two strains interbreed or hybridize easily; apparently the adaptation of the so-called body louse (as distinct from the head louse) is as recent as the invention of clothing and, like all recent or incomplete adaptations, has not yet led to genetic isolation of the new strain from the old.

The effects of lice on their human hosts have been very severe. Epidemic or louse typhus has depopulated parts of Europe, has changed the course of military history on many occasions by destroying armies, and still remains a threat wherever conditions are suitable for typhus to break out. Such conditions are (1) temperate climate with cold winters, (2) poverty and malnutrition, and (3) some degree of crowding. The typhus pathogen, *Rickettsia prowazeki*, already mentioned in connection with tick-borne rickettsial disease, multiplies in the intestine and tissues of the body louse. The pathogen is passed out with louse feces; infection of man occurs when such feces are rubbed or scratched into abraded skin. During times of social upheaval, war, and similar stress, people are ill-housed, undernourished,

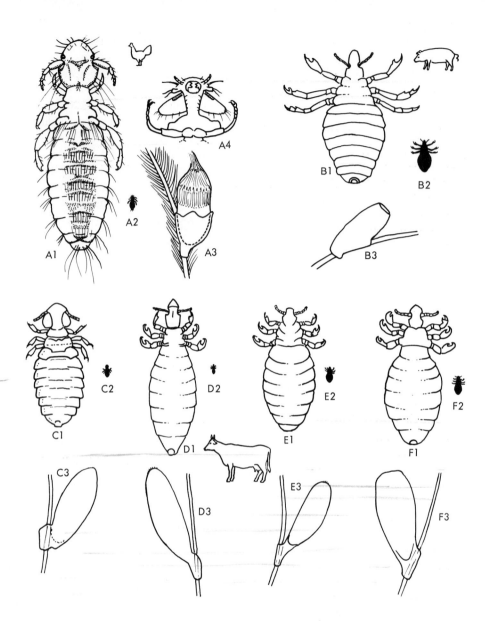

FIG. 23–1. Sucking and biting lice of farm animals. (A1) Chicken head louse, with egg attached to shaft of feather (A3) and ventral view of mouth, showing mandibles (A4). [After Metcalf, Flint, and Metcalf, 1962.] (B) Hog louse (*Haematopinus suis*), a sucking louse, with its egg cemented to a hair. [After Noble and Noble, 1964.] (C) Cattle-biting louse ("little red louse"), *Bovicola bovis*. [After Chandler and Read, 1961.] (D) Long-nosed cattle louse, a sucking louse, *Linognathus vituli*, with its egg. [After Metcalf, Flint, and Metcalf, 1962.] (E) Little blue cattle louse, a sucking louse, *Solenopotes capillatus*, with its egg. [After Metcalf, Flint, and Metcalf, 1962.] (F) Short-nosed cattle louse, a sucking louse, *Haematopinus eurysternus*, with its egg. [After Metcalf, Flint and Metcalf, 1962.] Silhouettes represent actual size.

and unwashed, and often move about from place to place as refugees or prisoners. Louse populations increase, and only a few human carriers of *Rickettsia prowazeki* are required to set in motion frightful epidemics. Human carriers of the pathogen occur in every community where typhus has been. Rickettsiae have been found in the blood of persons who had typhus many years earlier. Some persons suffer relapses (or recrudescences) as late as 40 years after an initial attack of typhus. Thus it appears that control of louse typhus depends chiefly on control of the human body louse, which, of course, depends on the human condition in general.

As was mentioned earlier, lice also transmit the relapsing fever caused by the spirochete *Borrelia recurrentis.* The other species of *Borrelia* are tick-transmitted.

The sucking lice of domestic animals are important pests, but only in swine have these lice been shown to transmit animal disease. Horses, swine, and cattle suffer greatly from sucking lice. Pain from the bites causes restlessness and scratching; loss of blood affects the animals' condition and resistance to disease. Conditions favoring the presence and multiplication of sucking lice are dirt, crowding, and neglect. The species involved are the horse sucking louse, *Haematopinus asini*, the swine louse (swine do not harbor Mallophaga), *Haematopinus suis,* and the cattle sucking lice, *Haematopinus eurysternus*, *H. vituli* and several others. (See Fig. 23–1.)

SUBORDER MALLOPHAGA

Biting lice, while superficially resembling sucking lice, are fundamentally quite different. The most obvious difference is the mouth parts; the biting lice have strong mandibles with which they chew hair, feathers, scales, and epidermis. They are common pests of poultry and mammals.

Poultry suffer from Mallophaga, which chew and destroy the feathers, eat the scales and epidermis of the skin, and constantly irritate and annoy the host. Indirectly these lice cause death, for chickens weakened and emaciated through loss of sleep are more susceptible to infection and suffer more from disease than fowl which are free from lice. Poultry lice are of several species; each is restricted in habitat to a particular part of the host. The chicken head louse (Fig. 23–1A), *Cuclotogaster heterographus,* is dark gray, large-headed, and about 2.5 mm long; these lice are found with head appressed to the base of feathers, and upon the skin they are constantly chewing or nibbling at the scales of skin. The irritation they cause in little chicks is often fatal. The eggs are cemented to down or small feathers; they hatch in 5 days and the lice are full grown 10 days thereafter. Such lice pass from mother hen to little chicks but otherwise seldom leave the host. They increase in numbers from generation to generation and are little controlled by the host, since the head region where these lice live can only be scratched, not pecked, by the irritated bird. Chicken body lice are of two common types: the species *Menacanthus stramineus* (Fig. 23–1A4) and *Menopon gallinae*. The former, called simply "chicken body louse," lives chiefly on chickens and turkeys, is a very rapid breeder, and is said to be quite injurious to its host. The second species, found on ducks, geese, guinea fowl, and even mammals kept near poultry, is not so completely host specific as the larger "chicken body louse." Other lice, less injurious than the above, live in the fluff under the vent or among the barbules of the wing feathers. Pigeons suffer occasionally from very severe infestations of the pigeon louse, *Columbicola columbae*. Poultry lice can be eliminated by dusting the fowl with insecticide and can be controlled by keeping chicken houses clean and airy, and providing clean dust baths for fowl; by dust bathing they effectively remove many of their ectoparasites. Artificial dust boxes to which rotenone or malathion has been added are very effective. Treatment of roosts and nests with residual insecticides is also good.

The horse biting lice, *Bovicola equi* and *Trichodectes pilosus*, behave on the horse's skin in very much the same way as poultry lice do among the feathers of a chicken; they run about chewing on epidermal scales and hair to cause great irritation. By scratching and rubbing against various objects, the animal damages its coat. The lice multiply well during the winter under conditions of crowding, neglect, and high humidity. The cattle biting louse, *Bovicola bovis* (Fig. 23–1C), is similar to the horse-biting louse in habits and effect. Other biting lice occur on sheep, goats, cats, dogs, and other mammals.

BUGS

The Hemiptera or true bugs contain two families of medically important insects: the Cimicidae (bedbugs) and the Reduviidae (assassin bugs). Members of both families have certain general hemipteran characteristics. They have a complex proboscis, which is able to penetrate prey and suck blood. Development is gradual. The young bugs (nymphs) have stink glands. They have well-developed compound eyes. There is a triangular scutum (dorsal plate) extending from the thorax back over part of the first abdominal segment. Wings are present in most bugs, but the bed bugs are wingless. In the other family mentioned, however, typical hemipteran wings occur; front wings, thick and leathery at the base, are folded diagonally across the abdomen of the adult bug and cover the more delicate and smaller hind wings. (The leathery proximal half of the front wings is, incidentally, the source of the name of the order, Hemiptera, or "half-winged.")

The bedbugs are of importance solely because of their bloodsucking abilities and their general unattractiveness. They stain their dwellings (mattresses, cracks in the wall, loose wallpaper) with black fecal matter; their nymphs exude a foul, sour odor which permeates a house, and at night they feed upon the blood of sleepers. They easily move, in clothing, from house to house; they may infect theatres, hotels, or public transportation waiting rooms. Their populations grow rapidly; a generation is completed in less than 40 days at moderate environmental temperatures (27°C). Although many of the dangerous blood-inhabiting pathogens of man can survive and multiply in bedbugs, the latter are not believed to be important vectors of any disease. It has been suggested that under the conditions of medieval Europe bed bugs were so numerous that in those days they may well have been important vectors of viruses, spirochetes, bacteria, and rickettsiae.

The assassin bugs feed on many vital liquids, including the blood of man. Some members of this family catch other insects, paralyze them with injected poison, and suck out their fluids with the powerful and efficient proboscis. A few of these predaceous bugs can bite severely when handled; *Arilus*, the wheel bug, is such a biter. Most of the reduviids which prey on man and other vertebrates belong to the genera *Panstrongylus*, *Triatoma* and *Rhodnius*.

Triatoma has over 100 species throughout the Americas. These rather colorful bugs are usually black with yellow or red abdominal markings. They hide in crevices near the nests of wild vertebrates. Coming out when the prey is quiet, they stealthily insert their long, jointed beaks with probing, piercing mouth-parts into subcutaneous blood vessels and pump blood into their intestine until the once flat insects are quite bloated. Their biting is nocturnal. Warm moist air, with a high carbon dioxide content, attracts them; consequently they often bite the face. They are called locally "kissing bugs" and "barbeiros." *Rhodnius prolixus*, a South American domestic species, is a greedy bloodsucker, which ingests over 0.5 gram of blood during its six moulting stages. Triatomas live by the thousands in thatch-roofed, earthen-walled homes in Brazil. They are present in similar houses in Colombia, Uruguay, Chile, and other South American countries.

Reduviid bugs transmit two kinds of trypanosomes: the dangerous *Schizotrypanum cruzi* and the harmless *Trypanosoma rangeli*. Since Chagas' disease has been discussed elsewhere (Chapter 4), and the role of reduviid vectors was brought out in Fig. 4–2, it is unnecessary to enlarge upon this important aspect of the Hemiptera.

FLEAS

Fleas, (order Siphonaptera) are highly specialized ectoparasites, as well adapted for life in the coverings (fur, feathers, clothing) of vertebrates as are the lice and acarines (ticks and mites). Unlike those other ectoparasitic arthropods, however, the fleas have a nonparasitic larval stage. Their development, a complete metamorphosis, reveals that they are related to the higher insects, such as Lepidoptera or Diptera; the latter order is believed by some to have given rise to the fleas at about the time of the origin of mammals. Fleas are important to mankind as pests of man and his domestic animals and as vectors of disease. At least one tapeworm, *Dipylidium caninum*, utilizes the flea as an intermediate host (see Chapter 11).

Adult fleas are wingless, laterally flattened, tough little insects. They have rows of backward pointing stout bristles at the edges of their body segments, and have strongly muscled hind legs. Their antennae fold into protective grooves above the simple eyes. The mouth parts (Fig. 21–2) consist of labium with palpi, three piercing stylets modified from epipharynx and maxillae, maxillary palpi, and

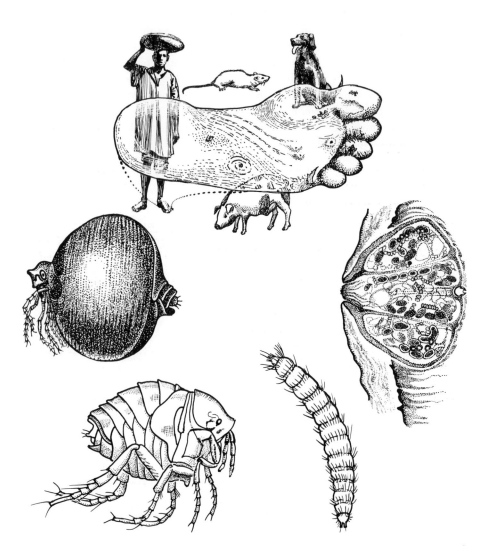

FIG. 23–2. *Tunga penetrans*, the tropical "jigger" or chigoe (not to be confused with the larvae of trombiculid mites; see Chapter 22). Larva (from soil), adult, and gravid female are shown. The latter has been removed from the skin of man, pig, or other mammal. [After Chandler and Read, 1961.]

labrum. The anatomy of the digestive tract of fleas has important implications in the spread of the plague bacillus. Between the midgut and the foregut is the proventriculus, which is lined with hundreds of spines and functions as a sort of filter or valve to prevent backflow from the midgut while blood is being digested. Contraction of this spiny region causes the spines to close in upon each other and blocks movement of material. Plague bacilli often form colonies among these spines. This formation blocks the flow of blood into the intestine and causes fleas to regurgitate frequently. Regurgitation of infectious material is an important factor in the rapid spread of plague. Larval fleas are segmented and hairy and have chewing mouth parts. When ready to pupate, the larvae spin a cocoon, within which the flattened pupa lies.

Because of the restrictions of the life cycle, fleas must live in or near nests, dens, or sleeping quarters of their hosts. Female fleas lay eggs at random (as many as 500 during an adult lifetime). The eggs fall among dust and debris and then hatch into larvae, which feed upon all sorts of organic waste; after four moults, they pupate. Among the food particles eaten by larvae are fecal pellets of dried blood deposited by adult fleas, fragments of hair and epidermis of the host, and bits of dung. The duration of larval or pupal stages varies with temperature, availability of larval food, and other factors. Larvae can develop slowly (taking up to 6 months) or rapidly (requiring only a few weeks). Likewise, pupae may "hatch" after a few days or may lie dormant in the cocoon for months. Emergence of the imago (adult flea) is sometimes triggered by vibration. Doubtless this adaptation greatly increases the chance that an individual flea will find a vertebrate host, since most vibrations near the breeding place of a flea would be caused by movements of the typical host animal.

Fleas are well known as pests. The human flea, *Pulex irritans*, was once extremely common in Europe when beds of straw were used by most people. Changes in human habits have greatly reduced the number of fleas, but they still exist as a cosmopolitan species and, where they are numerous, irritate by their lively hopping about as well as by their bite. In the absence of their normal hosts, dog fleas, cat fleas, rat fleas, and the fleas of various birds frequently attack man. The ability of most fleas to utilize abnormal hosts for blood meals implies that man may be the host or victim of almost any of the Siphonaptera.

A particular flea, the chigoe, *Tunga penetrans* (Fig. 23–2), causes more than mere annoyance. The adult chigoe sucks blood from man, swine, or other animals. The female burrows into the skin and then enlarges to a diameter of two or three mm as she becomes packed with eggs. The painful sore usually becomes infected and may cause lamenesss or other injury to domestic animals. Protection consists of avoiding chigoe-infested earth, where the eggs of this flea have hatched, larvae have grown upon a scavengers' diet, and adults have emerged from pupae.

The sticktight flea, *Echidnophaga gallinacea*, lives partly buried in the skin of chickens (Fig. 23–3) usually near the bill, eyes, etc. Occasionally these fleas invade the ears of cats and dogs. They often kill young fowl, and they injure older fowl by irritation and the sucking of blood. The larvae live and feed among the

FIG. 23–3. *Echidnophaga gallinacea*, the "sticktight flea," attached to the skin of the head of a chicken. The adult fleas are partially embedded and cannot easily be removed by scratching. Details of the anatomy of this flea are shown at right.

refuse of the chicken house, where they pupate on the soil. Clean quarters will prevent the spread of this flea, and treatment of infested birds with insecticide dusts will control it.

Control of fleas of all species involves not only treating the infested men and animals but also cleaning up the breeding places where eggs, larvae and pupae occur. Adult fleas must be killed, since they can survive away from their hosts for several months.

PLAGUE

Neither simple irritation nor the previously mentioned damage to skin and tissues is nearly as important as the disease-carrying propensities of fleas.

Plague is an ancient and terrible disease which has afflicted mankind periodically from the earliest historic times. A great pandemic, beginning in Asia Minor, struck the Mediterranean world in 165 A.D. Other outbreaks occurred in Europe in 542 A.D., in the 1640's and 1650's, and in the mid-18th century. With the great increase of world maritime trade in the 19th century, plague was "domesticated" in such major tropical ports of the world as Hong Kong, Calcutta, and Bombay. Entering the United States quite late, it was discovered on the West Coast about 1900.

The symptoms and effects of the disease depend, at least in part, upon the mode of transmission. Thus the pneumonic form is an acute, usually fatal, pulmonary infection which is spread by airborne particles. In bubonic plague (Black Plague) the death rate is also high, and the symptoms are swollen lymph nodes. A septicemic (blood) form of plague produces high fever with hemorrhages and is usually fatal. There is also a relatively benign minor form, in which the patient is not prostrated, and is able to recover. In mammals other than man, the disease is usually not extremely acute, but many animals die from it after illnesses of various durations.

Plague can be considered as a reservoir (Fig. 23–4) of disease which occasionally overflows its banks. Its ordinary location is among burrowing rodents of the semidesert areas of the world. There it exists in endemic form and is transmissible among these hosts in the same ways as among men, i.e., by contact of several sorts with the plague bacillus. When animal populations are large, epizootics of plague occur. Afterwards, because of reduced numbers of hosts, the plague foci shrink, and the disease is quiescent. Transmission during quiescence depends upon fleas. Thus the sylvatic reservoir itself fluctuates; it occasionally reaches large proportions but often shrinks into a few residual pools of rodent-flea-maintained infection. Overflow of the reservoir comes about in several ways. Historically, trade routes from desert foci have guided plague carried in the bodies of rats and rat fleas from its home in central Asia, for example, to one of the great Asiatic ports. There the plague spreads among rats infesting the city, and many rats die. Fleas, dispersing to seek new hosts, transmit the disease to men. And from man to man it spreads by contact, by contaminated food or water, by air, or by human fleas until an epidemic rages. The port city is the point from which a new dispersal begins;

FIG. 23–4. The reservoirs of plague. Central Asia is the historic home of plague, and ground rodents, such as the marmot, are reservoir hosts. Trade routes over land carried plague-infected rats to centers of population in several pandemics. Eventually the great port cities became foci of plague, and the lands near them became secondary reservoirs as local ground squirrels, etc., became infected. Modern plague control can remove the threat of pandemics, but the sylvatic reservoirs remain.

the disease may be spread by rat or by human traveller; a pandemic is under way. Then, after the world has suffered a "great plague," the susceptible persons have died, and only the survivors remain. The rats and their fleas are also fewer, and the disease seems to disappear. The reservoir has shrunken back to its original area—the lands where the pandemic began.

The above description is not entirely accurate, for the flood has left little pools of plague in new places. Thus when North America was first invaded by the plague bacillus in 1900, the fleas of the port rats of San Francisco attacked local rodents, including ground squirrels, and plague spread among these animals far into the mountains of California. Sylvatic reservoirs of plague now exist over much of Western United States, Canada, and Mexico. Similar events must have occurred elsewhere. Sylvatic plague exists in South America (in several distinctive reservoirs), in South Africa (in populations of a semidomestic mouse), and in Central Europe (in ground squirrels and marmots).

The danger of worldwide epidemics of plague has been eliminated by control of rats in port cities and by standard public health procedures which recognize and isolate human cases promptly. Insecticides and modern techniques for rat-proofing storehouses of grain have kept down both flea and rat populations. It is believed, however, that nothing can be done about the sylvatic reservoirs. These reservoirs always present a danger to explorers, travellers, and vacationers in areas where burrowing rodents and their fleas preserve the plague organism and prepare it for occasional excursions into human hosts.

MURINE TYPHUS

Murine typhus, another disease transmitted by fleas, is due to the organism *Rickettsia mooseri*. This organism was mentioned in the discussion of tick-borne disease, because ticks and mites form a part of the sylvatic reservoir of this typhus-like disease. Fleas, however, are important and effective vectors both for maintaining the pathogen in populations of wild mammals and for transmitting the disease to man. Murine typhus is seasonal in the temperate zones, for outbreaks occur in the summer when rodent and flea populations are high. Domestic reservoirs, in cities or houses, are established by rats and mice in association with fleas. Control of murine typhus depends on rodent control.

TRANSITION

No general discussion of the above insects will be attempted at this time. The major parts of the subject have been stated, and the medically and economically important lice and fleas have been briefly discussed. Other aspects of the problems posed by these insects and the diseases they carry will be considered in Chapter 25 (Control of Arthropods), when other arthropods—both the primitive ones discussed in Chapter 22 and the dipterous insects to be treated in the next chapter—can be included.

SUGGESTED READINGS

BURNET, F. M., *Natural History of Infectious Disease*, 3rd Ed. Cambridge University Press, Cambridge, 1962, pp. 202–225; 323–330.

CAMERON, T. W. M., *Parasites and Parasitism*. Methuen, London, 1962, pp. 237–252.

HERMS, W. B., and M. T. JAMES, *Medical Entomology*, 5th ed. Macmillan, New York, 1961.

HIRST, L. F., *The Conquest of Plague; a Study in the Evolution of Epidemiology*. Oxford University Press, London, 1953.

HOPKINS, G. H. E., and M. ROTHSCHILD, "Illustrated Catalogue of the Rothschild Collection of Fleas (Siphonaptera) in the British Museum of Natural History," Parts 1 and 2, British Museum, London, 1953–1956.

HORSFALL, W. R., *Medical Entomology*. Ronald Press, New York, 1962, pp. 65–106, 251–279.

LAPAGE, G., *Mönnig's Veterinary Helminthology and Entomology*, 5th ed. Williams and Wilkins, Baltimore, 1962.

METCALF, C-L., W. P. FLINT, and R. L. METCALF, *Destructive and Useful Insects, their Habits and Control*, 4th ed. McGraw-Hill, New York, 1962.

RODENWALT, E. (ed.), *World Atlas of Epidemic Diseases*. Falk-Verlag, Hamburg, Germany, 1952. (In German and English.)

CHAPTER 24

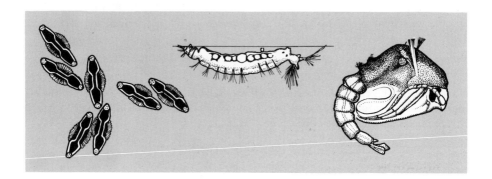

DIPTERA

INTRODUCTION

The members of the order Diptera are recognized as the most important of all the medically important arthropods. Many of these flies, gnats, and mosquitoes are predators on man and his domestic animals. They suck blood or lap up tissue fluids from punctures or abrasions made by their mouthparts. Others are parasites, which utilize the tissues of man and beast as shelter and food for the larval stage or, in some cases, for the adult stage as well. Still others, the house flies in particular, spread filth and disease by their feeding and defecation habits. Yet the most harmful aspect of the Diptera is their role as vectors of disease, for particular members of this order transmit malaria, African sleeping sickness, many virus diseases, and the filariases—surely some of the worst plagues of mankind.

Diptera have characteristic adults, larvae, and pupae. Adult Diptera have one pair of wings. The posterior wings are reduced to stalked knobs (halteres) which, in some species at least, have been shown to vibrate during flight. The oscillations are believed to provide sensory data on direction and inclination, much as a gyroscopic regulator is used in the control of ship or plane motion. Dipteran mouthparts are suctorial. Some consist of tubes with expanded, tonguelike structures (labella) at the distal end. Others are tubes within which movable, pointed stylets lie. Essentially, whether the mouthparts are adapted for lapping or for piercing, the food is liquid. A great many Diptera feed on the blood or tissue fluids of vertebrates. Larval Diptera are legless, although many aquatic forms have suspensory bristles and are able to "swim" by wriggling. The mouthparts of dipterous larvae are adapted for biting and cutting, and the food is quite varied. Some of these larvae are carnivorous, feeding on or in the flesh of larger animals. Many are predacious. Others ingest microorganisms in rotting material, or feed on living plants, or trap plankton in streams. Metamorphosis is complete in the diptera. There are 4 larval instars (forms) followed by a pupal stage. Pupae may be active (as in mosquitoes). Some pupae develop within the larval skin, which forms a tough covering called the puparium. Each kind of dipterous insect has its own peculiar combination of the above developmental stages.

CLASSIFICATION

Diptera are separated into suborders on the basis of certain larval characteristics, the type of pupa or puparium, and the shape and structure of the antennae of the adult. The suborders Nematocera (long, threadlike, flagellar tips on antennae) and Brachycera (with the distal part of the antenna short or compact) are distinguishable from the suborder Cyclorrhapha by the fact that in the latter group puparia are formed; in the puparia both a quiescent, nonfeeding fourth larval stage and the next stage, the pupa, are combined. Cyclorrhaphous Diptera escape from the puparium by a special organ on the head of the emergent imago. This organ, a distensible, bladderlike structure called the ptilinum, breaks the puparial skin and allows the imago to push its way out of the usually semiliquid mass in which these Diptera pupate. The suborder Nematocera includes four medically important families: the Psychodidae (sand flies, etc.), Ceratopogonidae (midges or punkies), Culicidae (mosquitoes), and Simuliidae (black flies). The suborder Brachycera includes the important Tabanidae (horse flies, deer flies) and the Rhagionidae (snipe flies). The suborder Cyclorrhapha includes the Chloropidae (eye gnats), Muscidae (house flies, stable flies, tsetse flies), Calliphoridae (blow flies, screwworms, flesh flies), Oestridae (cattle grubs, bots), and three families belonging to a special group (the Pupipara): the Hippoboscidae (sheep keds, louse flies) and the Nycteribiidae and Streblidae (flat flies of bats). For the most part, the families are groups of related flies, and the suborders may be natural, too (that is, groups of organisms related by descent). The above classification is a convenient way to arrange the following descriptive material.

SUBORDER NEMATOCERA, FAMILY PSYCHODIDAE

The Psychodidae are little hairy flies, some of which are called sewage gnats or moth flies. These cause no damage except by occasionally crowding the air with their bodies. Some of the Psychodidae are called sand flies and belong to the genus *Phlebotomus*. The latter are the vectors of leishmaniasis, a group of diseases of the skin, mucous membranes, and viscera (see Chapter 4), and of bartonellosis, a febrile disease occurring in the high valleys of the South American Andes. *Bartonella bacilliformis*, a very small, bacilluslike organism, lives in the red blood cells, the macrophages of the reticuloendothelial system, and the endothelial cells. It causes an anemia similar to malarious anemia, and is characterized by greatly reduced numbers of erythrocytes. Rather low fevers (99° to 102°F), lasting from days to months, accompany other symptoms such as dizziness, headache, thirst, and weakness. Skin eruptions appear chiefly over joints or bony prominences. These eruptions, verrucous (warty) in nature, may bleed or become infected. One of the common names for the disease, "verruga peruana," is derived from this symptom. Treatment with antibiotics (chloromycetin) and prevention of transmission by the destruction of sand flies with DDT can greatly reduce the incidence of bartonellosis in the high valleys of Peru, where the disease was first noticed, or in similar locations in Colombia and Ecuador.

Sand flies require three conditions of the environment: sufficient moisture in the soil, diurnal shelter, and a source of mammalian blood. The larvae develop in rather wet (but not soaked) soil containing a fair amount of nitrogenous waste. Rich soil near dwellings, soil in the burrows of rodents, and the moist places at the bottom of fissures during dry seasons are all good habitats for the larvae, which feed on organic debris and are killed by drying. Adults feed on blood and lymph during the early evening; they are weak fliers and are unable to withstand the buffeting of even the mild breeze which is apt to arise in the night. Heat and dryness are harmful to the adults, and they hide during the day in cracked earth, animal burrows, or crevices in buildings. Some species of *Phlebotomus* are domestic; *P. papatasii*, an important vector of visceral leishmaniasis, lives in and around low dwellings, breeds in rich, polluted soil, and feeds on human blood in the twilight and early evening. Many species are sylvatic; they breed in the lairs or burrows of animals, shelter under leaves or bark, and feed on a variety of mammals.

Transmission of leishmaniasis by sand flies was not easily demonstrated. Early workers (in the 1920's) discovered that *Phlebotomus* could serve as a host in which ingested *Leishmania donovani* would develop, but the erratic and inconsistent results of laboratory transmission experiments were puzzling. Then it was shown that the parasite does not grow well in sand flies fed exclusively on blood and that a diet of raisin juice permits rapid growth of flagellate forms (leptomonads) in the gut of the infected fly. These parasites actually block the fly's midintestine, cause the feeding fly to bite repeatedly and to regurgitate infective material into the same host many times. It is very likely that *Leishmania tropica* has a similar effect upon its *Phlebotomus* host.

As was seen in Chapter 4, the epidemiology of kala-azar and the cutaneous leishmaniases involves sylvatic and domestic reservoirs. In most parts of the range of kala-azar the parasite is maintained in a variety of sandfly-animal associations. In the Sudan, the sylvatic reservoir involves wild dogs. Because the sand flies of this reservoir are restricted to certain quite sharply limited "islands" of cracking soil, human infection is found only in and near such localities. In India, on the other hand, the disease is domestic and occurs wherever soil and humidity support populations of sand flies near human dwellings. The various forms of oriental sore are probably all derived from sylvatic reservoirs. In Asia, the two forms of sore, "wet" and "dry" (see Chapter 4), are of different epidemiological origin; the former is the result of a rodent-sandfly-human chain of transmission, and the latter is transmitted from human to human. Rodent association with sand flies is essential for the persistence of *Leishmania tropica* in Asia and North Africa, although other mammals, including man, may be temporary parts of a complex system of hosts. In Central and South America, the various strains of *Leishmania tropica* utilize similar reservoirs, yet climatic factors, especially forest ecology, make more complex mammal-sandfly associations possible. However, much remains to be learned about the transmission of cutaneous leishmaniasis. It has been suggested, for instance, that transmission by means other than biting should be investigated and that defecation by the flies near bites may be a common source of infection. The role of human "reservoirs" (asymptomatic carriers) should be studied, since there is evidence (in French Guiana) that such carriers are important. Above all, the ecology of the many species of *Phlebotomus* needs further investigation.

SUBORDER NEMATOCERA, FAMILY CULICIDAE

Mosquitoes are the most important vectors of human disease and are also serious pests. They belong to the family Culicidae, which contains two subdivisions, the gnats and the mosquitoes. There are about 30 genera and hundreds of species of mosquitoes, and these species are so widely distributed and so well diversified in ecological requirements that they are found throughout all the continents except Antarctica. Some thrive at high (12,000 feet) elevations, while others live deep in mines. Some require the cold water of melting snows for their larval development, while others can live in hot springs (up to 38°C). Every degree of water pollution is tolerated by the larvae of some particular species. The only habitat closed to them, besides the perpetual snows of mountain peaks or the south polar barrens, is the open sea and other deep waters, for they require quiet water from which to emerge as imagoes, and they need shallow water with some type of plant growth to provide the fine particles of organic material which the larvae feed on.

Although mosquitoes are noted for their diversity in habitat as mentioned above, they are rather similar to each other in form. These slender long-legged insects have bodies and wings clothed sparingly with scales similar to those of Lepidoptera. They have long antennae, which are slightly hairy in females and thickly bristled in males. There are two prominent compound eyes. The mouthparts are a pair of maxillary palps, plus the grooved, elongated labium wrapped around a bundle (fascicle) of six sharp, fine instruments of penetration, i.e., the maxillae, the mandibles, the hypopharynx, and the labrum-epipharynx. (See Fig. 21–2.) In female mosquitoes (but not in males) these structures are used for penetrating the skin, for injecting saliva as an anticoagulant, and for sucking blood from the lacerated capillary bed. The digestive tract consists of a very fine tube, the esophagus, which leads into one ventral and two dorsal food-storage sacs, or diverticula, which branch from the gut just in front of the proventriculus and midintestine. Opening by a separate duct into the hypopharynx are the salivary glands. In sucking blood, the mosquito inserts the bundle of sharp stylets beneath the epidermis, cuts into the capillaries of the dermis, injects salivary fluid into the wound (preventing the clotting of blood), and sucks blood through the canallike labrum. The blood is pumped into the esophagus and intestine by a muscular expansion of the pharynx, the pharyngeal pump.

CLASSIFICATION OF MOSQUITOES

The separation of genera depends on variations in egg-laying habit, appearance and behavior of larvae, as well as morphology of the adult. Separation of species within genera is a highly technical problem, in the solution of which minute differences in male copulatory apparatus happen to be useful. Since the structure of these organs in each species assures that only females of the same species can be fertilized, this particular taxonomic feature has true evolutionary and biological meaning; it is one of the best examples of a mechanism which restricts genetic material to its own "pool," the species. Of course the taxonomy of mosquitoes is an extremely important part of the research which is necessary in order for the control of mosquitoes and mosquito-borne disease to be possible.

The 30 genera of mosquitoes include about 10 which are of recognized medical importance. The genera *Anopheles*, *Aedes*, and *Culex* are the most important of the latter.

Anopheles are tropical and temperate zone insects. They are very common over the regions where broad plains and occasional or heavy rainfall provide permanent shallow water for breeding places. Eggs of *Anopheles* are laid singly on such water and float by means of lateral air bladders until they hatch. Depending on temperature, *Anopheles* larvae emerge less than a week after deposition and begin to feed actively on the surface film. One to two weeks may be required for larval growth to be complete; then pupation occurs. The pupa, like the larva, is an active, air-breathing form, which clings to the surface film, its source of oxygen. Two days to a week may be spent in the pupal stage before the imago emerges. The latter breaks open the pupal case along its back, rests briefly upon the shed, film-supported membrane, and then flies away. Adult anopheles hide by day in shady places and usually feed at night. The female, sucking blood, assumes a characteristic "head-down" position, one of the means by which mosquitoes of the genus *Anopheles* can be recognized. The importance of *Anopheles*, the vectors of malaria in all malarious lands, need not be discussed here (see Chapter 2).

Aedes is a mosquito of temperate and arctic regions. Unlike *Anopheles*, it is a floodwater breeder, and lays its eggs on dry ground, sticks, or plant stems at the

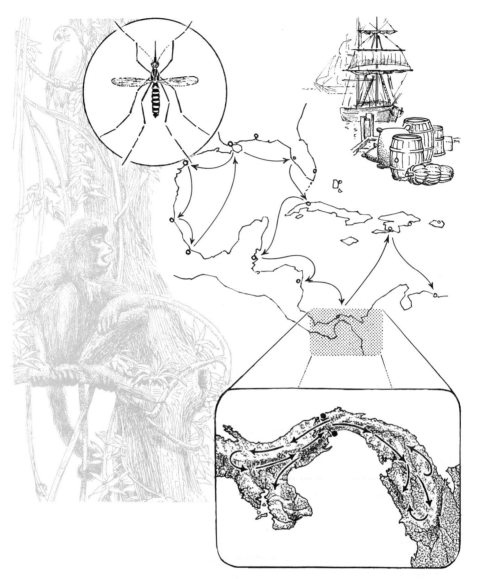

FIG. 24–1. The traveling virus of yellow fever. Until the 20th century, sailing vessels trading in the Gulf of Mexico and the Caribbean carried victims of yellow fever from port to port. *Aedes aegypti* also traveled and was found breeding in water barrels on the ships. This pattern of commerce allowed the virus to reach susceptible populations after other populations had become immune through epidemics, and it permitted time for new susceptibles to be born and grow up in formerly immune populations. Now that trade has speeded up and public health measures have controlled yellow fever, the virus cannot travel as it did among men. Yet it still moves from one population of howler monkeys to another in the rain forests of the Isthmus; the disease causes periodic epizootics among these animals and remains a threat to man in its sylvatic form.

YELLOW FEVER

Yellow fever was once a major disease of mankind. It struck port cities in rapid and severe epidemics, and spread from port to port, exactly as dengue did, by sailing ship, sick men, and *Aedes aegypti* (Fig. 24–1). Enough time had to elapse between epidemics for a susceptible generation to grow up, because yellow fever virus immunized everyone during an epidemic and the immunity was lasting. Yellow fever is no longer a cause of epidemics, for sailing ships no longer ply between the coastal cities of the world, and ships and sailors are inspected and, if necessary, quarantined. Nor can the disease be endemic in a human population because of the total immunization such a population would undergo. Yet yellow fever exists in sylvatic reservoirs. Such reservoirs curiously resemble the human "trade" reservoir of bygone days, for they probably consist of bands of monkeys usually isolated from each other in territories which they occupy and defend. The virus passes from one territory to the next through jungle canopy mosquitoes; it leaves sick but immunized animals as it passes, and moves on to susceptible populations. Thus by constantly traveling, the virus maintains itself just as it did during the age of sail among human populations. The sylvatic reservoir is, of course, a constant source of danger to man within range of its mosquitoes, and occasional outbreaks of yellow fever still occur in such regions.

From the above discussion of the Culicidae it should be obvious that these insects, which have been major forces of human destiny until quite recent times, are still a threat to human health and welfare. As will be pointed out later, the accumulated knowledge of mosquitoes—their ecology, physiology, behavior, and classification—is still not great enough to make control everywhere possible; mosquito-borne disease ranks very high in the list of human miseries.

SUBORDER NEMATOCERA, FAMILY CERATOPOGONIDAE

The midges ("punkies" or "no-see-ums") are members of the family Ceratopogonidae and of the genus *Culicoides*. They are so small that they can barely be seen, but their mouthparts tear and lacerate the skin savagely, so that they are truly significant pests of certain regions and seasons. Their larvae are carnivorous inhabitants of mud, tree-slime, and similar dense, wet materials. They are mentioned here because they are important as local pests, and because some of them transmit *Dipetalonema perstans* and *D. streptocerca*, filariids of man and chimpanzees of the African rain forest.

SUBORDER NEMATOCERA, FAMILY SIMULIIDAE

Black flies, or buffalo gnats, of the family Simuliidae, genus *Simulium*, are rather small, dark flies with large thorax ("hump"). As they attack man in bright light or sunlight, they pierce and cut the skin with sawlike mandibles and hooked maxillae, aided by a toothed labrum which anchors the fly against the wound during the feeding process. Blood, lymph, and fragments of abraded tissue are eaten. Black flies may feed on many kinds of hosts, including birds and mammals. They usually attack the lower parts of humans or the legs and belly of domestic animals. Black flies may appear in such numbers as to suffocate (by inhalation) large beasts. Their presence in some areas prevents or discourages human use of the region. Larval black flies occur in streams of various kinds. Some live on the surface of water-splashed rocks or grasses in upland streams. Others are riverine, and live on rocks at the bottoms of large rivers such as the Danube. All larvae feed by labral brushes which strain particles out of the running water while the larva lies attached by a suckerlike posterior appendage to the substrate. Pupae are protected by open silken cones attached to rocks or vegetation under water. They breathe by a branching mass of filamentous gills. Usually surrounded by air, the imago emerges and flies away when its bubble reaches the water surface. Black flies may travel for considerable distances from their breeding areas. The medical importance of black flies, the vectors of onchocerciasis, was emphasized in Chapter 14.

SUBORDER BRACHYCERA, FAMILY TABANIDAE

The largest of Diptera are included in the family Tabanidae—horse flies, deer flies, etc. The family is very extensive and is said to be in an active evolutionary state. Presumably this condition is reflected in the large numbers of subspecies reported among the more than 2500 known species. These big flies are important pests of livestock, and some of them transmit animal parasites.

Tabanid flies, the only important family in the suborder Brachycera, are characterized by the shortness of the flagellar portion of the antenna, sexual dimorphism with respect to the eyes (which in the males are contiguous along the dorsum of the head), and mouthparts adapted in the female for blood-sucking. (Males feed only on plant exudates, including the nectar of flowers.) The larvae are usually fusiform (tapering at each end). They breathe by means of a posterior airtube which can be retracted.

The eggs accumulate in masses on leaves, posts, etc., which overhang water or water-soaked soil or mud. From these, larvae drop onto the water or soil. Some larvae live in pasture soils, under cover of vegetation and humus. All tabanid larvae are carnivorous; they feed on other larvae and many invertebrates, as well as upon vertebrate tissue. They kill and macerate their prey by vertical blows of the hooklike mandibles. Pupation occurs in soil or mud which is drier than the larval habitat. A true pupa is formed, the walls of which outline the thorax, wing-buds, and abdomen of the imago. Development of some species is slow and may require a year per generation. Other species may have several broods per season. Temperate zone tabanids overwinter as larvae. Larvae of some tabanids can survive freezing, and it is thought that tropical forms aestivate during the dry season in conditions of great heat and environmental dryness.

The adult females usually feed in sunlight. Two kinds of feeding may occur: (1) lapping of blood oozing from the surface of wounds, or (2) sucking of blood as it flows from a deep wound made by the slashing or cutting action of the sharp-edged mandibles and the piercing action of the fascicle of stylets. Saliva of the flies is anticoagulant, and blood may flow for some time after feeding has stopped. Since tabanids are large flies and they feed voraciously, they may when numerous actually take enough blood to weaken their victims. The pain caused by their bites brings about much restlessness in cattle and other stock. While little is known about how tabanids find and respond to prey, there is an interesting African *Chrysops* which, while normally feeding on monkeys in the high canopy, follows smoke down to the cooking fires where men congregate. Whether the smoke itself is the stimulus or whether it coincides with some other nonapparent clue is not yet known.

The significance of tabanids in animal disease has been mentioned in Chapter 4 (the hemoflagellates) and elsewhere. Horses, camels, cattle, dogs, and other animals acquire the trypanosomes of surra and related diseases by the bites of tabanids. The tsetse-borne trypanosomes may be transmitted mechanically by tabanids and other blood-sucking flies. Tabanids transmit "el debab," a disease of horses and camels in Algeria. Tularemia, a bacterial disease often transmitted by fleas, is carried by a *Chrysops* in the western United States; rabbits are the commonly infected sylvatic reservoir, and the tabanids presumably feed on rabbits as well as larger animals, including man. It has been suggested (and in some cases demonstrated by laboratory tests) that tabanids and other blood-feeding flies can transmit a number of blood diseases—anaplasmosis, anthrax, and perhaps brucellosis—by infecting a susceptible host with pathogens which contaminate the fly proboscis. Interrupted feeding is a necessary part of such mechanical transmission, for normally the time between separate feedings to repletion would exceed the survival time of most pathogens. Loaiasis (see Chapter 14) is transmitted by *Chrysops* in Africa. As mentioned earlier, flies of this group feed on monkeys in the high treetops but descend to attack men around fires. Presumably *Loa loa* is transferred from monkey to man, and vice versa, through this feeding peculiarity of the vector. Tabanids not only transmit disease but also are vicious predators. Horse flies bite horses and mules repeatedly during the day to cause sharp pain and significant loss of blood. Work animals become unmanageable under these attacks.

SUBORDER CYCLORRHAPHA, FAMILY CHLOROPIDAE

Eye flies (family *Chloropidae*) are exceptionally troublesome members of a group of otherwise harmless small flies which feed on plants and plant exudates. *Hippelates* spp. cluster about the eyes, sores, and mucous membranes of man and his domestic animals. These flies transmit such eye diseases as "pinkeye," conjunctivitis, and trachoma in the New World. A related group of species, members of the genus *Siphunculina*, are similarly involved in disease transmission in Asia. Eye flies are very common in the tropics where yaws occurs, and are probably a significant mechanical vector of this painful, disfiguring, and once widespread spirochetal disease. Because the tiny flies are so numerous, and because they have peculiar

feeding habits—especially their way of ingesting, then regurgitating, then re-
ingesting semiliquid material—they are important in the transmission of any
pathogens found on the skin, in sores or scabs, or on the mucous membranes.
Larval eye flies develop in almost any moist, rotting material, including feces. They
swarm in moist, sandy areas, at the edges of forests, and in irrigated farmland.

SUBORDER CYCLORRHAPHA, FAMILY MUSCIDAE

The Muscidae are the family of Diptera to which most of the well-known medium-
sized "flies" belong. The family includes the house flies, stable flies, horn flies, face
flies, and tsetse flies—all quite important in human or animal disease.

House flies, including *Musca domestica*, a cosmopolitan and well-known species,
are common wherever garbage, organic litter, pit privies, or the excrement of
domestic animals, especially horses, are found. The maggots grow well in such
material; only about two weeks are required for a generation. Breeding is continuous during favorable conditions, and since a female *Musca* may lay over 1000
eggs in a few weeks, tremendous populations can build up. However, if breeding
places are not available or are altered by heat, moisture, etc., populations rather
rapidly diminish. House flies cannot overwinter in temperate zones in the out-of-
doors. The adults feed upon a great variety of foods, including most of the things
man eats. These flies soften food surfaces with saliva and then lap up the food
with the labella, which are fleshy lobes or pads covered by convergent channels
called pseudotracheae along which fluid food moves toward the labral groove into
the pharynx. After feeding, house flies usually rest while they slowly regurgitate a
droplet of food mass; they hold it on the end of the proboscis, and ingest it and
regurgitate repeatedly. Other species of *Musca* have similar breeding and feeding
habits. Their medical importance lies in their obligatory movement from privy or
garbage dump to kitchen or table, and in their contaminatory feeding habit, which
involves moistening food, walking upon it, and defecating frequently, thus liberally
collecting and scattering many pathogenic organisms. Among the more troublesome house flies are the face flies, *Musca vicina*, *M. sorbens*, and *M. autumnalis*,
which crawl persistently upon the skin. *Musca* species are known vectors of only
two metazoan parasites. *Habronema*, a spirurid nematode of horses (see Chapter
20), develops in fly maggots, then escapes from the labium of adult flies on warm,
wet surfaces—mouth, lips, etc.—of the vertebrate host, which licks and swallows
the parasite. *Thelazia*, eye worms of cattle and dogs, use *Musca* species as intermediate hosts, from which the infective larvae escape while the fly is feeding on or
around the eye of the vertebrate host.

Stomoxys calcitrans, the stable fly, is about the size of the common house fly, but
is distinguishable by its sharp, straight proboscis, with which it bites. This fly
annoys stock greatly and causes restlessness; the bites themselves, when numerous,
produce edema. Surra, anthrax, and other diseases are transmitted by the bites
of *Stomoxys*. Stable flies breed in manure and damp, soiled straw in shaded places
such as stalls and stables. The maggots grow and pupate in about a month, so that

several generations of flies are produced during a summer. This fly frequently attacks man, as it bites painfully and repeatedly about the legs.

Another expensive pest is the horn fly of cattle, *Haematobia irritans*. This, similar to the stable fly but only half as large, bites cattle, sheep, and goats; it gathers in swarms about the heads of these animals and often clusters at the bases of the horns. The animals are constantly restless and in pain. Eggs are laid in fresh dung, where the maggots grow and pupate. A complete generation occurs in two weeks. Some workers believe anthrax is transmitted by horn flies.

Tsetse flies, which have already been discussed (Chapter 4), are peculiarly important in the transmission of human and animal trypanosomes. However, little was said about their species or distribution, and these are important as an approach to control.

Tsetse flies (*Glossina*) occur as about 20 species which are distributed in equatorial Africa (and Southern Arabia). The species of *Glossina* fall into three natural groups, within which hybridization occurs; differing degrees of hybrid sterility maintain a rather precarious evolutionary isolation between one species and another. The natural groups are the *fusca*, with eleven species, the *palpalis*, with five species, and the *morsitans*, also with five species. Morphological differences among the groups, as well as within the groups, are matters of technical difficulty due to the above-mentioned genetic instability of the genus.

Each species has its characteristic habitat. Thus *Glossina morsitans* inhabits scrub woodland, with seasonal and sparse rainfall, and the vast extent of such woodland permits this species to occupy much of West, Central, and East Africa. Other types of forest can also support this fly. *Glossina palpalis* is restricted to grassy plains (savanna) where the plains meet thickets of low trees; this linear distribution is determined by the resting habits of the adults, which lie in wait in bushes, and by their larviposition needs, which require some sort of detritus or loose mould upon the ground. *Palpalis* thus has a distribution as wide as *morsitans*, but occupies a linear pattern instead of the general pattern provided by partly open woodlands. *Glossina tachinoides* requires a wetter environment than the above and occupies undergrowth near water. It can survive dry seasons, however. Its distribution can be described as convergent, following water courses from their sources to their confluence as rivers.

FIG. 24–2. Life cycle of the tsetse fly. Although tsetse flies of different species live in somewhat different habitats (some survive in relatively dry savanna, while others live along streams and near water holes), the female always rears its young, one at a time, within a uterus (C), into which a fertilized egg (B) passes from one or the other ovary (A). "Milk glands" secrete nourishment for the larva, which "pupariates" (D) (retaining its last larval exuvium as a covering for the pupa) after being deposited in litter or humus. Men, cattle, and wild game provide food for the adult tsetse (E). The larvae, of course, require no food other than their mother's milk. [Details of tsetse after Horsfall, 1962.]

Feeding habits of all *Glossina* species are similar; they bite large animals, and man is a suitable source of blood. But most species which have been carefully observed rely on certain kinds of animals for their regular food supply, and these animals have been called "dependable hosts." Animals occasionally fed upon are then called "casual hosts." Several species depend largely on pigs. *G. pallidipes* utilizes ruminants for about two thirds of its food supply. *G. brevipalpis* feeds on the hippopotamus, and *G. palpalis* feeds on large reptiles. *G. morsitans*, while subsisting mainly on pigs, is also a frequent feeder on man, and for that reason is of course a more important factor in human trypanosomiasis than are some other species.

Larval growth and pupation (Fig. 24–2) are unique in *Glossina* among free-living Diptera. Larvae grow and pupate within the uterus of the female adult and then, as puparia, are deposited in various sheltered places where, after about three weeks, the imago emerges. The pupal depositions characteristic of a few species are known, but more information is needed. It is likely that control depends on detailed knowledge about all aspects of the life history of each important species of tsetse.

SUBORDER CYCLORRHAPHA, FAMILY CALLIPHORIDAE

The Calliphoridae, the blow flies, are not at present very well defined. They include various carrion flies, as well as some genera having medical and veterinary importance. The larvae of the latter live in open wounds or decaying flesh, or actually invade the unbroken skin. One member of the family has larvae which puncture the skin and suck blood.

Screwworm flies (Fig. 24–3) deposit their eggs in masses at the edges of wounds. The larvae hatch and feed on living flesh. Obviously, this condition is very harmful to its victims—cattle, sheep, dogs, man, etc. An American screwworm epidemic between 1932 and 1934 resulted in over one million infestations of cattle and the death of over 200,000 animals. Losses have been estimated at $10,000,000 in some years. The chief species of screwworm flies are *Callitroga hominivorax*, the American screwworm, *Chrysomyia bezziana*, of Asia and Africa, and *Wohlfartia magnifica* of eastern Europe.

FIG. 24–3. Screwworms. The American screwworm fly, *Callitroga hominivorax*, deposits its eggs beside open wounds, which the hatched larvae invade. Control of this fly, eventually leading to its extinction in certain areas, is by means of the release of vast numbers of artificially sterilized males. Matings with these males prevent fertile matings and cause the population to sink below its "critical density." Predation then causes extinction.

The female screwworm fly is attracted to wounds, even such small ones as the bites of ticks or tabanids. Abrasions from fighting or scratches caused by barbed wire are particularly susceptible. The fly lays eggs in masses of 100–300; the eggs are cemented together at the edge of the lesion. One female *Callitroga* may lay a total of 3000 eggs in many such masses near a wound. The larvae hatch, invade the wound, and voraciously tear and consume the tissues. If untreated, the lesion enlarges, other flies, including carrion flies, are attracted to lay their eggs, and the severe myiasis results in death. Screwworm wounds may be treated with insecticidal ointments to remove and destroy the maggots. The fact that the maggots need air, even when they have deeply invaded the host's body, makes them susceptible to such treatment. In about 4 to 10 days the larvae mature and drop to the ground, where

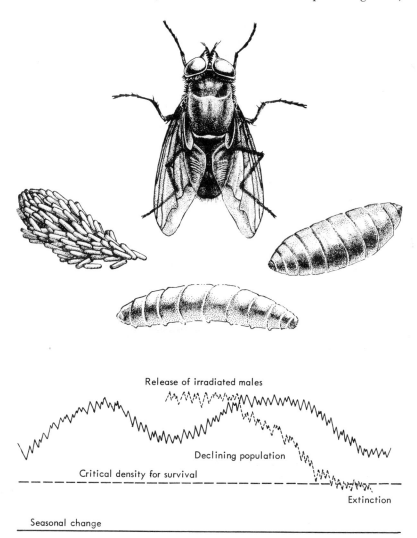

Release of irradiated males

Declining population

Critical density for survival

Extinction

Seasonal change

pupation requires another 3 to 14 days. In the southern United States, 8 to 10 generations of screwworm flies of the genus *Callitroga* develop in a single summer. The life cycles of *Chrysomyia bezziana* and *Wohlfartia magnifica* are similar to that of *Callitroga*.

Control of screwworm infestations classically consists of protection of animals, first, against wounds and abrasions where flies may lay their eggs and, second, against the damage caused by screwworm infestations once they occur. Since screwworms attack a great variety of animals, such measures for the benefit of domestic animals alone can do little to reduce the number of screwworms in nature. In 1960 the screwworm was eradicated from the state of Florida, a peninsula, by a unique ecological method, i.e., the sterilization and release of vast numbers of male *Callitroga*. Gamma radiation (from Cobalt 60 or waste Cesium 137) can sterilize male flies; the effective dose is about 2,500 roentgens, an amount of ionizing radiation which otherwise harms the insects very little. By rearing and releasing large numbers of sterile male screwworm flies, Knipling and others (see Bushland, 1959) caused nearly 100% of the egg masses sampled thereafter to be sterile, and the flies disappeared from the treated area. Since female *Callitroga* mate only once, the presence of a large number of sterile males ensured a high frequency of sterile matings. Predator pressure upon the relatively few successfully fertilized females undoubtedly brought about extinction. See Figure 24–3. Although a peninsula like Florida, or an island like Curacao (where experimental eradication by Knipling's method was first procured) is relatively convenient for application of the above technique, there seems to be no theoretical reason why the same method might not be used for other regions. Indeed, at the present time a screwworm control boundary is being moved south from Texas into Mexico. The plan is designed to eradicate *Callitroga* from more and more land until the fly has been squeezed into the Isthmus of Panama or to extinction. The same successful techniques described above are being used; financial support is international in origin and includes money from private ranchers. The success of such programs has stimulated efforts to apply the same principles to the control of other insects, such as mosquitoes, tabanids, or bots (see below), the males of which are harmless and would cause no damage if temporarily increased in number.

FIG. 24–4. The African floor maggot and the tumbu fly. In Central Africa, where people sleep on mats on the ground, two kinds of dipterous larvae may attack them. The adult flies look much alike. *Auchmeromyia luteola*, at left, lays eggs which hatch into predatory larvae. The larvae cut the skin of sleepers and then drink the blood which flows. *Cordylobia anthropophaga*, the tumbu fly, at right, deposits its eggs in sandy soil. The larvae invade the unbroken skin of man or beast in contact with the ground and live in boillike swellings for about nine days, after which the fully developed larvae emerge and drop to the ground to pupate. [After Faust and Russell, 1962, and Chandler and Read, 1961.]

The blow flies which cause "sheep-strike," an invasion of sheep wool and flesh by carrion-fly maggots, are not larval parasites in a true sense but are superficially similar, in their effects, to screwworm flies. Damp, soiled wool attracts flies which normally lay their eggs in putrefying carcasses. These flies, of the genera *Lucilia*, (important in Australia), *Chrysomyia* (in South Africa), *Calliphora* and *Lucilia* (in New Zealand), and *Phormia, Sarcophaga*, and *Callitroga* (not *hominivorax* but *macellaria*) (in the United States), cause extensive damage to sheep and wool during periods of the year when the flies are numerous. Since these flies breed normally in carrion, some control is possible through removal and disposal of carcasses by burying deeply or by burning.

Two blow flies of special interest are the African floor maggot, *Auchmeromyia luteola*, and the tumbu fly, *Cordylobia anthropophaga. Auchmeromyia* lives with man and swine in tropical Africa (Fig. 24–4). The adults find food and shelter in the dark shade of native huts; they feed as house flies do and lay their eggs in loose soil

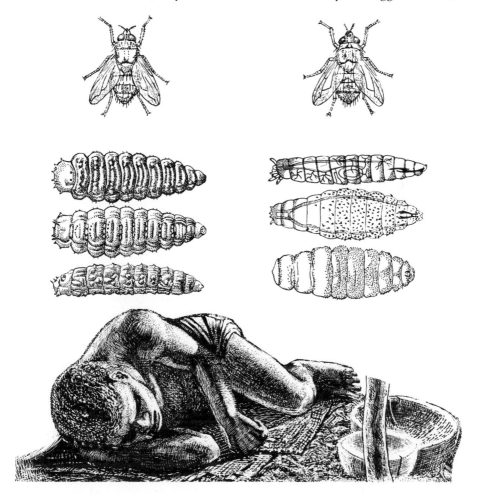

or debris of the floor. Larvae, hiding under sleeping mats, pierce the skin of sleeping persons or pigs and drink the oozing blood. The tumbu fly, *Cordylobia*, shares the range of *Auchmeromyia*. It lays its eggs in soil enriched with urine or feces; if such soil is found within a hut, the newly hatched larvae can invade the skin of a sleeping or reclining person. Within the skin the larvae grow; each one is surrounded by a tumor of inflamed tissue. After a little over a week, the larva leaves the host and pupates in the dirt floor of the hut.

SUBORDER CYCLORRHAPHA, FAMILY OESTRIDAE

"Bots" is a term used for a number of flies whose maggots develop within the bodies of vertebrate hosts; unlike the screwworms, the bots do not usually destroy the host while they develop.

In cattle and reindeer, bot larvae develop in hypodermal abscesses and damage the hide. Bots of horses develop in the stomach and esophagus, while those of sheep, as well as one of the horse bots, invade the sinus cavities. Although estimates of losses due to bots are not generally available, damage due to cattle bots in the United States is said to be about $160,000,000 a year.

The family Oestridae has only one group of species parasitic in man. This is the neotropical genus *Dermatobia*. The flies catch and hold large mosquitoes, and they cement a cluster of eggs to the latter (Fig. 26–5). The eggs incubate, and the larvae emerge only when stimulated by the warmth and moisture of nearby mammalian skin while the mosquito is feeding. Each larva burrows into the host to form eventually a hypodermal cavity, in which it grows in typical bot fashion. When grown, the larva emerges and drops to the soil for pupation.

Cattle bots (Fig. 24–5) include the heel fly, *Hypoderma lineatum* and the bomb fly, *Hypoderma bovis*, or "northern cattle grub," which is cosmopolitan in distribution. These large hairy flies are about 14 mm long and are banded with yellow and black stripes. In the southern species or heel fly the stripes are more orange than yellow and the fly is smaller. Adult bots are unable to feed. The females cement their eggs to the hairs of cattle and dart repeatedly at the host in the process. The southern cattle bot lays its eggs stealthily; it often deposits hundreds of eggs in small batches under the legs and belly of a resting animal. The northern bot is peculiarly aggressive in its egg laying as it strikes the host animal repeatedly. Although the flies cannot bite or sting the host, they cause extreme fright, so that animals attacked by egg-laying bot flies run wildly; the animals are frightened by the sound of the flies, and often injure themselves or become exhausted. The eggs hatch after approximately a week and first stage larvae penetrate the skin by means of the hair follicles.

After about two and one half months, larvae can be found in the walls of the esophagus and stomach, where they remain for about five months. Then migration toward the dorsal part of the body occurs, and the larvae, having moulted once, have reached a length of 3–13 mm; the size depends on their age. The larvae make holes in the skin of the dorsal region of the host and live beneath these holes, through which they breathe by means of a pair of spiracles located in their anal segment. As

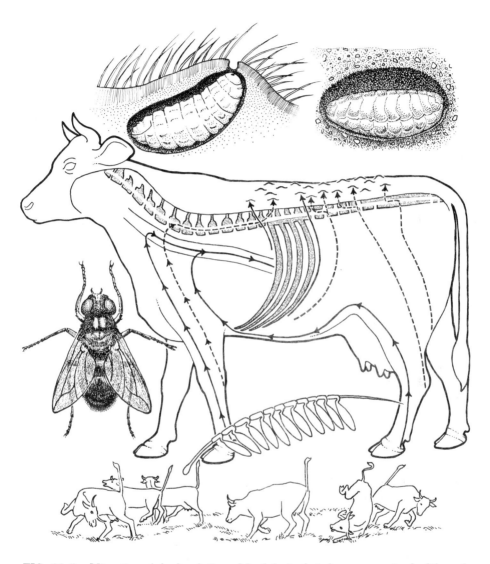

FIG. 24–5. Migration of the bomb fly and heel fly in their host; stages in the life cycle of the heel fly, *Hypoderma lineatum*. Eggs laid on the hair of the legs hatch and larvae penetrate the skin; the larvae move slowly through the tissues to eventually reach the back. There "bots" form; the larvae live under the skin and breathe through an opening in the skin. The larvae, when full grown, drop out of the hide onto the soil and pupate within the soil. Egg-laying adults of the bomb fly (especially) frighten cattle as they strike their sides to oviposit. (The heel fly is more subtle in its approach.) The development of these bots from egg to adult takes about two to three months, with an average of one generation per year. [After Metcalf, Flint, and Metcalf, 1962, and others.]

the larvae moult again, they enter the third stage, in which they grow larger (up to 20–25 mm) and turn darker in color (brown to black). The host surrounds the larvae with connective tissue capsules, within which blood and tissue exudates, often purulent, serve as food for the parasites. The bots remain in the skin for approximately two months. Thus by the time they mature as larvae and drop out of the host to pupate on the ground, the cattle bots have lived about eight months as parasites. Pupation takes up to two months, and the adult flies begin to appear on the range during warm weather and approximately ten months after the cattle have become infected. Since bot flies emerge, mate, and lay their eggs during the hot part of the summer and since an individual adult fly lives only a few days, there is considerable overlapping of developmental stages in any particular host; some animals acquire their bots in both early and late summer. The emergence of mature larvae at the end of winter and throughout the spring months provides a steady source of short-lived flies during the entire summer season.

Control of cattle bots is difficult. Since the adult flies emerge and live only a few days at widely scattered parts of the range over which infected cattle have wandered and since the puparia (hard capsules within which the flies pupate) are relatively resistant to insecticides, the only practical approach to control is the cattle themselves. The recent development of systemic insecticides (phosphatic compounds such as Ronnel or "Trolene" and Bayer 21/199 or "Co-ral") makes it possible to kill cattle grubs by feeding insecticide to the host or allowing the insecticide to penetrate the skin. Details of treatment are being worked out. Older methods consisted of laboriously rubbing insecticidal ointment into the cysts once these become palpable and of applying repellents to the bodies of cattle exposed to egg-laying adult flies. When herds are moved over considerable distances they may be protected from reinfection by simply moving the cattle far enough during the one to two months required for pupation of larvae on the ground so that emerging flies are effectively left behind.

Sheep, horses, and other animals are attacked by bots whose larvae develop in the sinuses of the head (Fig. 24–6) or the lumina of the upper digestive tract. These flies include the cosmopolitan species, *Oestrus ovis;* the old-world horse parasite, *Rhinoestrus purpureus;* a camel fly, *Cephalopina titillator;* and species of *Cephenemyia* attacking deer.

FIG. 24–6. Head bots of sheep and horses. (A) Head of sheep, dissected to show larvae of *Oestrus ovis,* the sheep botfly, in sinuses of skull. These larvae, when ready to pupate, leave the sinuses and are sneezed out. [After Herms, 1950, and Metcalf, Flint, and Metcalf, 1962.] (B) Head of horse (diagrammatic), showing larvae in sinuses and nasal passage. The adult fly, *Rhinoestrus purpureus,* is important in Eurasia and northern Africa. (C) Diagram indicating seasonal cycle of the head bots of horses and sheep. Maggots pupate in the ground in late summer, and larvae are deposited by adult flies in August and September. Maggots grow for about 10 months. [(B) and (C) after Ershov, 1956.]

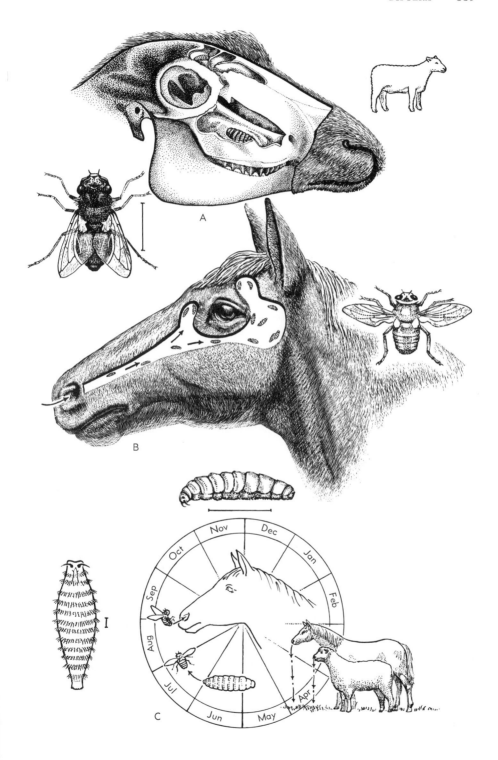

A

B

C

I

The sheep bot or nose fly, *Oestrus ovis*, is about 12 mm long and grayish in color. Like other bots, it is short-lived and unable to feed. These flies deposit active larvae, which have hatched in the body of the female fly; they inject these larvae in small drops of sticky fluid up the nostrils of sheep. The flies do not alight but merely dart past the sheep's nose. The maggots enter the mucous membranes to cause much local irritation, swelling, and discharge. The migrating larvae tunnel into sinuses, the bones of the head, and even the horns. They are said to invade the brain, but this is doubtful. Usually not more than 12 larvae are found in a sheep, but as many as 80 have been recovered. The larvae mature rapidly (in about 2 to 3 months) in lambs, but require almost a year in older animals. Since mature larvae fall out of the nose at various times (depending on the date of egg-laying and their own rate of growth), flies emerge upon the range during the entire summer. Control measures, until recently, consisted of smearing the noses of sheep with fly repellent, usually tar, and providing sheep with dark shelters into which they could crowd to escape the attack of nose flies, which are active only in bright light. Systemic insecticides are probably coming into use for treating infected animals, and surgery has been used to save especially valuable sheep.

The bots of horses (Fig. 24–7) belong to several species, all of which have similar habits. *Gasterophilus intestinalis*, *G. nasalis*, and *G. haemorrhoidalis* are cosmopolitan in distribution. In the United States, they are called the common horse bot, the lip or nose bot, and the chin fly or throat bot, respectively. These bots, like cattle bots, glue their eggs to the hair; some of the above common names of the flies refer to the regions of the body where the respective flies oviposit. Horses react violently to the egg-laying activity of these big hairy flies; yet, like all bots, these flies do not bite.

Eggs of the common bot, *Gasterophilus intestinalis*, cannot hatch without a thermal stimulus, which is applied when the horse licks hairs bearing ripe eggs. Eggs which have been present more than ten to fourteen days hatch immediately in the presence of sudden warmth; the maggots cling to the lips and tongue of the horse and enter the mouth, where they burrow into the mucous membrane of the tongue. They emerge from the tongue after 3 to 4 weeks, and pass to the stomach, where they live 9 or 10 months as pinkish maggots; eventually they pass out with feces to pupate on the ground. While in the stomach they cut and irritate the lining, feed on serous exudate, and interfere grossly with the function of the organ. By licking repeatedly the accumulated eggs of different ages, the host may acquire larvae constantly, and therefore can at one time suffer from symptoms due to all stages of larval invasion and establishment.

The lip bot (*G. haemorrhoidalis*) lays eggs near the mouth, chiefly upon the hairs of the lips. These eggs hatch when the first-stage larvae are fully developed (in about 2 to 6 days) and then the larvae migrate under the epidermis of the lips into the mouth. Eventually swallowed, the larvae live in the stomach. However, instead of being dropped with the feces, like the developed maggots of the common bot, the maggots of the lip bot after passing through the intestinal tract hang attached to the membranes of the rectum and anus to cause itching and scratching for some time before dropping to the ground.

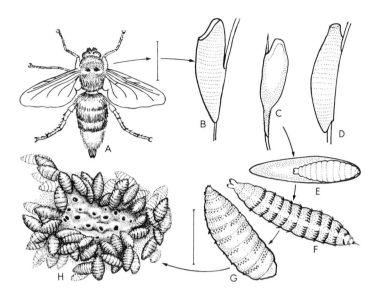

FIG. 24–7. Bot flies of horses. (A) *Gasterophilus haemorrhoidalis*, adult, the lip or nose bot fly. (B, C, D) Eggs of three common species: *G. intestinalis*, the common horse bot; *G. nasalis*, the chin or throat bot; and *G. haemorrhoidalis*, the lip bot [see (A) also]. (E, F, G) Stages in larval growth of horse bot flies. (H) Portion of stomach where numerous full grown larvae of *G. intestinalis* are attached. [(A) After Chandler and Read, 1961; (B–H) after Metcalf, Flint, and Metcalf, 1962.]

The chin fly or throat bot, *G. nasalis,* lays its eggs on the hairs of the throat and neck. These eggs hatch in 10 to 12 days, and the larvae migrate over the surface of the skin to enter the mouth. There they invade the tissues between and near the teeth, where they cause necrosis and infection of the gums and related membranes; they remain there three or four weeks until their first moult. Then they become attached to the walls of the pharynx and esophagus or move on to the duodenum; they are not commonly found in the stomach. Like the other horse bots, the chin fly causes digestive disturbances. Its presence in the pharynx damages the muscular walls of that organ and interferes with swallowing.

Control of horse bots has been centered around the protection of the animals from the egg-laying flies; special devices such as canvas throat and chin protectors, lip protectors, etc., have been devised. Smearing or sponging the legs and flanks of horses with warm (40°C) 2% phenol causes eggs of the common bot to hatch; the phenol then kills the larvae. Systemic insecticides, as well as "drenches" of carbon bisulfide, may be used to destroy maggots in the intestinal tract. Cooperative communitywide or nationwide efforts at control should substantially reduce the number of bot flies, since their population size depends directly upon the number of infected host animals.

SUBORDER CYCLORRHAPHA, FAMILY HIPPOBOSCIDAE

The sheep "ked" is a completely parasitic insect; it is often called a tick but is actually one of a group of flies of the family Hippoboscidae. It is wingless, flattened, and leathery, and is equipped with piercing and sucking mouthparts. The female ked produces one larva at a time; the larva reaches full growth within its parent's body and sticks by means of a "glue" to the hair of its host after being discharged from the uterus of its parent. The female produces, one at a time, about twelve of these larvae at intervals of about 8 days. Each larva requires about three weeks to pupate and emerge from the "nit," which the puparium is commonly called. (Note that "nit" is the common word for eggs of lice or other insects, and is inappropriately applied to the relatively large puparium of the sheep ked.) The complete cycle, from emergent parent to emergent offspring, takes at least six weeks. Heavily infested sheep are irritated by the crawling and biting of these flies, and suffer, of course, loss of blood. A U.S. estimate placed the yearly damage done by keds at 20–25 cents a head, or about 5 million dollars on 50 million sheep kept in the United States. Transmission by the sheep ked of the nonpathogenic sheep trypanosome, *Trypanosoma melophagium*, has been mentioned elsewhere. Control of sheep ked, involving spraying or dipping, is similar to tick control but is somewhat easier, because there is little contamination of pastures or pens by parasites, and survival time of accidentally detached keds or puparia does not exceed two months at mild temperatures. Sheared wool is a ready source of spread, however, and should be stored well away from animals, so that keds from the wool cannot reach new hosts.

Differing from the sheep ked, related hippoboscids possess wings yet are still parasitic throughout their life; they affect poultry, horses, and other domestic animals. These "louse flies" or "flat flies" have been studied in relation to numerous wild animals, birds, bats, etc. Their host relationships and evolution are very interesting.

TRANSITION IN LIEU OF CONCLUSION

The preceding discussion of the Diptera concludes the descriptive portion of a section of this book (Chapters 21–24) devoted to the medically and economically important arthropods. The chapter which follows will be an attempt to make some general statements about arthropods, human welfare, and the principles deducible from the data presented above.

SUGGESTED READINGS

BUSHLAND, R. C., A. W. LINDQUIST, and E. F. KNIPLING, "Eradication of Screwworms through Release of Sterilized Males," *Science*, 122:287–288, 1955.

Buxton, P. A., "The Natural History of Tsetse Flies: an Account of the Biology of the Genus *Glossina* (Diptera)," H. K. Lewis, London (London School of Tropical Medicine Memoirs 10), 1955.

Chandler, A. C., and C. P. Read, *Introduction to Parasitology*, 10th ed. Wiley, New York, 1961.

Dalmat, H. T., "The Black Flies (Diptera: Simuliidae) of Guatemala and their Role as Vectors of Onchocerciasis," *Smithsonian Misc. Coll.* 125 (1) :1–425.

Herms, W. B., and M. T. James, *Medical Entomology*, 5th ed. Macmillan, New York, 1961.

Horsfall, W. R., *Mosquitoes, Their Bionomics and Relation to Disease.* Ronald Press, New York, 1955.

LaPage, G., *Mönnig's Veterinary Helminthology and Entomology*, 5th ed. Williams and Wilkins, Baltimore, 1952.

MacDonald, G., *The Epidemiology and Control of Malaria.* Oxford University Press, London, 1957.

Metcalf, C. L., W. P. Flint, and R. L. Metcalf, *Destructive and Useful Insects, Their Habits and Control*, 4th ed. McGraw-Hill, New York, 1962.

West, L. S., *The House Fly.* Comstock, Ithaca, New York, 1951.

CHAPTER 25

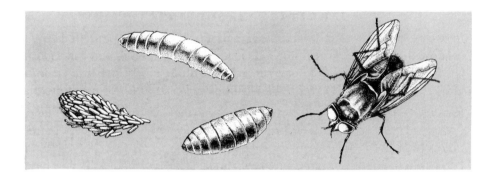

CONTROL OF ARTHROPODA

With a fair amount of data (on disease, life cycles, etc.) at hand, it will now be useful to review arthropods of medical and veterinary importance. Control, a highly important part of the subject, can be considered specifically, as it was treated, for example, in the chapter on malaria and in the section of Chapter 24 devoted to bot-flies, or it can be considered generally, as it will be discussed here. Following the discussion, there will be a condensed review of medical and veterinary entomology, i.e., an evaluation of knowledge.

Control of arthropods, as well as consequent prevention of both the direct damage and the many serious diseases caused or carried by them, succeeds only occasionally when certain key factors are both understood and acted upon. This statement may seem pessimistic, especially in the light of the apparent success of the worldwide campaigns against malaria. But the history of malaria (see Chapter 2) is a long history of research, education, and practical experience in solving a single problem —the problem of anopheline transmission. And while this problem is not really either single or simple (for malaria and anophelines exist in many different

344

climates, among peoples of greatly varying levels of education, economic strength, and political stability), still substantial success has been achieved. "Unstable malaria," which occurs in epidemic form on the fringes of regions where "stable malaria" is found, has yielded quite well to eradication programs. Eradication cannot work so well where malaria transmission occurs all the year round, where transportation and communication are so difficult or so undeveloped that some communities are inevitably missed by spraying and treatment units, or where the governmental structures are so weak, so poor, so capricious, or so unenlightened that the practice of public health administration has not yet become possible. Such regions, and such governments, exist, and where they exist, malaria is still an unsolved problem, a permanent part of the troubles of those regions. Thus even malaria, which we hailed as an example of man's victory over disease, is still a great problem. Other arthropod-borne diseases can be fought successfully by the same general strategy which has partially succeeded against malaria; but the vectors, the parasites, and the human situations are different for each disease, and different tactics must be discovered and put to use.

In relation to control, the diseases caused by arthropods may be put into two categories: simple and complex. "Simple," of course, does not mean simple in the usual sense, for none of these problems is easy to understand or to solve; instead, "simple" may be taken to mean that which involves relatively few factors. Malaria, for example, involves only man and anophelines. Plague, on the other hand, involves (1) a number of small mammals living in steppe-forest interfaces, (2) the fleas of these mammals, (3) transfer of the plague bacillus from the sylvatic hosts to rats living with or near man, and (4) the eventual epizootic decimation of rats, sending infected fleas questing for blood and resulting in human plague. Thus in malaria control is a simple problem, whereas in plague it is complex. Breaking the human chain of transmission in malaria is sufficient; abating the human form of plague leaves the sylvatic reservoir untouched and fails to prevent future outbreaks among men.

Some other examples of simple and complex problems follow.

In theory, the two major filariases of man are relatively simple to control. Mosquito-borne filariasis due to *Wuchereria bancrofti* probably has no nonhuman vertebrate reservoir. (Simian reservoirs are probably significant in brugiasis, however.) If it were possible to treat all human hosts in a region and thus prevent the infection of new vectors, wuchereriasis would vanish with the death of the last infected mosquito. Thus improvement in methods of treatment, effective discovery of cases, and careful public health surveillance should effectively eradicate this filariasis (see Chapter 14). Black fly filariasis (onchocerciasis) is also very likely man-vector limited. The chief difficulties in controlling this disease are lack of effective treatment (to destroy microfilariae and thus prevent transmission) and, alternatively, lack of effective means of destroying black flies. Progress can be expected, however, and the strictly man-limited filariases will disappear.

Another theoretically simple problem is gambian trypanosomiasis. Although sylvatic hosts of *T. gambiense* are believed to constitute a permanent reservoir, most human cases are derived from other human victims. Examination and treat-ment of man in some parts of Africa has greatly reduced the incidence of the above form of sleeping sickness and shows that the simple step of breaking the chain of human transmission can control this tsetse-borne disease.

A whole group of enteric diseases transmitted by houseflies can be controlled quite simply by reducing the number of flies or by protecting food from contamina-tion by flies. However, fly control itself is not simple, since if insecticides are not used, it requires storage and treatment of garbage, human excrement, and other fly-breeding materials; thus perhaps it would be necessary to attempt to change the whole human environment. If insecticides are used, other difficulties usually arise both from the danger to man and his stock due to wide use of poisons and from the well-known ability of insects, especially muscid flies, to develop resistance to or tolerance of insecticides.

Among the "complex" diseases, in addition to plague mentioned above, are the leishmaniases, Chagas' disease, the spirochetal "relapsing fevers," many virus diseases including yellow fever and dengue, and typhus and the other rickettsioses. In all these diseases sylvatic reservoirs exist, and therefore it would seem quite hopeless to attempt control. Actually, however, it is possible to control the complex diseases by protection measures once the nature and extent of the sylvatic reservoir are understood. Some sand fly vectors of *Leishmania tropica,* for example, live in the burrows of rodents far from human habitation, and only occasionally do they have the opportunity to bite man. Human chains of transmission depend on sand flies living near man; control of the local vector will greatly reduce the incidence of human sores. Forest strains of *Leishmania tropica,* moreover, may simply be escaped by avoiding forests or by using insect-repellent ointments when in dangerous regions. The same principles apply to visceral leishmaniasis. In some regions the problem is complicated by domestic reservoir hosts, such as the dog; detailed knowledge of local situations is therefore important. The virus diseases, yellow fever and dengue, have almost disappeared from cities where epidemics used to occur, although reservoirs of the virus exist in forest animals unaffected by human public health programs. In these diseases, as in the leishmaniases, protec-tion from contact with vectors from the sylvatic reservoir has been achieved, and thus yellow fever and dengue have been removed effectively from the list of impor-tant diseases. However, their potential importance must be recognized, for protection against mosquitoes can fail whenever private or public vigilance is relaxed. The epidemiologies of other virus diseases—encephalitis, the relapsing fevers, etc.—are so little understood and probably so complicated that no practical basis for control exists at present. These are examples of true zoonoses, diseases which only occa-sionally and unpredictably escape from their natural reservoir in wild birds, mammals, etc.; much ecological knowledge would be required for preventive programs. The rickettsial diseases are unusual in being maintained more signifi-cantly in vectors than in vertebrate hosts. As was stated earlier, many rickettsias are probably natural symbionts or parasites of ticks and mites, in which they multiply and pass from generation to generation. Their life in the blood of vertebrates may be considered as an unusual excursion from their normal habitat. The reservoir of such diseases as rickettsial pox, spotted fever, etc., is thus chiefly an arthropod reservoir. This fact implies that control should be based on reduction of contact between man and the naturally infected vectors. Typhus seems to be an exception to the above statements about rickettsial disease in general. Human typhus is transmitted by the louse, not by acarines, and *Rickettsia prowazeki* is not a symbiont but a harmful parasite in that insect. Modern sanitation has reduced human lice to a tiny fraction of their recent numbers, and epidemic typhus has likewise nearly vanished. Murine typhus continues to exist, however, in a mite-rodent domestic and sylvatic complex, where it remains an occasional but dangerous infective risk to man.

From the above rapid review of several arthropod-disease problems—simple (essentially man-limited) or complex (involving reservoirs)—it is obvious that there are many possible approaches to solving such problems.

Thus, where man himself is the chief vertebrate host and where treatment and cure are possible, control of the disease results from good medical practices, i.e., from the destruction of the pathogen in its human host and from the prevention of arthropod transmission while pathogens may still be present. In still other diseases, the conditions under which the disease is transmitted must be changed. Control of black flies in onchocerciasis is one such change in the environment; it is difficult because of the scattered breeding places in typical black-fly country, but it is necessary because of the slow, difficult, or ineffective treatment of human carriers. In parts of Africa, artificially changing the environment has effectively removed tsetse flies from the neighborhood of man; these flies pupate in underbrush along streams, trails, or at the edge of fields, and cutting such scrub forces the flies to move elsewhere. As mentioned above, improved personal sanitation—washing—has exiled the louse from many human localities where typhus was endemic.

But there are situations in which neither medical interruption of transmission nor change of the environmental factors in transmission is practical. Such "accidental" infections as cutaneous leishmaniasis, sylvatic rickettsial diseases, relapsing fevers, virus encephalitis, loaiasis, and Rhodesian trypanosomiasis must be prevented by avoiding contact with vectors. Screening, insect-repellent ointments, mite- or tick-proof clothing, and actual avoidance of regions where these diseases occur are the best means of protection. This kind of ecological prevention has an important place in tropical medicine, where so much remains to be learned in the control of arthropod-borne disease.

Control and protection against directly injurious arthropods—blood-sucking and biting insects, parasites, etc.—follow the ecological principles just mentioned. Change in the pest's environment is sometimes effective (as in control of muscid flies by garbage disposal, or the reduction of the number of mosquitoes by drainage, treatment of standing water, and planned variation in the level of reservoirs). Insecticides used to control the number of insects are often of great temporary benefit (as in preparing an area for some engineering development such as construction of a dam, in preventing an epidemic of typhus among a crowded wartime population, or in drastically reducing the number of mosquitoes during a campaign of malaria eradication). But it should be well known that insects, like all organisms, inevitably mutate, and that mutations toward resistance to insecticides repeatedly occur. Therefore attempts to reduce permanently the number of insects cannot be based on insecticides alone.

Indeed, it is clear that no single formula for the control of arthropods exists. It was my purpose in the foregoing discussion to illustrate the complexity of problems of arthropod control and to indicate the need for detailed information which alone can form a sound basis for the practice of control.

SUMMARY AND CONCLUSIONS

Medical and veterinary entomology omits those arthropods which have no relation to human and animal disease. Thus it neglects the arthropod parasites of plants and many other hosts, and it gives undue emphasis, perhaps, to some minor pests if they even occasionally or exceptionally affect human health. In the foregoing chapters the crustaceans, chilopods, millipedes, scorpions and spiders, as well as the curious pentastomes, were but briefly mentioned. Crustaceans are of medical interest as intermediate hosts of lung flukes, tapeworms, and guinea worms—parasitic helminths of man. Chilopods (centipedes) affect man by severe and painful "bites" (wounds made by the first pair of fang-like claws). Millipedes, harmless scavengers, offend and sometimes nauseate man by the defensive sprays, jets, or odors projected by repugnatorial glands in one order of this group. Arachnids of the primitive subclass known as scorpions sting severely, sometimes fatally, when disturbed. The large group of spiders has some venomous members, e.g., the "widows" and *Loxosceles* species of North and South America and certain large trap-door spiders of Australia. The pentastomes, possibly relatives of the acarine arachnids, were cited as potential dangers, since many animals, including man, may serve as hosts to larvae, the adults of which parasitize reptiles or carnivorous mammals. Two large groups of arthropods, the Acarina and the Insecta, are medically much more prominent than the above. The acarine arachnids, so-called ticks and mites, are important agents of disease and are also parasites themselves. They irritate and injure man and his domestic animals by feeding on skin, feathers, scabs, exudates, and even tissue and blood. They are the vectors of several important diseases, while they serve as components of sylvatic reservoirs of many others. Protozoa, spirochetes, rickettsial organisms, bacteria, and viruses are transmitted by acarines. Viruses and rickettsias, in particular, are actually symbionts of ticks and mites; multiplying within the host tissues, they invade the ovaries and are transmitted to succeeding generations without the need for vertebrate passage. The insects, of course, received the most attention. Among the nondipterous insects, pests (such as cockroaches and bedbugs) and dangerous disease vectors (such as true bugs, lice, and fleas) are responsible for much annoyance as well as the spread of such diseases as South American trypanosomiasis, epidemic typhus, and plague.

From the medical and veterinarian point of view, however, the Diptera appear to be the most important of all groups of arthropods. The sand flies (*Phlebotomus*) transmit the often fatal visceral leishmaniasis, a cosmopolitan disease of the near tropics, as well as the disfiguring cutaneous leishmaniases of the Old and New Worlds. Mosquito-borne disease includes malaria, filariasis, yellow fever, dengue, and virus encephalitis. Mosquitoes even serve as phoretic hosts of a painful skin bot, *Dermatobia*. In addition, mosquitoes often become numerous enough to repel man from certain areas at certain seasons because of their persistent biting. Control of mosquito-borne disease is among the urgent unfinished tasks of entomologists. Black flies are responsible for onchocerciasis; they present a peculiarly difficult problem of control. Tabanid flies, large biting bloodsuckers of man and beast, are responsible for mechanical transmission of some trypanosomes and biological transmission of others. An African tabanid, *Chrysops*, transmits a filarial worm. Eye flies are minor pests, but have some potential and actual importance as vectors of various skin and eye infections; they have been implicated in the transmission of yaws. Muscid flies include the house fly and the tsetse. The former carries practically every enteric disease and is undoubtedly responsible for much infant mortality and adult sickness. The latter has until quite recently denied almost a third of the great continent of Africa to agricultural development and human settlement by its transmission of human sleeping sickness and animal trypanosomiasis. Blow flies are frequent invaders of the flesh of domestic animals; they cause severe injury to babies in parts of Europe. These screwworm flies deposit eggs or larvae which develop within living tissue. Also within this group of flies are the Congo floor maggot, a predatory larva which bites and sucks the blood of sleepers, and the tumbu fly, whose larva seeks and enters skin in contact with the soil; the larva then lives like a bot in human tissue. True bot flies affect man very little but are important parasites of cattle, horses, etc. *Dermatobia*, mentioned above as utilizing mosquito phoronts, does affect man as well as other mammals. In short, the Diptera have many harmful species which serve as both vectors and parasites.

From the above resumé of the chapters on arthropods, it should be obvious that a taxonomic arrangement is only one of several reasonable possibilities. For instance, important arthropods can be listed (1) in terms of their way of life—free-living or parasitic as either larvae or adults, etc.—or (2) in order of their importance to man—a difficult method but one which has the advantage of emphasizing significance, with the *Anopheles* mosquito at the top of the list, tsetse, lice, triatomid bugs, cattle bots and certain ticks and mites in intermediate positions, and certain spiders, tabanids, or fleas perhaps near the bottom; such a list, however, would find little acceptance. Another arrangement would be based on the kinds of arthropod-borne pathogens—helminths, protozoa, spirochetes, bacteria, rickettsias, viruses, etc. Geographical, epidemiological, or even historical arrangements might be made. The student is invited to try his own hand.

The question of control of arthropods had to be treated very briefly. Consideration of general ecology, engineering, the biology and evolution of arthropods, the safety and long-term efficacy of insecticides, the diversity of control measures, and the uniqueness of each problem requires a discussion in much greater detail than an introductory text could give. Yet such considerations are perhaps the most important problems which a modern parasitologist or medical entomologist faces.

SUGGESTED READINGS

Bushland, R. C., A. W. Lindquist, and E. F. Knipling, "Eradication of Screwworms through Release of Sterilized Males," *Science*, 122:287–288, 1955.

Herms, W. B. and M. T. James, *Medical Entomology*, 5th ed. Macmillan, New York, 1961.

MacDonald, G., *The Epidemiology and Control of Malaria*. Oxford University Press, London, 1957.

Metcalf, C. L., W. P. Flint, and R. L. Metcalf, *Destructive and Useful Insects, Their Habits, and Control*, 4th ed. McGraw-Hill, New York, 1962.

West, L. S., *The House Fly*. Comstock, Ithaca, N.Y., 1951.

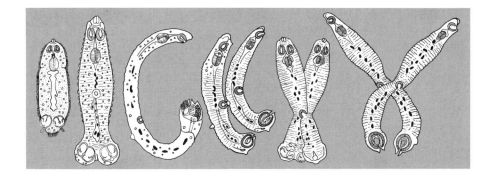

CHAPTER 26

HOST-PARASITE RELATIONSHIPS

INTRODUCTION TO PARASITOLOGY

In keeping with the purpose of this book—to inform and to teach—I have tried to avoid on the one hand encyclopedic indigestibility and on the other the tendency toward cryptic generalization that comes from unsuccessful attempts to simplify. The "introduction" has been postponed until late in the text. The general principles of parasitology are not simple; in order to understand them one has to have some background of specific knowledge, such as that which has been obtained from the first 25 chapters. Thus an "introduction to parasitology" appears near the end! This location, in my opinion, emphasizes both the intricate complex nature of the subject and the natural or "feedback" way of learning, which is to observe particulars, then to generalize, and then, let us hope, to continue observing with the particulars in mind and with the mind open.

The final chapters should be read with such open-mindedness, for they concern frontiers in parasitology. Knowledge of immunology, pathology, and evolution of

351

parasites (that is, of host-parasite relationships in space and time) is considerably less complete than the information already given concerning individual parasites and the diseases they cause. Likewise, world public health, the subject of my concluding chapter, is a quite recently developed concept and a field in which great progress is being made today. It is hoped that the reader will share with all active parasitologists the speculative and exploratory spirit which enlivens today's investigations in these problem-filled areas.

This chapter, after the above necessary apology, begins with a classification of the relationships among living things and emphasizes the facts that parasitism is not sharply or clearly set off from other ways of life and that it can be understood best as a set of ecological dependencies. Next, the parasitic relationship is examined in detail, with examples (some familiar, some new) of various host-parasite associations. Finally, the theory of parasitism is examined (perhaps with too little emphasis on general ecology and evolution, which are handled in other chapters). The conflicting notions of parasitism, as degeneration in one view and as opportunism in another, are reconciled by contrasting parasitic dependency with the evolutionary compensations for such dependency—abundant food, ingenious cycles, shelter, long individual life.

In effect, the chapter is a bridge between individual problems of parasitology (like malaria and schistosomiasis) and the subject of parasitology as a whole. But it also points out that this subject cannot be isolated from biology in general and that parasitology, a complex of interacting interrelated organisms itself, is part of the study of the interactions and interrelationships of all living things.

A CLASSIFICATION OF RELATIONSHIPS

The relationships among organisms may be arranged in order of dependency (of parasite upon host, for example) or of intimacy (the extent of physiological interaction between two organisms). Both dependency and intimacy are involved in true parasitism. Degrees of each, however, may be found in other relationships.

The autotrophic relationship avoids both dependency and intimacy. Autotrophs are organisms which utilize only inorganic sources of food and energy. Needless to say, true or complete autotrophs are rare on this planet; iron and sulfur bacteria, which metabolize inorganic compounds of these elements, are examples. The autotrophism of the green plants is far from true or complete. These plants, while utilizing radiant energy (sunlight) in a truly autotrophic way, depend on the activities and products of other organisms. They require nitrogenous compounds which derive from living processes, they utilize carbon dioxide, and they depend upon the complex physical conditions of soil (a biological product) for their growth. Green plants exist as parts of biological communities. Yet it is true that green plants (and the other photosynthetic organisms) are the energy sources of the organic world; they alone furnish the food for all other kinds of organisms.

Therefore the series of relationships involving dependency begins with the green plants.

The herbivores, animals directly dependent on plants, are food for predators. The predators, therefore, depend upon the herbivores. Even the most intricately dependent parasite is not more dependent, nutritionally, upon its host than is a beast of prey, such as the tiger, upon its herbivorous victim. It is well known that the populations of predators suffer extreme fluctuations in some parts of the world; such fluctuations follow and often exceed increases and decreases in the number of herbivorous prey. Dependency, a feature of nearly all relationships among living things, thus is seen to increase with the degree by which a way of life differs from autotrophism or from the pseudoautotrophism of the green plants. Dependency is exaggerated in predatism and parasitism alike.

The degree of intimacy is the best means of arranging or classifying parasites and of setting them apart, wherever possible, from predators. Thus an organism which inhabits the body of another, and even feeds on or eventually destroys the other, is a parasite, not a predator. The many examples in this book of harmful and intimate relationships should make this fact perfectly clear.

But there are many forms of intimate relationships which are not obviously harmful to either participant. Such are the relationships called phoresis and commensalism, and the intimate relationship beneficial to both participants which is called mutualism. These relationships, which may all be considered forms of parasitism, will be examined next.

Before a discussion of phoresis, commensalism, and mutualism, however, a digression seems appropriate. There is a social analogue to parasitism. In human affairs, there is a well-recognized cycle of producers and consumers, similar to the green plant and animal chain of dependency. In a very loose sense, one can also recognize in human affairs analogues of primary and secondary consumers, "herbivores" and "predators." These analogues may be individuals within a village economy, or they may be large units, such as states, corporations, nations, or supranational entities. Even "parasites" may be identified in the human community. It would be unfair to name any of these. Obviously, the successful human relationship which emerges from the darkness of a world torn by strife is one not of predatism, parasitism, or commensalism, but of mutualism. The practice of brotherhood and cooperation is a concept taught by all the great religions; but it is a concept also taught by the world of nature, where balance and security depend on an exchange of benefits. In the world of nature predation and harmful parasitism are marginal occupations which are not central to the economy.

PARASITISM IN DETAIL

Phoresis is a term used for a parasitic relationship which emphasizes habitation or transportation (phoresis) rather than nourishment. Any community rich in numbers and kinds of associates, such as the plant community called rain forest, or an animal community such as a coral reef or shellfish bed, has such a profusion of growth that some organisms are attached to, or grow upon, others. Oysters, for instance, bear sponges and anemones upon their shells, and often are attached upon

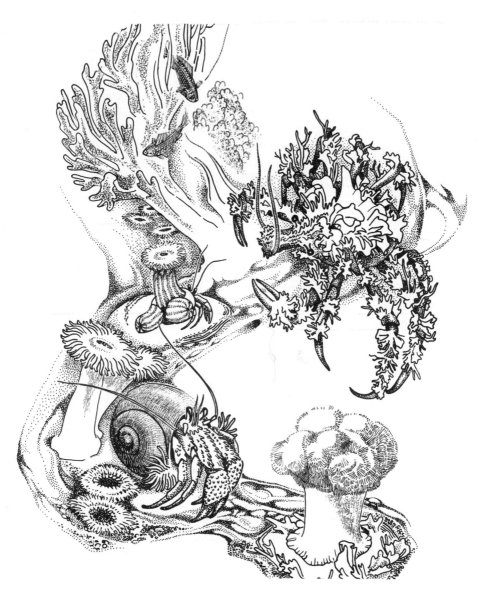

FIG. 26–1. Camouflage and phoresis in the sea. Three crabs are seen. At lower left is a hermit crab, *Eupagurus prideauxi*, with the obligate commensal *Adamsia palliata* attached under its shell; the anemone's mouth is adjacent to that of the crab. At left of center is another hermit crab, *Pagurus arrosor*, which bears on its shell an adventitious phoront, *Sagartia parasitica*; the latter is capable of living alone. At right of center is a small spider crab with an assortment of sponges growing all over it. The protective or mutualist significance of these phoretic associations is obvious. [After Baer, 1951, and others.]

a base of other oysters, living or dead. Crabs in such an environment may also bear sponges or coelenterates upon their shells. Such associations may prove beneficial to all members in only a slightly more specific way than the benefit accruing to all members of a community from the existence of the community itself. Phoresis is a more permanent relationship than those just mentioned. In well-established phoresis, the smaller of the pair, i.e., the animal carried or sheltered by its host, benefits in an obvious way. Thus the sedentary sponge or coelenterate gains the advantage of a wide environment by being attached to a wandering crustacean. A protozoan inhabitant of the gill chambers of a clam, a flatworm living in the body folds of a crayfish, and a fish inhabiting the cloaca of a holothurian all receive shelter and protection. In phoresis, the host does not benefit so obviously. Hosts bearing parasites (phoronts) on their surfaces (crabs bearing anemones, for example) probably receive some protective coloration or disguise from the parasite's presence; hosts with internal parasites probably derive little or no benefit. Yet the benefit to both parasite and host is such that through evolutionary trial and error (the origin and multiplication of mutations strengthening the relationship, and the failure of nonadapted phoronts to maintain a chance establishment), many well-established cases of phoresis can be recognized.

An interesting example is the hermit-crab–anemone relationship. (See Fig. 26–1.) Hermit crabs (marine decapod crustacea having a soft abdomen and utilizing the shells of gastropod molluscs as coverings) must select new shells (homes) as they outgrow old ones. Anemones (coelenterates with fleshy bodies surmounted by a mouth and with tentacles bearing stinging threads) live attached to pebbles, shells, or other sea-bottom objects. If a hermit crab chooses a shell which supports an anemone, the latter may remain attached (as a "temporary phoront") or may separate itself from the now mobile shell. Various species of hermit crabs form such loose associations with various species of anemones. But a particular species of crab (*Pagurus arrosor* Herbst) and the anemone (*Sagartia parasitica* Gosse) have a characteristically permanent relationship. Whenever a crab of the above species selects a shell bearing *Sagartia* for its home, the anemone remains attached. And when the crab outgrows its home and finds a new one, the crab removes the anemone and places it upon the new shell. This behavior must reflect evolutionary changes in both species favoring phoresis. Yet the phoresis in this case does not seem to be complete, since either *Pagurus* or *Sagartia* may live without the other.

Commensalism, like phoresis, is a rather broad term for various host-parasite relationships. Commensalism involves the sharing of food. This sharing is seldom mutual; the parasite eats the food of the host, not *vice versa*. Probably any form of parasitism in which food and shelter are acquired by the parasite with no demonstrable harm to the host may be called commensalism. The relationship may even be defined by degree or extent of harm caused. Commensalism which significantly affects the host's nutrition is not commensalism, but should be called true or strict parasitism; parasitism from which the harm is slight, as in some very light infestations of intestinal helminths, can be called commensalism. Commensalism is

best defined, however, by an example of another hermit crab sea anemone association. *Eupagurus prideauxi* Leach, the crab, always uses a shell which is too small
to protect its abdomen. On this shell the anemone, *Adamsia palliata* Forbes, is attached in such a way that its body protects the exposed portion of the crab. The
crab cannot survive without its partner, and the latter cannot live without the crab,
whose food, chopped up by the crab's pincers, is easily seized by the anemone. The
relationship is actually physiological both in the nutritional dependence of the
anemone upon the crab and in the curious fact that the crab's blood contains antibodies which can protect the crab against the toxins produced by the particular
anemone. Thus commensalism seems to be only a few steps removed from phoresis,
yet it also shows, in this case, a mutual benefit that suggests another relationship,
mutualism.

Mutualism is an extremely intimate form of the relationship most characteristic
of living things—interdependence. The host and parasite which live in such mutual
dependence resemble, in a simplified form, the animal-plant community, the socioeconomic world, perhaps the universe of life itself. For example, cattle and other
ruminants harbor millions of ciliate protozoa in the rumen (the stomach in which
cellulose fermentation goes on). These protozoa never occur outside of a host
except as cysts. They reproduce very fast, with a doubling of the population each
day. Of course this doubling is actually replacement, for the amount of ciliates in
a cow, for example, remains constant at about six pounds. Therefore the dying
ciliates must provide their host with a significant amount of protein. Incidentally,
some of the ciliates, but not all, produce cellulose-splitting enzymes and thus are
able to utilize directly the food ingested by their host. Termites and the more
primitive insect, *Cryptocercus,* harbor a remarkable fauna of flagellates. The
flagellates secrete enzymes which digest wood particles ingested by their host. This
relationship is not understood in so much detail as the mutualism of ruminants and
their ciliates, but termites and *Cryptocercus* cannot survive when defaunated, and
their protozoa have neither habitat nor food outside their hosts. The stability of
mutualism is apparently very great. In the case of termite and *Cryptocercus*
protozoa, the same genera of protozoa can be found in both groups of insects. (See
Fig. 29–4.) Yet the insects have had separate evolutionary histories since the
Carboniferous period, about three hundred million years ago. Therefore this
mutualism must have begun at least that long ago and must have persisted without
substantial modification of the protozoa. Significant mutualisms have been found
among plants, of course. Lichens, which are intimate algal-fungal structures, and
legumes, which usually depend on root-nodule nitrogen-fixing bacteria, should be
mentioned. A curious drought-resistant mutualism is said to result from a root-
nematode desert-plant association, in which the modifications of root structure
caused by nematodes permit some storage of water.

Mutualism, like parasitism itself, has interesting analogues and extensions. Thus
the interdependency of man and the giant grasses, while not intimate like parasitism,
is a mutualistic relationship of the greatest importance. When prehistoric men
(probably women) chose the ancestors of wheat, maize, barley, rice, etc., for

special care, they started these plants in the direction of dependency on man. Now most of these grasses cannot seed themselves, but must be planted, protected, and carefully nourished by man. Without man's agriculture, the cereals would become extinct. And man himself became dependent on the cereals. Until cereal agriculture made possible the feeding of large populations, man existed in very small numbers. It is estimated that the entire preagricultural population of the world did not exceed twenty millions. Today, of course, the huge and expanding population depends absolutely on the cereals. Man and his giant grasses are locked in mutual need.

Parasitism, in the restricted sense used generally, is a relationship which harms one partner, the host. Parasitic organisms live on or within host organisms, deriving shelter and nourishment at the host's expense. Since the forms such a relationship may take are numerous and may be very complex, it is helpful to arrange parasites in several ecological classes.

ECTOPARASITISM

Ectoparasites live on the external surfaces of their host or occasionally within cavities open to the exterior. Among protozoa, true ectoparasites are difficult to distinguish from phoronts. The remarkable ciliate *Spirophrya* (Fig. 26-2) forms reproductive cysts on the surface of colonial coelenterates. These cysts liberate small forms which become attached to copepods. When the copepods are caught and ingested by the polyps of the coelenterate, the protozoa feed upon the partially digested copepod. As the protozoa grow, they are expelled, along with the remnants of the copepod, from the coelenterate and then become attached as reproductive cysts to the same or other polyps. *Spirophrya* thus seems at some stages an ecto-parasite, at others an endoparasite (stealing food from its host) and at still another stage a free-living organism. Metazoan ectoparasites include representatives of several phyla. Among the best known are various flatworms, arthropods, and annelids. The monogenetic trematodes, for example, include many species of organisms parasitic on the gills of fishes. The parasites cling to the gills by various hooks and suckers, feed actively on the membranes of the host, and reproduce simply (hence the name Monogenea), usually producing a few eggs, one at a time, which are attached to the host by cement or filaments. Except for size and sexual maturity the larvae of these trematodes resemble adults.

The largest group of ectoparasitic arthropods is probably the Acarina, or mites and ticks (see Chapter 22). Since this group includes many free-living organisms, it is not surprising that degrees of parasitism are found among them. For instance, ticks require a blood meal before moulting or egg production, and to obtain such a meal they must find suitable vertebrate hosts. But the hosts need not be of a single species of vertebrate, and the time during which the tick lives on the host may be only long enough for engorgement. Such a relationship seems closer to predatism than to parasitism. On the other hand, many mites live their entire lives feeding on the epidermis, fat, or body fluids of their hosts. They are modified

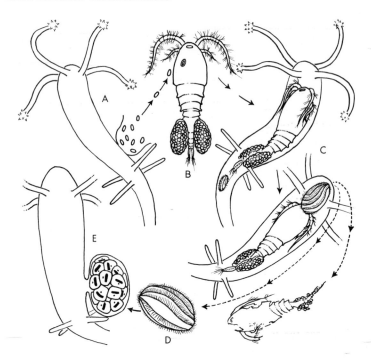

FIG. 26–2. The ciliate, *Spirophrya subparasitica,* and its relationship to two hosts, a hydroid and a copepod. The hydroid becomes a resting place for a cyst which liberates small, ciliated "tomites" (A). The latter attach themselves to copepods (B). If these are ingested by a second hydroid (C), the encysted tomite emerges and grows by feeding on the disintegrating copepod within the hydroid's gastrovascular cavity. Then the copepod remnants (indigestible portions) are expelled by the hydroid, and the fully grown "trophont" emerges (D) to encyst upon the surface of a hydroid and form many "tomites" (E). [(D) and (E) are enlarged somewhat more than the other figures.] [After Baer, 1951.]

variously for clinging to hairs or feathers, they burrow under the epidermis or scales, and they cut or pierce protective surfaces with their mouthparts. Other ectoparasitic arthropods are crustacea belonging to several orders, as well as various insects.

Fleas and lice illustrate somewhat different forms of ectoparasitism. Fleas, while intricately adapted for concealment, effective movement, and escape from attempts by the host to catch and kill them, do not spend their entire lives on the host, but lay eggs which hatch among the organic debris of the host's nest, lair, or customary environment. The larvae feed on materials such as hair, fecal matter, their own shed exoskeletons, etc. Only the adults are parasites. And the various species of fleas, while they "prefer" one or a few species of hosts, may obtain blood and shelter from accidental or "foreign" hosts. Thus fleas suggest in their life habit not only ectoparasitism but also predatism and free existence. Lice, on the other hand, usually stay on one host throughout life, and may be transferred successfully only to other hosts of the same species. Some lice, such as the pubic, head, and body louse of man, occupy only certain parts of the host. Lice are specifically adapted ectoparasites; they show morphological and physiological adjustments to particular environments. The body louse, by living in the clothing of its host and by depositing its eggs on cloth fibers, illustrates extremely well both the evolutionary opportunism of parasites (since it has evolved along with the recent culture of man) and the completeness with which an ectoparasite can become adapted to its host.

ENDOPARASITISM

Endoparasitism is, in general, a more complete or perfect relationship than ectoparasitism. Where the examples of ectoparasitism may resemble phoresis, commensalism, or predatism, endoparasitism nearly always involves both the intimacy and dependency associated with true parasitism. But different degrees of dependency and intimacy may be found among the internal parasites as well as the external, and it is convenient to classify the endoparasites in several categories according to habitat.

First, there are the lumen dwellers. These include most of the worms, many of the protozoa, and a few arthropods. The amoebas of man (see Chapter 6), living in the mouth and lower intestine, show complete dependency (since none of them can grow outside their host). But their parasitic habit may or may not harm the host. *Entamoeba gingivalis* of the mouth is apparently a commensal and not a harmful parasite, since it feeds on bacteria. But it occasionally ingests blood cells of its host and is associated at times with inflammations or infections of the mouth. The harmfulness of *E. gingivalis* has not been established. *Entamoeba coli,* one of the two common intestinal amoebas, is considered harmless (hence a commensal, not a true parasite) by most workers; but sometimes infections with this organism have been associated with gastrointestinal symptoms. *Entamoeba histolytica* is, of course, an important pathogen which causes death to some hosts and severe pathological lesions in many. Yet even this organism is usually so mild that it is able to live unsuspected in its host. Many factors influence the degree of harm which a parasite may cause; in the case of *E. histolytica,* the strain or variety of the parasite (there being two recognized races: the small cyst, mild form; and the large cyst, virulent form) is important in limiting the effect, and so is the condition of the host. As was pointed out earlier (see Chapter 2), the very existence of a parasitic infection may, by inducing "premunition," protect the host against more virulent strains of the same organism. Therefore, *E. histolytica* not only illustrates one extreme in a series of lumen-dwelling amoebas (ranging from supposed commensals, *E. gingivalis* and *E. coli,* through avirulent and virulent strains of the dangerous pathogen, *E. histolytica*) but also suggests the intricate and complex nature of the endoparasitic relationship. Other lumen-dwelling parasites, when adequately studied, have shown a similar complexity of host relationship.

Still more intimate and complex relationships are involved when parasites live in the tissues and cells of their host. Examples are the larvae of *Trichinella spiralis*, a nematode (see Chapter 17), and the important protozoan pathogen, *Leishmania donovani* (see Chapter 4).

The first example, *Trichinella spiralis*, involves the well-known migration of larval nematodes from the walls of the host intestine to the skeletal muscles. (See Fig. 17–1.) This migration and subsequent establishment is intercellular. Among the muscle fibers, the parasites become trapped and are enclosed within cystlike capsules deposited by the host connective tissues. During and after the migration of the parasites, various host reactions occur. (The symptoms evidencing these are the varied aspects of trichinosis, the disease.) Depending on the relative number of larval worms present and probably also depending upon the individual condition (allergic or otherwise) of the host, the relationship between host and parasite will become stabilized, or the host will die. If not killed by the ravages of the acute condition, the untreated human host eventually outlives its encysted larvae, which die and become calcified.

The second example, a protozoan parasite of cells, *Leishmania donovani*, causes the important disease, visceral leishmaniasis or kala-azar. Introduced into the human host by the bites of infected sand flies, the flagellates are carried in the lymph to the blood system and are then distributed to the organs where the characteristic lesions develop. (See Fig. 4–7.) The parasites invade or are phagocytosed by macrophages of the liver and spleen; in these cells the protozoa multiply and eventually infect most of the cells of the reticuloendothelial system of the above organs. Since this system is the major source of antibody production, the disease exposes the body to other pathogens, while it severely damages organs essential for life. Without treatment, visceral leishmaniasis is usually fatal. Thus this cell-inhabiting parasite causes extreme harm to its host by destroying essential cells.

The above examples of tissue and cell parasites, respectively, illustrate only two of the many ways in which such parasites interact with their hosts. These interactions, like malaria and many other diseases, are relationships which are complex and intimate, and are of great practical and theoretical importance.

THEORY OF PARASITISM

All parasitic relationships involve lack of competence by the parasite. This statement is not meant to be anthropocentric; nor is it meant to imply that parasites are in some way degenerate or inferior. Emotional attitudes toward parasites may be as absurd as the implication sometimes made that parasites themselves have attitudes or aims. "Lack of competence" is merely a useful concept by which to understand the several kinds of host specificity, the frequency and prevalence of parasites, and the effects of various parasites upon their hosts. In fact, the concept "lack of competence" may be applied to the supposed origin of animal life itself.

In the beginning (the beginning, that is, of life upon this planet) organisms were simple, and their environment was complex. Modern biochemistry combines with geophysics to inform us of the probable states under which life arose. Both theory and experiment indicate that relatively complex chemical compounds were once available. Metabolic units (the first living things) need not have been more complex than virus particles. Their physiological competence need not have been great, because their environment was rich in amino acids and even proteins. Yet existing life eventually used up available food; a new competence was forced upon living things. As Horowitz has explained, small mutational steps which had the effect of improving the enzyme systems of the primitive organism made possible the increased utilization of environmental resources, even as these resources became more and more depleted. Thus the evolution of photosynthesis occurred to prevent extinction of life from the exhaustion of energy sources. Not all organisms, of course, developed photosynthesis. Many were less competent than the new "plants," and fed upon the latter or their products. Thus began the dichotomy of living things on earth: the plant world (energy-binding and productive) and the animal world (energy-consuming and wasteful). In the broadest sense, therefore, animal life is parasitic upon plant life. And this parasitism results from lack of competence.

In parasitology, lack of competence implies dependence of a parasite upon a host. The closeness and exclusiveness of this dependency may be examined under the concept of "host specificity," which means the restriction of certain parasites to particular kinds or species of hosts.

Baer (1951) has analyzed host specificity and has defined two kinds, physiological and ecological. In physiological host specificity, the parasite lacks certain enzymes or enzyme systems which the host possesses, and certain products of the host metabolism supply what is missing in the parasite's metabolism. Detailed explanations of particular examples are not easily found, because not much work has been done upon parasite physiology. But the absence of a digestive tract in several groups of parasites should be ample evidence of lack of digestive competence, and should lead to the reasonable conclusion that such parasites depend upon their hosts for utilizable food. Likewise, as immunological studies have verified, tissue and cell parasites must be in extremely close physiological union with their host. When details of parasite metabolism are worked out in the future, it is probable that the exact nature of many host-parasite physiological relationships will be described. From the fact that many parasites can grow in only one species of host, it is clear that physiological host specificity exists.

Ecological host specificity also exists. It may be defined as a specificity determined by the environment of host and parasite rather than by their physiological interdependency. Ecological, or environmental, host specificity is difficult to distinguish from physiological specificity without complete knowledge of the life cycle of the particular organisms involved. For example, certain groups of digenetic trematodes are specific to certain groups of final hosts. In 1938, Dubois showed that the strigeid trematodes occur in aquatic hosts and do not occur in hosts unassociated with water. The fact that the hosts of strigeids include representatives of all the major groups of vertebrates except fishes and amphibia indicates that there is little or no physiological dependence of these worms on particular kinds of hosts. It is not surprising, therefore, that in the laboratory strigeids have been reared successfully in several "foreign" hosts, such as chicks, doves, rodents, dogs, and cats. Because their larvae require aquatic intermediate hosts, these flukes are restricted in nature to animals that prey upon these hosts. But this specificity is ecological only, and the strigeids can develop in many kinds of physiological environments.

Another example of ecological specificity and physiological nonspecificity is the digenetic trematode *Cryptocotyle lingua*. In 1930, Stunkard obtained naturally infected snails, which he kept with fish until the latter contained many encysted metacercariae. After feeding these cysts to a variety of hosts (including rats, guinea pigs, and cats), he was able to recover adult parasites from these hosts. *Cryptocotyle lingua* has been found naturally infesting only fish-eating birds and mammals.

A third example may be remembered from the discussion of *Schistosoma japonicum* in Chapter 9. This blood fluke infects in nature several different kinds of mammals—presumably almost any mammal which comes in contact with cercaria-infested water. Yet the molluscan hosts of *S. japonicum* can be only the members of a particular species, indeed a particular strain, of snail in each area where the blood fluke occurs. In the Philippines, only *Oncomelania nosophora* is readily infected by the miracidium of *S. japonicum*. (See Fig. 26–3.)

Thus the occurrence of strigeids in many kinds of hosts which have in common only an ecological connection with water, the laboratory experiments which show that certain parasites, such as *Cryptocotyle lingua,* can exist readily in unnatural hosts, and the occurrence of the blood fluke *Schistosoma japonicum* in many water-loving mammals, serve to demonstrate the fact of ecological host specificity. It is worth noting that in the trematode stages which are found in snails or other molluscs (the sporocyst and redia stages) physiological rather than ecological host specificity occurs, probably because the larval trematode is a tissue parasite, which is unable to feed, except by absorption, and hence is biochemically locked to the metabolism of its particular species of snail. Thus different phases of a parasite's life cycle may involve different kinds of host specificity. In the case of most parasites, it is probable, of course, that both kinds of specificity interact; ecology is predominant in some, and physiology, in others.

An example of such interaction between ecological and physiological factors may be seen in the Mallophaga, or biting lice. Some of these insects show definite preferences for particular parts of the host integument, for example, epidermal cells, hair, feathers, or even particular feather groups; thus physiological factors are clearly involved in the specialization of these insects. Yet the Mallophaga also show a high degree of host specificity which can only be based on the ecological isolation of the various host species; the speciation of hosts is accompanied by speciation of their biting lice. Interestingly, in situations where the hosts of Mallophaga are brought together, and their interspecific isolation is ended (as in the zoo or laboratory, or on the Galapagos Islands, where, on the shelterless land, birds of several species and subspecies must live together), the parasitic insects of one species of host often infest other species, and the host specificity due to ecological factors breaks down.

FIG. 26–3. Ecological and physiological host specificity in *Schistosoma japonicum*. The vertebrate hosts of this blood fluke are practically all mammals which enter the water where infective cercariae occur. Hence the adult schistosome is limited as to its final host chiefly by ecological factors, that is, contact by the host with infested waters. On the other hand, the invertebrate hosts (for example, snails of the genus *Oncomelania* in the Philippine Islands) are limited to a certain species (*O. nosophora*) and even to particular strains or varieties of this species of snail. This specificity is physiological and depends on nutritional, immunological, and perhaps other intrinsic factors of snail or parasite. Miracidia of *S. japonicum* are not ecologically host specific for they may come in contact with many kinds of snails but can develop in only one.

From the above discussion of loss of competence, and the consideration of the interacting physiological and ecological factors, which stabilize parasitic dependency (or host specificity) in its many forms, one might conclude that parasitism as a way of life is futureless and without compensation. Because this conclusion would be false, however plausible it may seem, I shall now point out, in terms of evolutionary advance or survival, some of the values of the parasitic way.

The first test of evolutionary success is, of course, survival itself. In parasites, this means their continual establishment and reestablishment in hosts, a distribution and frequency which do not risk the danger of extinction, and the achievement of a reasonable balance (discussed in Chapter 29) between the health and survival of the host and the growth and reproduction of the parasite. All the well-known parasites are, in the above sense, successful. In a purely logical way, it must be true that fantastically high rates of egg production (as in the larger tapeworms of man, for instance) indicate high improbability of survival of any one egg. On the other hand, relatively low reproductivity, which is characteristic of many ecto-parasites, indicates by the same reasoning that chances for individual establishment and transmission are high. It has been pointed out that time or duration, as well as number, may be important. If the total number of eggs laid is in each case the same, the short life of one egg-laying parasite may compensate for the long life of another kind of parasite. Also, duration of reproductive life may in a practical sense spread the opportunities for transfer over a variety of environmental changes; a single reproductive burst may not coincide with the ideal season for transmission. Reproductive balance, it seems, is a universal characteristic of successful parasites, as it is of all successful species.

Perhaps the most fascinating object of study in parasitology is a second factor in parasite success—the ecological ingenuity of parasites, as shown by their life cycles.

Some life cycles are almost incredibly ingenious, so that the evolutionary basis of such specialization seems questionable. (Actually, as will be seen in Chapter 29, evolutionary hypotheses are strengthened and illustrated by examples of parasitic adaptation; however, the immediate impact of such examples is to startle and surprise.) The skin bot of tropical America, *Dermatobia hominis,* requires for its larval development the subcutaneous tissue of a warm-blooded animal, such as man. Female botflies, when ready to oviposit, catch a large, female mosquito and, without harming it, lay several eggs on the under side of the mosquito's abdomen, where the eggs stick tight by a cement which covers them. Released by its captor, the mosquito continues its life and seeks blood from time to time. When the eggs which it carries are ripe, and while the mosquito is biting a warm-blooded host, the eggs hatch, and the first-stage maggots crawl out and drop to the warm skin of their next host (Fig. 26–4). They penetrate the skin and then develop into characteristic bot maggots within several months; they live within a swollen inflamed cyst and feed on the host tissue. When ready to pupate, the maggots leave the host and pupate on the ground.

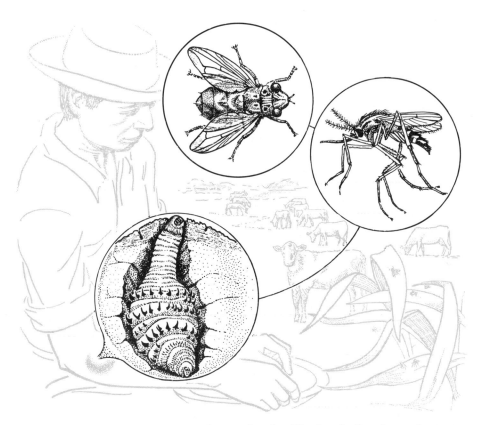

FIG. 26–4. *Dermatobia hominis*, a human bot fly. The female fly seizes a large mosquito, on the abdomen of which it cements a few of its eggs. These hatch and release active maggots while the mosquito is resting on warm skin during the act of biting. The maggots penetrate the human host and grow in the subcutaneous tissue of an eroded, boillike cavity. The maggot breathes through a small opening in the cavity. The full grown larva leaves the cavity, drops to the ground, and pupates. [After Baer, 1951, and others.]

The nematode *Strongyloides stercoralis* (Fig. 26–5), the life cycle of which was discussed in Chapter 20, probably represents a unique compromise between fecund dependency and the free-living state, since this worm manages to pursue four different patterns of reproduction. Although the strictly parasitic cycle may result in vastly increased numbers of individuals through autoinfection (Cycle 1 of the figure), the parasites would end their lives with the death of the host unless a transmission phase (Cycle 2, involving brief larval sojourn in the soil) is added. Further opportunity for growth and reproduction comes through Cycle 3, in which larvae from the host mature, mate, and produce new infective larvae in the soil, as well as through Cycle 4, which, potentially at least, transforms the parasite into a free-living animal, capable of reproducing indefinitely without the food or shelter of a host. Most parasites, however, retain little of such ancestral independence.

Another interesting cycle is that of *Diplozoon paradoxum*, a monogenetic trematode of the gills of fish. The egg of this animal becomes attached to the host gill by means of a long polar filament. The larva hatching from the egg bears cilia and two suckers. It attaches itself to the gill surface and develops into a sexually immature stage called "diporpa" (which was the name of a genus once erected in the belief that such larvae were adults). The diporpa larva may remain unchanged for months; but if another diporpa larva becomes attached to it, development proceeds to the adult stage. The two diporpas are attached to each other by a small ventral sucker and a small dorsal lump or papilla which each larva bears. The sucker of each seizes the papilla of the other, so that the bodies become twisted around each other to form an X. As they develop further, the vagina of one lies opposite the sperm duct of the other, so that cross-fertilization seems assured. At the points of attachment the organisms become permanently fused together. It is difficult to see how such a morphological adaptation could arise; the exact matching of sucker and protuberance seems inexplicable in terms of the gradual (ineffective) development of either part of the interlocking mechanism. One may safely assume, however, that the peculiar requirement of *Diplozoon* for cross-fertilization has provided the selective stimulus for this bizarre life cycle. The self-fertile forms (if any existed) have been suppressed to a sterile, immature diporpa, and sexual development itself depends on the presence of two "mates."

Evolutionary opportunism, as exemplified above and as manifested throughout the parasitic way of life, may be considered a third factor (after survival and the ingenuity shown in complex life cycles) which compensates for the restrictions

FIG. 26–5. Life cycle of *Strongyloides stercoralis*, a nematode parasite of man and other mammals. Essentially four kinds of cycles can occur. In Cycle 1, adult parasitic females in the mucosa of the intestine deposit rhabditiform larvae which emerge into the lumen of the intestine, moult into filariform or infective larvae, penetrate the mucosa, and are carried by the blood to the lungs, where they develop into adolescent males and females. The females may mate but probably after reaching the intestine will produce larvae by parthenogenesis. In Cycle 2, rhabditiform larvae in the intestine reach the soil with feces, moult into the filariform stage, and penetrate the skin as hookworm larvae do; they reach the lungs by the blood stream. In Cycle 3, rhabditiform larvae in the soil become rhabditiform (free-living) adults, which mate. Females deposit larvae which moult to the infective stage. The larvae enter the host by the skin as in Cycle 2. In Cycle 4, the free-living adults may produce larvae which mature into a second generation of free-living forms and thus may avoid any parasitic stage whatsoever. [Details of worms after Faust and Russell, 1964.]

imposed on parasites by dependence on hosts. The fact that all groups of animals have parasites is evidence that rates of evolution among parasites must at least equal the evolutionary rates of hosts. The remarkable fact that parasites in general belong to more ancient or primitive taxonomic categories than their hosts seems to indicate that parasites must mutate (vary or adapt) at a faster rate than contemporary free-living organisms, i.e., their hosts. (These ideas will be discussed and critically examined in Chapter 29.) Direct evidence of the adaptability of some parasites is available in the records of drug resistance among pathogenic bacteria

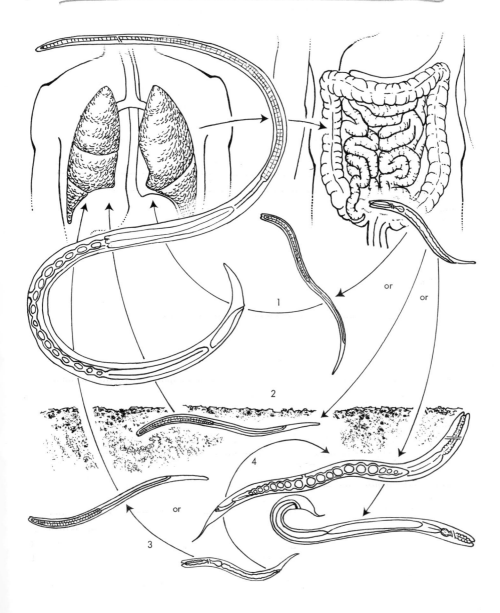

and protozoa, in the marked ecological diversification of the biting lice, mites, and parasitic diptera, and in the existence of varied and intricate life cycles among the parasitic worms, whose free-living relatives (Turbellaria, for instance) have comparatively simple and direct cycles of development.

CONCLUSION

Perhaps the remarkable trematode, *Diplozoon paradoxum*, discussed above, can be used as a general symbol of parasitism. The interlocking sucker and papilla are mechanical analogues of the interlocking factors in the biological world. Relationships which work and are successful become fixed. The establishment of such relationships suppresses other potential developments. A close look at ecology gives such an impression of interdependence, of permanence, of specificity not just between males and females of a species, between parent and young, between matched diporpa larvae, or between parasite and host, but also between prey and predator, animal (consumer) and plant (producer), living organisms and the inorganic planet itself. And such a view of the conservative complexity of nature is valuable and true. But the very existence of such interlocking systems of life, like the existence of the sucker and papilla in *Diplozoon*, implies evolution. Hence the appearance of permanent adaptations must not be the whole truth about nature. Change is as real as order. And when the order is disturbed by man or by time, then changes will occur in organisms until a new appearance of order is determined. Parasitism represents only a special case of order resulting from change.

SUGGESTED READINGS

BAER, J. G., *The Ecology of Animal Parasites*. University of Illinois Press, Urbana, Ill., 1951.

BLUM, H. F., *Time's Arrow and Evolution*, 2nd ed. Princeton University Press, Princeton, N.J., 1955.

DUBOIS, G., "La specificite de fait chez les Strigeida (Trematoda)," *Premier Symp. sur la spec. parasit. des paras. de Vertebres. Neuchatel*, 1957, pp. 213–227.

HOROWITZ, N. H., "On the Evolution of Biochemical Syntheses," *Proc. Nat. Acad. Sci.* (U.S.), 31:153–7, 1945.

ROGERS, W. P., *The Nature of Parasitism*. Academic Press, New York, 1962.

STUNKARD, H. W., "The Life History of *Cryptocotyle lingua* (Creplin) with Notes on the Physiology of the Metacercariae," *Jour. Morphology*, 50:143–191, 1930.

WALD, G., "The Origin of Life," *Sci. American*, 191:45–53, 1954.

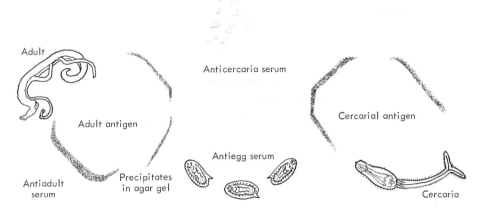

CHAPTER 27

Adult

Anticercaria serum

Adult antigen

Cercarial antigen

Antiegg serum

Antiadult
serum

Precipitates
in agar gel

Cercaria

RESPONSES

INTRODUCTION

The host-parasite relationship, which in the form of examples was introduced in the first 25 chapters of this book and which was defined in very broad terms in Chapter 26, must now be visible to the reader as an intricate and marvelous inter-action among living things. This relationship is, moreover, very old—as old, in principle, as animal life itself. It is therefore one of the most persistently surviving products of evolution; and, while ancient, it is still alive with the capacity for ingenious growth.

The present chapter is a continuation of the preceding chapter, for here we shall begin to consider special aspects of the host-parasite relationship. A student of science asks what mechanism is involved in each mystery being investigated; he is not satisfied with a broad or general description, but wants to use, if possible, operational terms, i.e., terms which tell what will happen if certain acts are carried out. What are the specific defenses which protect some hosts or render some parasites harmless? What are the mechanisms by which a parasite is able to invade and to utilize a particular host? These two questions will be approached in this chapter and the next. The present chapter considers the results in some cases

369

when the host resists, destroys, or tolerates potential or actual parasites. The next chapter will consider the way in which parasites injure or kill their hosts. Since it is human to think first of the host member of the host-parasite relationship, we may call this and the next chapter discussions of success and failure, respectively, in host defenses against parasitism.

RESISTANCE

Resistance to parasites is a term which includes a complicated group of factors opposing the establishment of potential parasites.

The latter, however, are not those parasites which cannot conceivably invade human hosts because of ecological barriers; resistance of man to the hookworms (*Uncinaria*) of sea lions (worms transmitted by mother to offspring through milk) does not need to be considered, since these worms can never test man's resistance to them. However, the ecological barriers which occasionally break down ought to be mentioned. For example, among the zoonoses is "visceral larva migrans," the larva of cat or dog ascarid worms which can cause liver damage and other pathological effects in humans who ingest eggs of these ascarids. Hydatid cyst (see Chapter 11), a serious disease of man, is properly a zoonosis, since the normal life cycle of this cestode is in herbivores and dogs or wolves. Also *Trichinella* (Chapter 17), maintained in nature by various food chains, reaches man in a sense accidentally when he eats infected meat, and of course it can maintain itself in man only by the now quite rare custom of cannibalism. Although the zoonoses are important and interesting, there are more than enough well-established human parasitisms to fill our need in discussing the resistance of hosts to parasitic invasion and damage.

An important form of resistance is due to surface barriers which hinder or prevent the entry of parasites. The skin is such a barrier. Some parasites enter their hosts through the skin, but in many cases the entry is through some tear or break in this barrier, or, if not, entry must be aided by unusual conditions or by invasive enzymes produced by the parasite itself. Thus, for example, skin puncture by insect vectors is required for the entry of *Plasmodium*, the filarioid worms, and the hemoflagellates and leishmanias; screwworm-fly larvae (see Chapter 24) hatch from eggs laid near open wounds. The intact skin is a barrier to all of the above. Needless to say, the skin is also completely impervious to all the delicate organisms which flourish in the digestive canal. The human skin fails to protect against hookworm larvae, *Strongyloides* larvae, and the cercariae of blood flukes. The latter, however, are aided in penetration if a drying water film holds them against the host, for they require a certain time during which their proteolytic enzymes work to soften the host's skin (see Chapter 9). Barriers other than skin are the mucous membranes which line the nasopharynx, urogenital tract, and digestive canal. It seems to be established, for instance, that the ulceration caused by *Entamoeba histolytica* is stimulated or, more likely, permitted by the action of enteric bacteria or other irritants which injure the lining of the intestine. Amoebas invade the tissues through minute tears or rents in the mucous membrane. Without such opportunity, *Entamoeba histolytica* possibly remains a harmless commensal. A dramatic demonstration of the protective role of mucous membranes occurs when ionizing radiation destroys the highly sensitive interstitial cells of the intestine, i.e., cells which normally replace sloughed lining cells. All sorts of ordinary intestinal bacteria can then invade the body to result in septicemia. Radiation sickness and death are due largely to damage to the repair function in the lining of the intestine. The above examples suggest that skin and mucous membranes protect the host from many potential invaders.

Once a parasite successfully enters a host, other barriers oppose its spread or growth. These may be called "environmental" in the sense that the chemical and physical environment provided for the parasite by the host's tissues or cavities must be suitable; if such an environment is not suitable, it is a barrier to parasitic success. While little is known about the factors which cause or permit the hatching of helminth eggs or the excystment of protozoa in the host intestine, it has been shown that the eggs of certain worms pass unhatched through the intestine of various "unsuitable" hosts. Sewage-feeding birds in England, for instance, carry the eggs of the beef tapeworm, *Taeniarhynchus saginatus* (see Chapter 11), far from the sewage-disposal drying flats and scatter these eggs in pastures with their droppings. The eggs of various ascarid worms are similarly scattered by being passed unhatched through the intestine of coprophagous hosts. (It should be noted that the above examples are not merely demonstrations of environmental barriers to hatching but are also cases of parasitic "ingenuity," for the dispersion of a parasite population in space is accomplished quite effectively by "transport," i.e., unsuitable hosts.) The acidity of the stomach of most mammals, with the protease pepsin of the stomach, destroys the parasites which pass through this region unless they are protected by egg shell or cyst wall. Thus while many animals, including man, drink water containing the infective cercariae of the schistosomes, only ruminants, which have an alkaline stomach content, normally acquire blood flukes by ingestion. Presumably the larvae are destroyed in an acid or pepsin-containing environment. Another environmental barrier is the movement of fluids in the various cavities of the body. Hookworm larvae, having completed their third moult, are carried with tracheal mucus up to the pharynx. They must be swallowed, however, to complete their development, and great numbers of them may be lost by expectoration. The therapeutic value of chewing some expectorant—tobacco or betel, for instance—has been pointed out by hookworm specialists. Similarly, the emptying time of the intestine may be too short to permit excystation or hatching of certain parasites. Another environmental barrier is the various tissues through which migrating larvae must pass. The dog hookworm larva, for instance, causes creeping eruption in man, presumably because it cannot escape from the germinative layer of the epidermis; it therefore wanders in the epidermis and does not complete its development by lung and intestinal stages as it would in its normal host. Swimmers' itch, caused by cercariae of blood flukes found in nonhuman hosts, is a similar condition; the parasites wandering in the skin are unable to penetrate underlying barriers. Various degrees of inhibition of larvae may occur. The larvae of *Toxocara* and *Toxascaris*, ascarids of dogs, cats, and other carnivores, have been found in the liver and other tissues of children; hepatitis, eosinophilia, and lung involvement may result. These worms, called "visceral larva migrans," are prevented from reaching the intestine and completing their normal development by some (unknown) factor in the human tissues. In this case the protective quality of the environmental barrier may be questioned, since the wandering larvae are more harmful to the human host than they would be if they pursued a normal pathway to the intestine. The role of the human host of "visceral larva migrans" can perhaps be understood in terms of the normal life cycle of the ascarids of carnivores. Some of these worms, while capable of developing from egg to adult in one host, are commonly acquired by the definitive host as third-stage larvae (already suited for an intestinal habitat), which have developed in the tissues of some small animal which is the frequent prey of carnivores. The abnormal host (man) is thus usurping the role of an intermediate host which aids both the survival and the dispersion of the parasite. Thus some environmental barriers, when examined ecologically, may turn out to be adaptations toward effective parasite-host relationships.

Certain barriers other than the above mentioned mechanical (surface) and environmental (tissue and gut) obstructions to successful invasion and development exist, but they are hard to classify. Time is one of these. If the biological life span of a parasite is greater than that of an otherwise suitable host, successful parasitism is impossible. Temperature is another significant factor. Both time and temperature affect the malaria organism in its mosquito host. If the temperature of the mosquito is relatively high (as it would be in the tropics), sporogony may occur in as short a time as 10 days. If the temperature is lower, up to three weeks may be required. At still lower temperatures (characteristic of the nonmalarious ranges of *Anopheles*), the parasite fails entirely to reach a step infective to man. Thus temperature affects both the time of development and the limits of range of *Plasmodium*.

Finally, in discussing resistance, it should be emphasized that not every host of the same species is equally resistant to invasion by a particular kind of parasite and that parasites of the same species are far from identical in their ability to invade a host. That is to say that variation occurs. An individual host may change in condition or state of health from day to day; thus he may acquire, when in bad condition, parasites which would be unable to enter or survive if the host were healthier. Also, individual hosts of the same species differ in many ways, environmental as well as genetic, and may thus be expected to differ in respect to their responses to invasive parasitism. Parasites also vary. Most well-studied pathogens have strains ranging from almost harmless to virulent; invasiveness is a part of resistance, and it varies among strains of parasite. Therefore resistance is, like the host-parasite relationship, an interaction and not a quality of the host alone.

TOLERANCE

Another such interaction is the response of a host after invasion; such a response may involve drastic reaction (immune response, which will be discussed later) or disease (pathogenic effect), but characteristically it is a sort of adjustment between host and parasite, an adjustment which, from the host's point of view, can be called tolerance.

The bases of tolerance are varied and for the most part unknown. Commensalism (see the preceding chapter) illustrates complete tolerance. For example, *Enterobius*, the pinworm of man, living attached to the mucosa of the human large intestine, excites no demonstrable response from the host beyond *pruritus ani*, the mild and transient itching induced by the parasite's eggs and by the perianal migrations of the female worms. This itching, however, is varied and is nearly intolerable for some sensitive individuals. *Entamoeba coli*, another commensal, causes no response on the host's part. Human *Ascaris* was believed by ignorant people to be a natural, indeed beneficial, part of living; some went so far as to call the white worms which sometimes escaped from a child's mouth or anus "guardian spirits." Apparently the response to ascarid infection was so mild that it escaped notice until medical parasitology became a scientific discipline. Now it is known that *Ascaris* is dangerous both during its tissue phase and while it is in the intestine. Yet in most cases the body hardly acknowledges the presence of the adults of this worm. Tolerance to *Ascaris* is not universal, of course; some persons become hypersensitive to the worm and its products, while the undernourished can ill afford even the small demands for food made by this commensal. Other forms of tolerance may occur in true parasitism or even in disease. The well-known relationship between the sickling hemoglobin gene and tolerance of *falciparum* malaria is an adaptive tolerance which probably results from long association between man and malignant malaria (see Chapter 29). Only the races in which *falciparum* malaria is stable and endemic show a high frequency of the sickling gene, and only these people can tolerate this particularly severe disease. In some diseases, the number of parasites present may greatly affect tolerance. Thus chickens which acquire only a few oocysts of *Eimeria* suffer correspondingly few lesions in the affected organs and usually survive to become immune. Other chickens, during the course of an outbreak of coccidiosis, may ingest millions of infective oocysts and are overwhelmed by a fatal infection. Liver flukes in cattle or man, and blood flukes as well, produce few symptoms unless many flukes are present; numbers here affect tolerance. The scab mite, *Sarcoptes*, is tolerated in small numbers. Another mite, *Demodex*, is found in the skin of most persons but causes few or no symptoms in man. In dogs, demodectic mange (a nontolerant response) occurs only in animals on poor diet or in bad general condition, yet these follicle mites must be extremely common. Still another basis of tolerance is the behavior of the parasites themselves. *Schistosoma japonicum*, the most harmful of the human blood flukes, lays many eggs, whereas *S. mansoni* and *S. haematobium* lay very few. Some hemoflagellates (*Trypanosoma theileri* of cattle, *T. melophagium* of sheep, and *T. rangeli* of man) are believed to produce no symptoms. Since most trypanosomes do cause symptoms, it should be concluded that the above trypanosomes have become mild (which is to say that the hosts, perhaps, have become tolerant) over the ages during which the above associations have persisted. One may speculate that adaptive change in the parasite from the virulent to a mild form may be one explanation of host tolerance. Therefore, tolerance, like resistance and other aspects of host-parasite relationships, is probably not one-sided but involves adaptive behavior by both parasite and host.

EXAMPLES OF IMMUNITY

Apart from the ready-made or preexisting conditions mentioned above, the most important mechanisms for achieving tolerance are probably the immune responses. Before discussing some of the theories of immune response, however, it would be best to give some examples.

Examples of immunity are familiar to all educated people. Before the time of Jenner or Pasteur, it was well known in the Middle East, India, and other regions that certain disfiguring sores, once healed, never recurred. Thus it was said that a visitor to Calcutta could be recognized by his sore (implying that all long-time residents had once had such sores, had recovered, and were henceforth immune). Cutaneous leishmaniasis became a classic example of immunization when an English woman, the wife of a diplomat to Turkey, wrote to Jenner, the inventor of vaccination, that mothers in the Middle East deliberately inoculated their daughters with matter from active sores in order to prevent disfigurement by accidentally acquired sores on the face. Many bacterial and virus diseases confer immunity with recovery. Diptheria, measles, poliomyelitis, and typhoid fever are among such diseases. As the Turkish mothers first demonstrated, it is possible to produce immunity artificially by using the disease agent itself in a controlled way. Or, as Jenner showed, a similar agent (cowpox virus instead of smallpox) can produce immunity (a cross-immunity). Pasteur used attenuated virus in his classic inoculation against rabies, and some of the modern antipolio methods follow the latter principle. In all the above examples, the inoculated host responds by producing its own protective substances. However, it is possible to utilize serum from an immune host to protect a nonimmune host temporarily against a pathogen; the use of antitoxin against diptheria or botulism is an example of this method of protection called "passively acquired immunity." Such immunity is strictly temporary and is superseded by an "actively acquired immunity" resulting from the patient's response to the actual pathogens. Before proceeding to discuss immunity to animal parasites, it is well to point out that immunities may be life-long or very brief and that the amount of protection they afford may be absolute or may be so little as to be practically undetectable. It should also be mentioned that some immunity is far from helpful to the responding organism; hyperimmunity or sensitization frequently occurs, and the disease organism, foreign protein, or substance which causes such sensitivity then becomes more irritating, more damaging, and less well tolerated than it was on first contact. Thus it appears that immunity is a very complex reaction. Some further examples from animal parasitology will help to make this clearer.

Malaria, as we have seen in Chapter 2, confers no lasting immunity. Yet the course of the disease itself is a demonstration of immune response. (See Fig. 27–1.) First, as at each paroxysm in which toxins, metabolic wastes, protein detritus, and parasites themselves are released into the blood, strong reactions (chills, fever) occur which simulate shock reactions in some types of allergy or hypersensitivity. As the disease progresses, the patient is threatened with an overwhelming number of parasites, which increase in geometric progression. Obviously, the liver and spleen must respond to the crisis by increased phagocytosis, or the parasites would destroy all the red cells in the body. This increased phagocytosis is evidenced by the accumulation of parasite-produced pigment in liver and spleen as well as in the final result, i.e., recovery of the patient and the disappearance of demonstrable parasites from the blood. Unfortunately, recovery does not confer lasting immunity

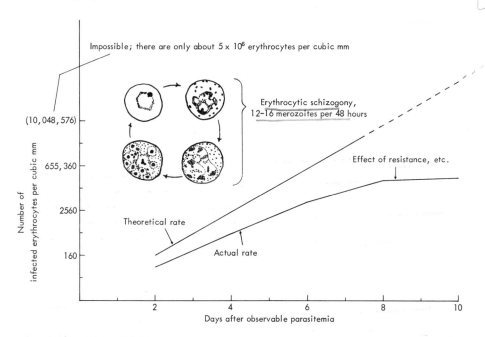

FIG. 27–1. Theoretical and actual growth of a population of an infectious agent (*Plasmodium*). The vertical axis is marked in a logarithmic scale to show numbers of parasitized blood cells arising by a 12- to 16-fold reproductive factor per 48-hour generation, as in *P. vivax*. The horizontal axis tells time after onset. In about eight days every red blood cell would be infected, theoretically, but actually even in very heavy infections the number of infected cells does not go over 1 or 2 per 100 erythrocytes. The effect of acquired resistance is thus seen in the reduction of parasite populations far below their theoretical potential.

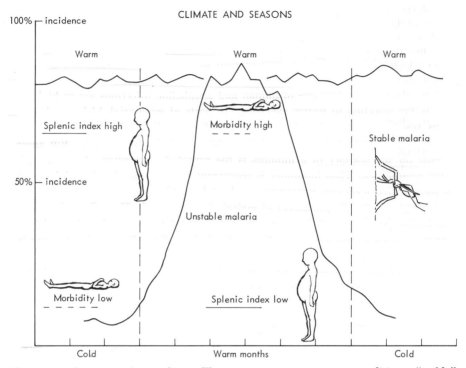

CLIMATE AND SEASONS

100% ⌐ incidence

Warm   Warm   Warm

Splenic index high   Morbidity high

                                                    Stable malaria

50% ⊢ incidence

                        Unstable malaria

Morbidity low   Splenic index low

Cold   Warm months   Cold

FIG. 27–2. Premunition in malaria. The two curves represent two conditions: "stable" malaria (the horizontal line, top) and "unstable" malaria. In the tropics, transmission rates are constant throughout the year, and infection levels remain high. There is little mortality because frequent reinfection maintains a degree of protective immunity. The high splenic index (frequency of persons with enlarged spleens) is evidence of the high infection rate and probably of immune response also. "Unstable" malaria, seasonal and epidemic, is characteristic of temperate zones where transmission fails during cold weather. Morbidity is high during epidemics, while splenic index remains low. Premunition is a name for the protection afforded by constant reinfection in regions where malaria is stable.

upon the malaria patient. The relapsing malarias, *vivax* and *malariae*, probably remain dormant as liver schizonts for the variable length of time between recovery and relapse. The early relapse is often called "recrudescence," because the symptomless period is so short that recovery does not seem to have occurred. Some relapses occur so long after the last disease episode that one must suppose there is a rather lasting protective immunity. Relapse may be considered (oversimply) as a multiplication of blood forms when the phagocytic response has diminished. The rate of diminution of that response varies with many factors; it varies particularly with the physiological condition of the host.

In addition to the temporary condition emphasized above, malaria illustrates another aspect of immunity. Repeated reinfections play an important role in maintaining this temporary immunity at a protective level. (See Fig. 27–2.) "Stable malaria" (see Chapter 2) has been defined as a condition in which control and eradication are either not contemplated or too difficult at present; it is also a condition in which the frequencies of human and anopheline infections are so high as to ensure the infection of nearly everyone. In parts of equatorial Africa such conditions result in universal malaria but little disease. Part of this tolerance is no doubt due to the high frequency of the sickling trait in these people. But another reason for the scarcity of symptoms is the repeated acquisition of new parasites from mosquitoes. Thus immunity is constantly being restimulated, and it remains at such a high level that parasitemia is kept low. Since low parasitemia produces few symptoms, the disease aspect of malaria is suppressed. This effect of continued reinfection has been called "premunition," a useful term meaning "armed in advance" against pathogens. It applies particularly well to diseases due to parasites, but it should not be considered a special kind of immunity. It is the kind of immunity that protects over a long period of time against those parasites which do not, by a single attack, provoke a lasting immune response.

Immune responses to several helminths have been studied. One of the first worms used in such studies was *Hydatigera taeniaeformis*, a cat cestode which develops into pea-sized cysts in the liver of its intermediate host, the rat. The presence of developing cysts partly inhibits the establishment of new infections; secondarily invading larvae usually die after only slight growth. Yet the antibodies which are present are ineffective against the continued growth of cysts already established. Field collections of rats usually show a rather low number of *Hydatigera* cysts in individual livers, although laboratory animals can be easily infected with hundreds of cysts. Since wild rats undoubtedly have opportunities to ingest many *Hydatigera* eggs over a considerable period of time, their average low infection rates must be due to the immunity mentioned above. This example of immunity coexisting with established infection and terminating with the loss of the established forms is probably a good case of classical premunition—protection depending upon infection. Natural immunity to the lungworm of cattle, the nematode *Dictyocaulus viviparus*, is known to occur. Young calves infect themselves by ingesting larvae with contaminated hay, pasture grass, or water, and sometimes become very ill with this parasite. Recovery occurs with increasing immunity, a condition which not only prevents further infection but reduces the reproductive abilities of worms already present. A practical protection against lungworm has been provided by the use of living larvae which have been sterilized by exposure to large doses of ionizing radiation. (See Chapter 20.) These larvae, injected into cattle, migrate and grow as do normal lungworms, but they cannot reproduce. They cause immunity, however, and thus they protect calves against symptom-causing levels of infection. Similar techniques are being tested with other helminths. Such elaborate and difficult methods of inducing immunity as the above use of sterile but living inoculum have been necessary for preventive immunization, because immunity to helminths seems to depend upon the growth and metabolic activity, not merely upon the substance, of the parasites.

A form of immune response called hyperimmunity or sensitization may occur in many kinds of parasitism, since many foreign substances, including the proteinaceous exudates of various worms and arthropods, can stimulate certain individuals to produce abundant antibodies of a more harmful than helpful nature. In the form of filariasis (see Chapter 14), caused by *Onchocerca volvulus*, for instance, one symptom of the disease is swollen sensitive skin with itching and redness. This symptom is characteristic of many allergies, of course. The hypersensitive state in onchocerciasis is particularly dangerous, because unwise chemotherapy can kill all at once the millions of microfilariae in the skin and thus can release enough toxic antigen to cause even fatal shock in the patient. In scabies (see Chapter 22) there is little itching at first, although many mites may be tunneling in the skin. Then the host begins to react with antibodies which combine, presumably, with the mite feces and other wastes to cause the very severe pruritus commonly seen in developed scabies. Even after the mite population has been greatly reduced by the positive aspects of the host's immune response, the hypersensitivity to mite antigen persists, and itching continues.

A few more examples of immunity to animal parasites should be mentioned. It has been suggested that malarial relapse might be the result of the release of mutant merozoites (among many to which the host was immune) by persistent exoerythrocytic schizonts. While this explanation cannot be supported by evidence from malaria, a similar situation has been analyzed in the hemoflagellate of rats, *Trypanosoma lewisi*. The course of infection in the normal host is predictable; there is at first a rapid increase of trypanosomes in the blood, and then the population reaches a brief stable period, after which it slowly and steadily declines until the parasites can no longer be found. By contrast, in guinea pigs infected with *Trypanosoma rhodesiensis*, the parasite population curve is quite different. After the rapid rise in parasite level with fever and other host symptoms, there is a rapid decrease in parasites and a cessation of symptoms. Within a short time, another paroxysm of disease occurs; it is like the first but is often more severe. Similar episodes follow one another until the host is finally overwhelmed. It has been

FIG. 27–3. Ablastin, an antibody that prevents adaptation. Taliaferro's data, here reproduced in combined form, show that *Trypanosoma lewisi* in rats (broken line) rises rapidly in numbers at first and then diminishes as it gradually disappears from the blood. Coincident with the pattern of rise and fall in population is a change in frequency of growing and dividing forms. The coefficient of variation (solid line) is high during the population rise and then falls abruptly, so that even while the population is still very large there are very few dividing or growing forms present. It is known that ablastin, an antibody produced by the rat in response to the antigenic stimulus of *T. lewisi*, inhibits reproduction of the latter and prevents the variation from which new antibody-resistant strains of the parasite might arise. [After Raffel, 1961.]

shown that while the trypanosomes in the rat are reaching their highest number the host is producing a unique antibody (called "ablastin" by its discoverer), which suppresses further reproduction of the trypanosomes. The stable population then is gradually consumed by phagocytes. (Figure 27–3 shows these events graphically.) The guinea pig, however, does not produce ablastin, although it produces other antibodies against trypanosomes which destroy most of the parasites during each

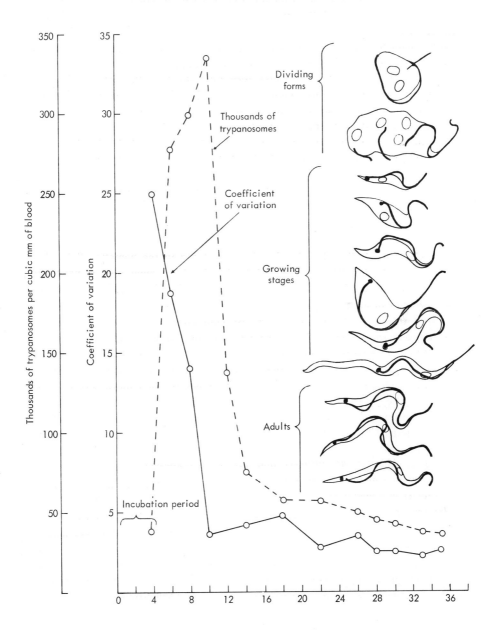

episode or paroxysm. A very reasonable explanation of the failure of guinea pig antibodies to save the animal is that these destroy most, but not all, of the parasites; some parasites at each episode of disease are inherently able to withstand antibody attack and create a new resistant population as they continue to reproduce. Being genetically variable, the parasite thus avoids each attempt by the host to destroy it and finally destroys the much weakened host. In rats, on the other hand, ablastin prevents reproduction of the parasite once an original population has grown; thereafter no further variation becomes possible, and phagocytosis slowly but effectively destroys the parasites. Thus there is evidence that variation (i.e. evolution) is involved in immune responses and that the ability of a host to destroy its parasites is affected by the variability of the parasites in much the same way that the ability of man to control such diseases as malaria and sleeping sickness is affected by the variability (insecticide resistance, drug resistance) of vector and pathogen.

Some parasites are able to make effective use of the host's immune response. *Leishmania donovani*, the pathogen which causes kala-azar (visceral leishmaniasis), is transmitted to man and other mammals by the bite of the sand fly (see Chapter 4). By way of the lymph and blood vessels of its vertebrate host this parasite reaches the usual biological "filter," i.e., the phagocyte-lined sinuses of bone marrow, spleen, and liver. There the tissue phagocytes ingest the foreign organisms, but the latter, instead of dying, multiply rapidly within the host cell and spread to others until nearly the entire reticulo-endothelial (macrophage) system may be severely damaged. Not only does this parasite fail to elicit a defensive immune response but also it so injures the antibody-producing system of the host that the host becomes easy prey to various other pathogens. Another parasite may benefit from a curious immune response. The "CHR" (Cercarienhüllen Reaktion) phenomenon in schistosomes seems to be a protective reaction on the part of the parasite in the presence of the host antibody. When a cercaria of one of the schistosomes is placed in the serum of an animal which has harbored blood flukes, the cercaria forms a delicate membrane or "hull" around itself; the hull prevents other substances (a cercaricidal antibody, for instance) from reaching the parasites. This reaction thus seems to be a protective adaptation, and the host antibody which elicits the reaction is an aid, not a hindrance, to the parasite. The actual working of this reaction *in vivo* (that is, as the cercaria penetrates an immune host) has not yet been described; it would be difficult indeed to observe. But it is not surprising that some parasites, such as *Leishmania donovani* and certain schistosomes, may have responded through evolution to the immune responses of their hosts. The relationship, we are reminded, is dynamic.

Finally, in this group of examples of the immune response to animal parasites we should consider those responses which are neither protective nor harmful but are merely detectable. Antibodies of many kinds are produced in response to many kinds of antigens, that is, foreign substances in the body. The examples given so far do not include the reactions which are merely recognizable, although these reactions are quite useful and informative. For example, serological tests can be used to diagnose various diseases caused by parasites. Such tests may involve the mixing of suspected antibody (from the serum of the patient under diagnostic observation) with known antigen (extracted or excreted fractions of the suspected parasite) in order to see whether some visible reaction such as precipitation might occur. Some of these tests are very delicate, specific, and reliable. Other uses of serology include identification of particular strains of parasite. Various intraspecies strains of *Leishmania*, both *donovani* and *tropica*, have been identified; in fact, were it not for these differences, as well as the differences in pathogenicity, host preferences, etc., mentioned in Chapter 4, all the members of the genus *Leishmania* would be indistinguishable. Thus in parasitology antigen-antibody reactions may be used to differentiate, define, and describe organisms.

IMMUNOLOGICAL THEORY

The above examples of immunity to animal parasites are far from self-explanatory. Together with the very numerous studies of immune responses to other agents, the examples present a picture which is more puzzling than informative. A broad explanation for these puzzling facts, i.e. the evolutionary explanation, has already been suggested. Beyond question both the immune responses and the counteradaptations of parasites to these host reactions are the results of mutation and selection and are examples of the dynamic balance which exists throughout the living world. But this explanation is not helpful, of course, in particular cases. It is such a general theory that it cannot be used for experiment except in rather simple situations like the familiar studies of populations and genes in *Drosophila*. What is required to explain immunological phenomena is an immunological theory. Naturally, workers in the field of immunology have tried to construct such a theory. In the following pages of this chapter, the outline of this theory will be sketched. Hopefully, parasitological studies may help modify and extend ideas which have come mainly from biochemists, microbiologists, and cell physiologists.

Over the past fifty years, certain important elements of a theory of immunity have been developed. These elements include knowledge of the substance from which antibodies form, the cells and tissues instrumental in the production of antibodies, and the patterns in time and titer (level) of the rise and fall of antibodies. In addition, something is known of the chemistry of the interactions between antigens, antibody-producing cells, and antibodies. Research on all these matters continues.

The material basis for the immune response is gamma globulin, a protein soluble in saline but not in water; it forms a substantial part of the serum fraction of normal blood. Several types of gamma globulin have been distinguished by the newer separation techniques (electrophoresis, diffusion, chromatography), but the essential material is a protein which presumably can be altered in specific ways in the synthesis of antibody. The site of origin of both gamma globulin and antibody is now known to be the lymphoid cells, called plasma cells by the histologists. Lymphoid cells occur in the spleen and liver, in the lymph nodes and nodules throughout the body, and as a normal part of the circulatory system. Because these cells are part of the network (reticulum) which filters the lymph and because they are associated with the lining (endothelium) of various blood sinuses (as in the liver, for instance), the system of phagocytes and lymphoid cells is called the reticuloendothelial system. (This is the system mentioned in the discussion of visceral leishmaniasis above and is also the system which removes malaria organisms from the blood of convalescent patients.) Part of the system is a population of plasma cells recognizable by the stronger staining reaction of their cytoplasm as compared with lymphocytes; this reaction is due to the active state of the cytoplasmic organelles of these plasma cells, i.e., the active state of globulin and antibody synthesis. Moreover, the plasma cells do not all respond alike or at the same time to the stimulus of infection. Small groups of cells become active when the body is injected with a particular foreign protein (antigen); the responding cells multiply until enough antibody is produced to destroy or otherwise remove the antigen. A theory of immunity must allow for the above facts plus a generally available substance, gamma globulin, and the specifically reactive subpopulations of lymphoid cells.

The pattern of the immune response is also well established. When a foreign protein (few nonproteinaceous antigens are known) has entered the body, three things occur in sequence. These happenings are detectable by measuring the amounts of antibody in the serum; whether or not immunity is effective (that is, protective) is immaterial, for the rise of antibody is the operational definition of response.

First, there is a period of latency during which no rise in antibody can be detected. Regarding latency there is some question. Is this delay between the entry of an infectious agent and the production of antibody real, or are the earliest amounts of antibody simply too small to be measured? That the latter is probably true was recently shown by some ingenious experiments in which different amounts of antigen were injected into similar laboratory animals. Latency varied directly with the amount of antigen injected; it was greatest (longest delay) for the larger amounts and smallest (shortest delay) for the smaller amounts of antigen. This experiment strongly suggests that the antigen combines with the first antibody produced in response to it, and thus antibody cannot be observed until there is an excess. The time required to produce an excess of antibody determines the duration of latency.

Second, there is a period of effective antibody synthesis. This period may be long-lasting—for years or even for life. It may also be as short as a few weeks. The latter situation is probably what is seen in several parasitic diseases, such as malaria. Antibody level is maintained only by repeated reinfection similar to that which occurs in regions of "stable malaria." It is very likely that the concept of premunition mentioned above is a special case of short-lived antibody synthesis.

Third, there is an establishment of immunological memory or anamnesis. This means that the responding organism can react promptly and effectively to a second inoculation with the antigen which had produced an earlier response. The host "remembers" its invader. This pattern is of great practical importance. Anamnesis

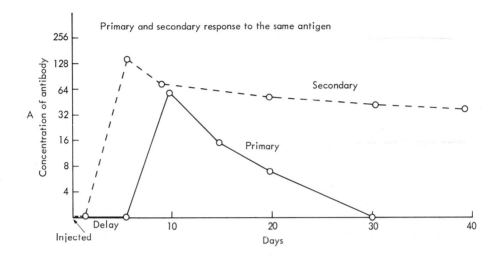

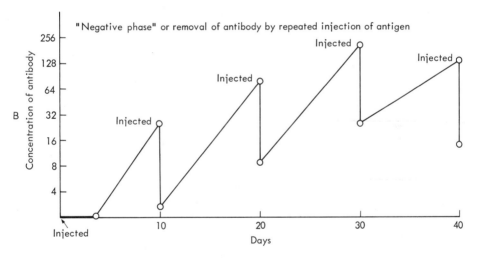

FIG. 27–4. Basic patterns of the immune response. In both figures, "titer" (level or concentration of antibody) is plotted against time after injections of antigen into a suitable animal. (A) The primary response involves a substantial delay in the production of antibody and a relatively rapid disappearance of antibody after it is formed. The secondary response is both prompter and more lasting than the primary, a phenomenon called "anamnesis" (nonforgetting). (B) When antigen is injected in sufficient quantities after an immune response has begun, the rise in antibody titer is abruptly stopped, and the amount of antibody actually falls sharply. It is believed that injected antigen removes antibody from the circulation by combining with it. A parasite may, by producing antigen, cause such a negative phase and induce a state of tolerance or chronicity. [After Raffel, 1961.]

explains why immunity can be protective even when antibodies against a particular disease are present in only trace amounts. Reinfection stimulates a production of antibodies which is much prompter and higher than the production induced by the first infection. Also, the anamnestic response is usually broader in scope; the twice-elicited antibodies are apt to be more effective against both the inducing antigen and related antigens. The familiar protection against *Variola* (small-pox) by inoculation with *Vaccinia* (cow-pox) virus is an example of a broadened anamnestic response.

In summary, the immune response goes through (1) a period of latency when antibody is very low or absent; (2) an initial response, with a relatively slow rise in antibody, destruction or other disposition of the eliciting antigen, and a rate of disappearance of antibody which varies with the particular antigen-antibody reaction involved; and (3) an indefinite period during which an anamnestic response—prompt and effective synthesis of antibody in the presence of reinfection—can occur. (Figure 27-4 represents these patterns as curves on a graph.) It should be emphasized that each phase of the above pattern depends on the particular antigen and host or responding organism involved; an outstanding characteristic of immunity is its heterogeneity.

If the above descriptive outline of the immune response can be called a theory, it is only in an operational sense. Research over the past decade or so has been aimed at discovering the mechanisms which underly the response. This research is biochemical and cytological in nature, for the immune response is essentially a molecular phenomenon at the cellular level. Some important facts have been revealed. The origin of the subpopulations of lymphoid cells is now known to be in the thymus, a once mysterious organ which is prominent in the growing mammal but retrogresses or atrophies in the adult. The thymus has been shown to be a center of origin of cells which come to lie in many other organs—the lymphatic tissues—of the body. Within these tissues the cells form colonies. It is believed that these colonies from the thymus are the primary centers of immune response. These and related cells, including the resident phagocytes (fixed macrophages) of the reticuloendothelial system, respond specifically to antigens. The response is believed to be a combination of ingestion (engulfment of foreign molecules), processing (some kind of molecular alteration of these molecules or perhaps their inclusion in a model for synthesis), and synthesis (the production of many new molecules related to the antigen in structure). (See Fig. 27-5.) This group of events may set in motion a chain reaction, which will stimulate cells of the responsive type to divide and will thus increase geometrically the amount of specific antibody. Studies of the molecular properties of antigens and antibodies have also been carried out. Light has been shed upon the structural and reactive features of these substances.

Eventually the immune response (heterogeneous, complex, and deeply hidden, as it now seems, below the resolving power of microscopes and beyond current biochemical and cytological knowledge) will be fully described. When such a description is made, we shall have a comprehensive theory and, in addition, a means of controlling, augmenting, and even creating particular defenses against

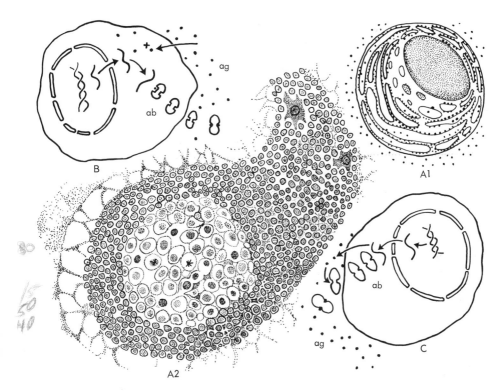

FIG. 27–5. Theory of cellular response to antigenic stimuli. Cells (A1) of lymphoid tissue (A2) are the sites of antibody production. How this occurs is the subject of two theories. The instruction theory (B) states that antigen, ag, enters the cell, where RNA molecules manufactured at a DNA chromosomal template combine with it to make a new molecule, the specific antibody, ab, which now is capable of reacting with antigen outside the cell. The clonal selection theory (C) states that the presence of antigen at the cell membrane stimulates certain cells (a selected clone or group) to manufacture specific antibody and to grow and multiply. At the same time the amounts of useful antibody and the number of cells capable of producing it are increased. Neither theory can be said to be either complete or exclusive; the role of intermediates such as free phagocytes and mast cells is proving to be quite important in the immune response.

pathogens. Some optimism regarding the realization of these specific possibilities comes from the history of genetics. This recently developed science, which began with the amazing and paradoxical data from plant and animal breeding, reached a set of operational definitions (Mendel's laws) before the discovery of chromosomes, developed rigor and precision with the mapping of genes in *Drosophila*, and eventually unfolded the helix of the DNA molecule to reveal the genetic code, upon which the very structure of heredity is based.

SPECIAL ASPECTS OF IMMUNITY

Before concluding this chapter, there are some aspects of immunology which seem
particularly applicable to host-parasite relationships. These are such nonprotective
responses as hypersensitivity, suppression of immunity, and homeostatic immunity.
Hypersensitivity is definable in operational terms as a condition in which the
immune response causes harmful symptoms. These symptoms are familiar under
another term—allergy. They may include local inflammation with swelling, itching,
vesiculation, or even necrosis. They also may be general and may involve severe
"shock" reactions, such as interference with breathing and circulation. In most
cases, hypersensitivity is due to the presence of a foreign substance (antigen),
usually protein, in the tissues of the host. In combination with antibody, this antigen causes the release of one or several kinds of substances which damage or
destroy the normal cells of the host. One such harmful substance has been identified
as histamine; this chemical is normally present in certain cells of the intestinal
epithelium, in the lymphoid cells including leucocytes, and, in high concentration,
in the granules of mast cells, characteristic cells of the loose connective tissues.
When histamine is artificially injected or is released by mechanical injury, the
surrounding tissues are damaged in the same way as by hypersensitivity. Also,
antihistamine, a neutralizing substance, can be used to treat allergy and hypersensitivity. Certain types of the latter, however, do not respond well to antihistamine.
This fact supports the belief that more than one kind of harmful substance is released by the antigen-antibody reaction. The details of hypersensitivity effects are
well known. One type of reaction, the so-called Arthus effect, is readily observed
in the walls of small arteries; the endothelium appears "sticky," the walls contract,
and blood cells pile up to block the vessels. This reaction is due to precipitation
of substances from the serum. Its results are, of course, severe damage to the
tissues whose blood-supply has been affected. Histamine and histaminelike substances are known to cause spasm in smooth muscle; this effect explains both
asthma (the restriction of breathing by spastic contraction of the bronchial rings)
and changes in blood pressure which may occur with hypersensitivity reactions.
Other more subtle effects have been observed. One of these, common in so-called
"delayed hypersensitivity," is death and lysis of certain cells, with breakdown of
collagen (connective tissue substance). A secondary effect of blood-vessel damage
is edema, swelling of the tissues with fluid released from the injured vessels. The
many aspects of allergy—involvement of skin, blood system, lungs, etc.—are traceable to the operation of the above pathological changes in different parts of the
body. A still more serious condition, anaphylactic shock, can occur when large
amounts of antigen-antibody are present; this condition is often fatal. Accompanying hypersensitivity is a rise in the number of eosinophile leucocytes. No explanation of this seems to be available at present. The fact that eosinophilia is also a
very common accompaniment of helminth parasitism suggests, of course, that such
infections may actually produce hypersensitive states. As we have seen elsewhere,
several parasites (especially the nematodes *Onchocerca* and *Trichinella,* the various

blood flukes causing swimmers' itch, and the itch-mite, *Sarcoptes*) do induce hypersensitivity, which must be taken into account during treatment. It is obvious that hypersensitivity, a harmful aspect of immunity, has particular importance in parasitic disease.

Suppression of immunity, sometimes called a "negative phase" in the immune response, occurs when antibody is present in the system in such large amounts that it combines with all available antigen. The result is actually an inhibition of antibody production, because the latter depends on the stimulus of free antigen. Negative phase response (or lack of response) is characteristic of certain chronic infections. In fact, it may explain why such infections persist; a sort of balance is reached between antigen (released by the parasite) and antibody (manufactured by the host) so that the combining of these substances removes both from active participation in any new reaction. Most parasitisms (see Chapter 26) are long-lasting chronic conditions. It seems reasonable to assume that part of the evolutionary adjustment of host to parasite, and *vice versa*, has been accomplished by the development of just such an antigen-antibody balance as the negative phase reaction.

Perhaps the above balance should be considered as a homeostatic mechanism. Homeostasis is the sum of all factors which contribute to the maintenance of living things in a "steady state." Well-known examples from physiology are the feedback mechanisms, such as the thyroid-pituitary interaction. Thyroxin, a hormone which stimulates increased rates of metabolism in the body, affects all organs in which blood circulates. Thus thyroxin stimulates metabolic activity in the pituitary body or hypophysis. This endocrine gland responds by producing a thyrotrophic hormone; this hormone is released into the blood stream, reaches the thyroid gland, and inhibits the production of thyroxine. Thus thyroxin, by stimulating its own inhibition, is self-regulating. Such a mechanism, whether in a living organism, an ecological community, or a man-made activity such as heating or manufacturing, is called "negative feedback," because part of the energy produced by the system is "fed back" into the system to restrict production. Antigen-antibody relationships are known to involve negative feedback (especially in the two aspects of immunity just discussed—hyperimmunization and negative phase response). Thus in cases of chronic disease, such as tuberculosis, syphilis, and helminth infections, one would expect antibody production to continue to an unlimited extent until the host becomes dangerously hyperimmune. Yet such is seldom the case. Some factor holds antibody levels down. Probably this factor is the antibodies themselves, which, as was just explained, combine with new antigen as it is formed to "bind" it (in the immunologist's term) so that it is not antigenic and cannot elicit new and dangerous amounts of antibody. This is an example of regulatory feedback or a homeostatic mechanism. What is interesting about it is the fact that two organisms are involved. The parasite produces antigens (excretions, secretions, exuvia, etc.) which cause the production of antibody or a response on the part of the host. Thus the parasite in a sense limits its own activity (growth, invasion, fertility, etc.) by the negative feedback of its own products. Likewise, the host, producing antibodies which reduce or neutralize the antigenic properties of the parasite, limits its own response also by negative feedback. The net result of these two regulatory feedbacks is a mutual homeostasis, i.e., the steady state of tolerance and low virulence which characterizes most well-established host-parasite relationships.

SUGGESTED READINGS

ALLISON, A. C., "Protection Afforded by Sickle-Cell Trait against Subtertian Malarial Infection," *British Med. J.*, 290–294, Feb. 6, 1954.

COHEN, S., and I. A. McGREGOR, "Symposium on Immunity to Protozoa," British Soc. Immunology, London, 1962.

MASON-BAHR, P. E. C., "Immunity in Kala-azar," *Trans. R. Soc. Trop. Med. Hyg.*, 55:550–555, 1961.

NOSSAL, G. J. V., "How Cells Make Antibodies," *Sci. American*, 211(6):106–115, 1964.

RAFFEL, S., *Immunity*, 2nd ed. Appleton-Century-Crofts, New York, 1961, pp. 1–240.

SEN GUPTA, P. S. and S. L. ADHIKARI, "Observation on the Complement Fixation Test for Kala-azar," *J. Indian Med. Assn.*, 23:89–93, 1952.

SPIERS, R. S., "How Cells Attack Antigens," *Sci. American*, 210(2):58–64, 1964.

TALIAFERRO, W. H., "Ablastic and Trypanocidal Antibodies against *Trypanosoma duttoni*," *J. Immunol.*, 35:303–328, 1938.

UHR, J. W., "The Heterogeneity of the Immune Response," *Science*, 144:457–464, 1964.

VOLLER, A., "Fluorescent Antibody Studies on Malaria Parasites," *Bull. WHO*, 27:283–287, 1962.

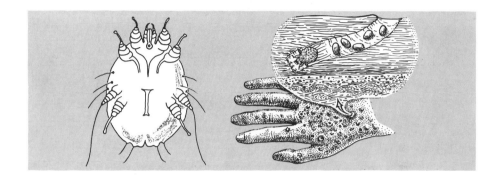

CHAPTER 28

PATHOLOGY

INTRODUCTION

Continuing the discussion of parasitism begun in Chapter 26 and pursued in Chapter 27, we may at last consider the host, from our own natural point of view, as ourselves, victims of insidious and persistent attack by foreign organisms. We need not try to admire the ingenuity by which parasites compensate for their loss of independence through all sorts of evolutionary tricks of life cycle and super-fecundity, nor must we struggle with the difficult concepts of host response, which through the immune mechanism strikes a bargain with many parasites, a "live-and-let-live" arrangement for slowing the activity and keeping down the numbers of an invading species while allowing the parasite some shelter, some nourishment, and some means of propagating its kind. Now we can frankly view the parasite as our enemy, and we can consider the many ways in which he attacks us and our domestic animals and does us harm.

INJURY BY ENTRY

Some parasites harm us by the very act of entering our bodies. *Entamoeba histolytica* should not be considered as being truly within the body unless it is more than a mere inhabitant of the lumen of the bowel. When it invades the mucosa,

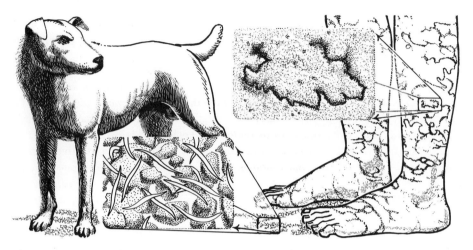

FIG. 28–1.   Hookworm-induced creeping eruption. A pattern of sinuous and anastomosing lines on the feet and legs can result from invasion of the skin and migration by such hookworms as *Ancylostoma braziliense* and *A. caninum*, two common parasites of dogs. The upper insert shows the pathway of "cutaneous larva migrans" under magnification; the lower shows filariform larvae in coarse, moist soil. [After Faust and Russell, 1964.]

whether by its own tissue-destroying (histolytic) enzymes or by the aid of bacterial lesions (see Chapter 6), it begins its destruction of bowel tissue. It can be said that intestinal amoebiasis is the result of active invasion of host tissue. No other protozoa are believed to act so aggressively, but many helminths do. As we have seen, the schistosomes penetrate human skin as cercariae. Usually they cause a transitory dermatitis, but sometimes they produce a severe and long-lasting rash (swimmers' itch) if the victim has been sensitized by prior exposure to species of blood fluke normally parasitic in other vertebrates. The liver flukes, *Fasciola*, *Clonorchis*, and *Dicrocoelium*, cause some injury by the very acts of leaving the intestine for their sites in the bile ducts. The young *Fasciola* moves through the intestinal wall, penetrates the surface of the liver, and wanders about in that organ; while the other two move into the small bile ducts from the common duct. From the beginning of their entry, all of these worms cause irritation. This damage, compared with subsequent harm done by the adult flukes, is probably slight, yet in a list of pathological invasions, the liver flukes should be mentioned. Two nematodes, the hookworms (as we have seen in Chapter 15) and *Strongyloides* (see Chapters 20 and 26), enter the human body through the intact skin. Both of these actively penetrate the feet to cause local lesions and itching. The dog hookworms, *Ancylostoma caninum* and *A. braziliense*, cause a severe rash or creeping eruption in man (Fig. 28–1), since their larvae cannot get below the skin in the human (abnormal) host. Other nematodes, notoriously *Ascaris* and the strongyles and trichostrongyles, attack their hosts from within; the larvae, upon hatching or being freed of larval exuvia in the host intestine, immediately bore through the intestinal mucosa to begin various migrations within the host. The extent of larval damage in these cases depends upon the number of larvae invading at a given time as well as the immune (sensitive) condition of the host. As seen in Chapter 11, some tapeworms enter the tissues of man, hogs, or cattle in the oncosphere (six-hooked embryo) stage, but they probably have negligible effects during their brief journeys to regions of encystment. In general, the same is true of most actively invading parasites; entry is made by a relatively small stage in the life cycle, and the lesion is minor compared with later pathological effects. Parasites that enter the body through lesions produced by vectors are even less harmful than the above in the act of entering. *Plasmodium,* the hemoflagellates, and the filarial worms do not appear to injure their hosts while creeping or being injected into a lesion. It is the migration, growth, and reproduction of these forms which cause harm.

INJURY BY MIGRATION

After invasion comes migration usually to some site of development. Parasitic migration by larvae is serious in the following cases. In onchocerciasis (Chapter 14) the microfilariae which are produced in subcutaneous nodules move in vast numbers throughout the dermal layer of the skin; they even invade the orbit and the optic nerve. They cause increasingly severe allergic reactions; the latter are the chief source of pathological changes in the victim. The skin may become edematous, red, and very sensitive. In the African form of onchocerciasis, thickening and scaly texture of the skin develop. Other nematode larvae are also harmful. The pathway of hookworms, *Strongyloides,* and *Ascaris* through the body leads to the lungs; large numbers of larvae of any of these cause local lesions in the alveoli and irritation of the pleura. Symptoms of pneumonitis are observed, and sometimes such pulmonary involvement is quite dangerous. However, the lesions heal, and the symptoms disappear with the completion of tissue migration. Unable to find the lungs of a human host, larvae of the ascarids of carnivores migrate actively among the organs of the body (hence the term "visceral larva migrans"), where they destroy and injure tissue in their path. Hepatic symptoms in children have been attributed to the larvae of dog or cat ascarids acquired by ingestion of eggs from dirty yards or sandpiles. There is evidence from occasional reports of unidentified nematode parasites in the tissues of man that perhaps a number of nematodes are potential invaders and wanderers in the human host. Among flatworm larvae, those of several trematodes and cestodes are significant. *Clonorchis* and *Opisthorchis,* the liver flukes of man, probably cause little damage as they move from the intestine to the small bile ducts by way of the common bile duct, but once in the liver the young flukes actively feed upon host tissue and fluid to cause the inflammatory reaction and fibrosis which eventually lead to liver malfunction and the progressive symptoms of clonorchiasis. However, the sheep liver fluke, *Fasciola*, moves through tissue—intestinal wall, peritoneum, liver parenchyma—to reach its destination, and it causes injury along the way. *Paragonimus* (see Chapter 10) also migrates dangerously; it is especially dangerous when the young flukes lose their way to the lungs. Amphistomes, reaching their sites in the abomasum of ruminants, also cause damage as migrating larvae. Cestodes which migrate in the human host are the pork tapeworm and *Echinococcus*. Although the results of migration (i.e., establishment of cysts in various parts of the body) may be quite damaging, the cestode embryos in their migratory phase are too small and usually too few to cause discernible damage or reaction. Occasionally cestode larvae do have very serious effects on man. The form of *Echinococcus* cyst, called multilocular or budding (probably the larva of a particular species restricted to sub-Arctic regions), produces daughter cysts which are carried to many parts of the body; the cysts eventually overwhelm the victim with many parasites. Another cestode larva, the young form of various species of *Spirometra* (a cestode of fish-eating carnivores), may occasionally enter human tissue, where it moves about rather freely for a long time. One kind of sparganum (the name given to such spirometrid larvae in human or other mammalian hosts) proliferates in the same manner as the budding cyst of *Echinococcus*, and of course the danger is equally as great. No treatment is known for such proliferating cysts or larvae, for surgery is ineffective once proliferation and metastasis (scattering of larval material) has started. Larval activity is probably most harmful in the case of the myiases (see Chapter 24). Maggots of the screwworm fly enter open wounds; they excavate and destroy healthy tissue, and actually kill many domestic animals. Human infections are rare, but not unheard of. Other maggots—the bots of cattle, sheep, and horses —migrate in various ways through their hosts, as they feed upon tissue exudates and cause irritation and physical destruction of material in their path. Although most parasitic worms remain in one organ or tissue after completing their larval migration, a few organisms migrate as adults to cause typical lesions and symptoms. *Ascaris*, the large intestinal roundworm (see Chapter 16), is powerful enough to penetrate tissues, including the intestinal wall. Perhaps impelled by some irritant in the intestine, *Ascaris* adults sometimes leave the intestine either directly or via the bile duct; such wandering by a worm 30 cm or more in length can be very harmful to the health of a host. A final example of wandering or migration is the behavior of the mange mite, *Sarcoptes scabiei*. This pest burrows in the germinative layer of the epidermis and forms tunnels in which it oviposits and defecates. In sensitive hosts, the result is scabies, an itching, burning syndrome, which causes local inflammation and edema as well as secondary tissue damage and infection through scratching. Other kinds of mites (see Chapter 22) affect various domestic animals in similar ways. The above instances of larval and adult migration within a host show that the movements of parasites are pathogenic whether the parasites are the tiny larvae of hookworm, the much larger larvae of flies, or various adult organisms.

INJURY BY RESIDENCE

Residence and growth of parasites can also cause pathological effects either by mechanical injury—pressure, erosion, abrasion—or by the actual displacement and destruction of tissue.

Mechanical injury is evident in a number of helminth infestations. (See Figs. 28–2 and 28–3.) The liver flukes, *Clonorchis* and *Opisthorchis*, are relatively large animals. They live tightly packed within the middle-sized bile ducts; they stretch and abrade the epithelium of these tubes, actually feed on tissue fragments, and interfere with the normal functions of the whole liver. While the eventual injury from clonorchiasis is fibrosis due to host reactions, the initial damage is largely mechanical. Likewise, schistosomes cause mechanical injury; however, the approach is different, for the eggs laid in the small veins of the intestine or bladder block these vessels to cause local stasis of blood flow and the death by starvation and oxygen deficiency of certain cells. Of course, the damage is not merely mechanical, since the miracidia of blood flukes secrete histolytic enzymes which further injure the tissues in egg-infested regions of the body. Acanthocephala (Chapter 12) are

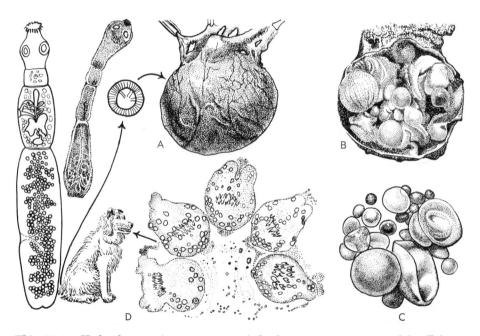

FIG. 28–2. Hydatid cyst. An entire cyst of the human omentum caused by *Echinococcus granulosus* is shown (A) as it was removed by the surgeon and then opened (B) to show daughter cysts within the main cavity. (C) Brood capsules and daughter cysts are shown enlarged, and (D) scoleces from a brood capsule are illustrated. (The latter are similar to those diagrammed in Fig. 11–2.) [After Bohrod in *Medical Radiography and Photography*, 1951.]

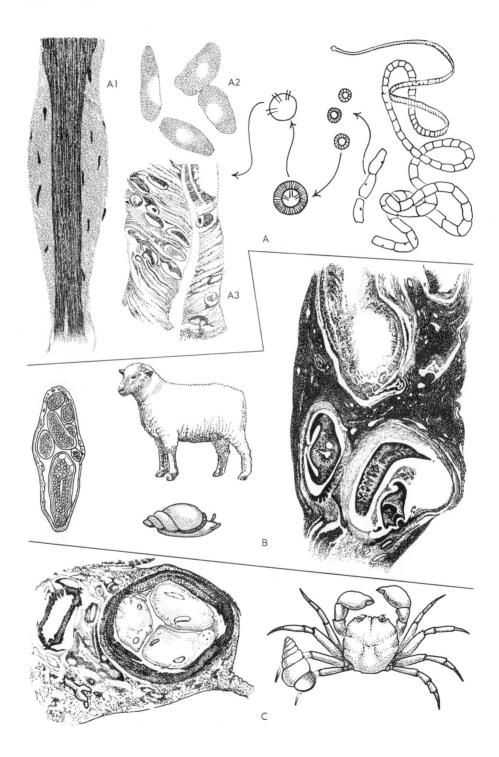

A1

A2

A3

A

B

C

FIG. 28–3.   Some parasites *in situ*, causing lesions in their hosts. (A) Cysticerci of *Taenia solium* as x-ray shadows in human thigh (A1) and seen enlarged as whole objects (A2) removed from muscle tissue, and as cysts (A3) in a portion of a muscle. (B) Section of liver of sheep showing parts of two or more specimens of *Fasciola hepatica* near a grossly enlarged bile duct. (C) Section of a cyst containing lung flukes, *Paragonimus westermani*. The flukes are enclosed in a fibrous capsule deposited by the host. [After Gradwohl and Kouri, 1948, and Chandler and Read, 1961.]

very well equipped for mechanical injury; the armed proboscis of *Prosthenorchis*, for example, frequently perforates the delicate intestinal wall of small monkeys or marmosets and causes peritonitis and death to these interesting and scientifically valuable animals. Although human infestation with acanthocephala is rare because probably people do not often eat the arthropod intermediate hosts of thorny-headed worms, such parasites would be quite dangerous. Similarly, certain cestodes have a spiny rostellum capable of digging into the mucosa; a fowl tapeworm, *Raillietina echinobothrida*, sometimes injures infected birds fatally. The mechanical damage done by adult nematodes is widely evident. Among the strongyles and trichostrongyles of ruminants are nodular worms in sheep. These small strongyles cover the lining of the stomach with purulent lumps in which they live. They feed on blood, cause malfunction of the organs infested, and seriously weaken the host animal. Certain trichostrongyles can form a furry lining in the host intestine, as each one of the thousands of slender worms cuts into the tissues and drinks blood. The human hookworms are similar in their feeding habits. Perhaps the most spectacular of mechanical attacks is the well-known destruction caused by the huge nematode *Dioctophyma* (10 mm thick by 100 cm long) which literally devours one kidney of its carnivore host and leaves a mere shell in place of the devastated organ. Less spectacular but more important mechanical damage is caused by the bots which ruin the hides of cattle by living in dermal cavities on the backs and sides of infested animals; they rasp and excavate these holes as they grow, until they drop out and pupate. *Dermatobia*, a bot which attacks man, is a painful and horrifying parasite; its ingenuity in utilizing a large mosquito to transport its eggs nevertheless (see Chapter 26) excites one's admiration. Nearly all parasites cause some mechanical harm to their host; the above examples are illustrations of the general fact.

CHEMICAL AND PHYSIOLOGICAL INJURY

Somewhat less obvious—indeed, sometimes obscure and mysterious—is the damage caused by the chemical and physiological actions of parasites. The growth of *Plasmodium* in the erythrocytes obviously destroys individual cells, but not so obvious is the manner in which subjective chills and measurable fevers are elicited. Loss of blood cells causes the weakness, emaciation, and anemia due to malaria; but the dramatic paroxysms of the disease are due to causes other than mere erythrocyte destruction. Likewise, it is obvious that amebic ulcers can explain some of the discomfort associated with chronic amebiasis, and that by invading and disrupting liver or lung, amebic abscesses can cause symptoms typical of the location of the abscess. However, the vague symptoms often referred to amebiasis are due to very complicated causes. The fact that diet can be used to control amebiasis shows that the effect of the parasites may be largely physiological. As was stated earlier, schistosomes cause several kinds of damage in the tissues where the eggs are deposited. The early stage of mechanical blockage of blood vessels and enzymatic softening of tissue permits the passage of eggs into the lumen of the bowel or bladder. This mechanical and chemical attack is followed by a reactive state in which damaged tissue is replaced with fibrous scar material. Hence the structure (and function) of the organ involved is drastically altered. Thus the primary attack by the parasite again sets in motion a chain of complex events; some are due to the parasite itself, while others are due to peculiar host reactions. Lung flukes (see Chapter 10) become established in pairs in cystlike recesses of the lungs of crustacean-eating mammals, including man. The flukes irritate the lung tissues and cause symptoms which mimic those of pneumonia and tuberculosis. Probably the irritation is due to metabolic wastes which pass with the eggs of these flukes out of the cysts into the air chambers and ducts of the lungs. As suggested above, liver flukes cause so much tissue destruction that liver parenchyma is replaced by fibrous tissue, which in cases of very heavy infestation may block hepatic function completely and result in death. The cause of death here may be not only the worms but also the host's reactive processes of reconstruction, isolation, and repair. (In this entire discussion of attack and harm by parasites, the reader should keep in mind that these concepts are oversimplifications and that host-parasite inter-actions occur even in the most obvious examples of parasitic aggression.) The behavior of *Leishmania* in host tissue is not simple aggression, probably because host phagocytes are important in the pathogenic process. These defensive cells engulf the *Leishmania* bodies, but instead of destroying the bodies, the cells permit them to multiply. As we have seen in Chapter 4, the pervasive sickness called kala-azar affects the reticuloendothelial system of macrophages. Possibly, without phagocytosis, *Leishmania donovani* could not become established; but, of course, phagocytosis is essential to host defense against other pathogens. Dermal and mucocutaneous leishmaniases are similar in pathology to kala-azar. The difference is that, in the former diseases, monocytes of the skin or nasopharyngeal tissues are affected, while the deeper parts of the system of phagocytes are not parasitized. Perhaps a true contest between invasive forces (the parasites) and defensive cells (the phagocytes) can be seen in oriental sores. The phagocytes, while harboring the parasites and for a while nourishing them, first limit and then reduce the area of invasion until at last the sore disappears and permanent, probably serological, defenses are established. However, in Chagas' disease, another of the hemoflagellate diseases, phagocytosis is not important, and the parasites actively invade such cells as the heart myocytes. The ability of *Schizotrypanum cruzi* to multiply in such tissues in intracellular habitats greatly limits the effectiveness of host defenses and may explain why Chagas' disease has such a long, progressive, and generally unfavorable course. Only in its early stages, when blood forms have been introduced with the feces of the vector bug, does a victim of Chagas' disease exhibit symptoms of immune response (edema, fever). Later, when the heart or other organ shelters the growing leishmania forms, the disease, depending on the tissues affected, resembles various bodily malfunctions such as chronic heart failure. The physical presence of the parasite causes such symptoms through destruction of host cells. Chagas' disease is thus an example of the direct effect of parasitic invasion upon the host's health.

Indirect effects of parasites upon health also occur. Veterinarians recognize as an important effect of parasites the "unthriftiness" of affected farm animals. A similar effect is observed in human victims of parasitism. We call this condition "poor health," and we recognize its manifestations in reduced resistance to disease, lowered levels of energy available for work, and changed attitudes, such as apathy, listlessness, irresponsibility, irritability, etc. None of these vague manifestations can be called a symptom, and in the above general sense it is impossible to point to pathological changes as the direct cause of "poor health." Yet attempts may be made to relate "unthriftiness" of people to a few specific kinds of parasitism. *Ascaris*, for example, while causing certain specific pathological damage during its invasive migration and its occasional wandering, is usually considered as a mere commensal in its ordinary life in the human intestine. It is known, however, that swine heavily infested with *Ascaris* are able to utilize only 90% of their food; this is a serious economic factor in the production of pork. By analogy it must be true that human hosts of *Ascaris* fail to utilize all their food, for they share with their "guests" an amount proportional to the worm burden. Studies in Japan and elsewhere have confirmed this fact. In the many parts of the world where nutrition is insufficient in quality and quantity for normal human growth and development, the share of the food each person loses to hundreds of large and voracious ascarid worms might well be critical and may contribute substantially to that person's nutritional deficiency. Thus, the indirect harm caused by *Ascaris* is probably much greater than the worm's direct injury to the tissues. Similarly, indirect effects of other parasites may be recognized. The blood is affected by three distinct mechanisms in the unrelated diseases caused by *Plasmodium*, hookworms, and *Dibothriocephalus latus* (the fish tapeworm of man). The first organism destroys a large number of red blood cells and causes a malarious anemia characterized largely by iron deficiency. The second, by biting the intestinal lining and drinking blood, causes a hookworm anemia characterized by loss of both blood cells and plasma. The results are deficiencies of both iron and protein and very serious effects on the oxygen supply to the tissues and on immunity mechanisms. The third, a very large cestode, extracts from the intestinal lining of its host and presumably from the usual nutritional sources vitamin B_{12}, which is essential to the manufacture of hemoglobin. In persons unusually sensitive to vitamin B_{12} deficiency, pernicious anemia develops. Although this development is rare (observed in about one out of one thousand *Dibothriocephalus* infections), the incidence is much higher than the incidence of pernicious anemia in the general population, and it must be considered an important effect of the fish tapeworm. Moreover, there is probably a subclinical vitamin B_{12} deficiency in many, if not all, victims of this cestode. *Trichuris trichiura*, the whipworm (see Chapter 16), is a very important parasite of children in tropical and warm temperate regions where epidemiological conditions permit heavy infestations to occur. By irritating the lining of the colon and actually feeding on blood, this worm causes abdominal pain, vomiting, and diarrhea. Emaciation, stunting of growth, and anemia resembling hookworm anemia may occur. This combination of effects is due to a complex of causes, only some of which are direct pathological changes. A final example of indirect effect is the familiar kala-azar, or visceral leishmaniasis, the disease in which the cellular system of immune defenses (the monocyte and macrophage system) is the habitat of the parasites. This disease does not directly cause death, because the affected tissues are not essential to ordinary metabolism; the cause of death, which may occur months or years after infection, is any of a number of common diseases against which the victim of kala-azar is powerless to develop antibodies. The above set of examples of parasitisms which have serious effects in addition to their obvious damage to specific cells, tissues, organs, or systems can be extended by the student of parasitology to include practically all parasites. It is characteristic of parasitic disease that the parasite becomes part of the host and affects in some way nearly every aspect of the host's well-being.

INJURY DUE TO HOST REACTION

The above list of harmful effects which parasites have upon their hosts emphasizes the importance of parasitic invasion, migration, residence, and growth in causing physical damage to the host's tissues and organs. Indirect or general effects were also noted. However, it is improper to attribute all damage to the parasites themselves. The following paragraph will mention some ways in which the host cooperates to its own disadvantage.

In the preceding chapter hypersensitivity or allergy was mentioned as a harmful aspect of immunity. This is only one of many injurious host reactions to parasitism. Another is the process of repairing damage caused by parasites. For example, both of the organisms which damage the wall of the intestine—the schistosomes and *Entamoeba histolytica*—cause extensive lesions which eventually heal. Yet the healing process is not a restoration of tissue as it was before the pathological change took place; the flexible, vascular, muscle-layered wall of the colon, capable of metabolic movement and absorption of water, becomes infiltrated with fibrous connective tissue; the wall is inflexible, incapable of movement, and quite useless in carrying out the functions of the colon. Permanent impairment of function results from the repair process itself. In clonorchiasis, the damage done in the liver is repaired by fibrous tissue not by restoration of host hepatic cells, and there follows a condition—fibrosis—which is pathological in its own right. Probably some permanent harm results from all tissue invasions, since in most cases the body is not able to make perfect restoration or substitution for lost parts. And in the inflammatory responses which initiate most repair processes, of course, immune reactions, even hyperimmune responses, also play a part.

REVIEW AND CONCLUSION

In conclusion, the reader should be reminded that this chapter was actually a continuation of the discussion carried on in the two preceding chapters, i.e., a discussion of host-parasite relationships. Thus in Chapter 26 the concept of parasitism was introduced. Different in only one aspect from other relationships where dependency is involved, the parasitic relationship is intimate. In other ways it may resemble predatism (the predator being dependent upon its prey), or community or social mutualism (the members of a plant-animal association being dependent in many ways upon one another), or even the "great chain of life" in which each living part may be considered as a member of the universal whole. But the intimacy of the host-parasite relationship is its distinguishing feature. Nevertheless, there are two other features which are so characteristic of the parasitic way of life that they also serve to define it. These are loss of competence (or dependency) and the compensating gain in fecundity and ingenuity (or opportunism in the utilization of infecting and transmitting mechanisms). Also implicit in parasitism is the fact that both host and parasite are undergoing organic evolution, which, here as elsewhere in the living world, maintains various kinds of dynamic balance.

As the discussion continued, the role of each member (host and parasite) in maintaining this balance was then appraised—first in terms of host responses, then in terms of parasite attack. The host response is chiefly immunological. Although certain hosts resist even the entry of parasites and others are able to live reasonably well with parasites in or upon them, neither resistance nor tolerance is so well understood as the defensive response called immunity. Therefore immunity to animal parasites was outlined briefly, and the accepted principles of immunity— sites of production of antibody, curves of antibody production and persistence, material (gamma globulin) out of which antibody is formed, types of antibody action— were mentioned. Transient immunity (probably the commonest kind in animal parasitism) was shown to be a plausible explanation of "premunition," the protection against severe infection afforded by persisting or frequently repeated infection with particular parasites. Harmful aspects of immunity (i.e., hypersensitivity at one extreme, and "negative phase immunity," or antibody binding by parasites, at the other) were shown to be significant forms of the host response to parasites. It was explained that hypersensitivity or allergy is one of the dangerous features of such chronic infections as helminthiasis, and that "negative phase" or suppressed immunity is an avoidance of the allergic state by suppression of active antibody altogether. Finally, the aggressive activities of parasites were organized into categories—invasion, migration, residence, and growth—in order to show by example the kinds of damage resulting from such activities. After such rather fragmentary approaches to the host-parasite relationship, let us now consider the relationship as a whole.

A well-known parasite, the hookworm, can be used to illustrate nearly all the above aspects of parasitism. Intimacy and dependency are self-evident. The invasive larvae live in tissue (blood and lung) before entering the intestine, where, as lumen parasites, the adults are not quite so intimately involved as in the tissue phase. All stages of the hookworm found in its host are dependent on that host for food and shelter, but the eggs are not so dependent nor are the first stage larvae. The latter are truly free-living organisms which feed and grow in the soil. On the other hand, the infective larvae, while being able to survive and being capable of responding to sensory stimuli, are unable to grow or even replenish their dwindling store of energy, for they have no functional mouth. They are potentially parasites, and they have no other future. Through these larvae, the hookworm shows its parasitic ingenuity. A simple thermotropic response, plus stylet and histolytic salivary glands, enables these larvae to become active when in contact with warm skin and to penetrate that skin. What further guiding responses these larvae have, we do not know. Perhaps, after being carried passively to the lungs, they move toward the cool alveolar spaces by a reversal of their former thermotropism; perhaps they respond to oxygen gradients; or possibly no response occurs, and they are stopped by the pulmonary capillaries or the first passageways which they would encounter as being too small to permit transit. The system of larval behavior is sufficient to account for transmission of hookworms in numbers limited only by soil conditions, temperature, and host susceptibility.

Hookworms reveal through their host specificity a parasite's evolutionary relationship with its hosts. In Africa, the probable center of origin of human hookworm, the indigenous Negro race shows marked resistance to clinical infection. This "racial immunity" is undoubtedly the product of long interaction between host and parasite; the host acquires genetic factors protecting him from heavy infection, and the parasite perhaps acquires genes predisposing it to mildness in relation to at least one kind of human host. Individually, each host of hookworm is capable of developing antibodies which inhibit the parasite in various ways. Inflammatory responses at the level of the skin may prevent further entry, or deeper tissue responses may prevent development in the lungs to the preintestinal stage. One effect of immunity is to lower the number of eggs produced daily by adult female worms; in this case a general inhibitory effect of host blood upon the parasite is evident. But these protective responses depend upon the availability of gamma globulin, the protein precursor of antibody, and this in turn depends upon dietary protein. Because the latter must be sufficient to overcome the constant drain of protein through the blood used and wasted by the hookworm, ordinary diets do not suffice to support immunity, and in impoverished communities immunity against hookworm usually fails to develop. (See Fig. 28–4.) As we have seen

(Chapter 15), hookworms, by sucking blood, prevent the development of immunity and in a sense attack the defenses of the host. More direct damage, of course, comes from invasion of tissues by larvae and from the hemorrhage caused by adults. Indirect damage includes all those effects of anemia, especially an anemia in which whole blood and not just the oxygen carrying red blood cells is lost. (Of course, no one parasite can cause all the kinds of damage listed in this chapter, but hookworms have remarkably varied effects.)

FIG. 28–4. The hookworm "facies," or expression and appearance upon which a clinical diagnosis may partially rest. Puffy, listless face, stunted body, evidences of allergy, anemia, and long-term deprivation due to hookworm. [After Faust, 1949.]

Finally, hookworms show how host populations and parasite populations interact. Thus hookworm thrives where poverty, filth, mild temperatures, and moist loose soils exist; in more fortunate regions, where education has followed economic success and public health measures are wise and effective, hookworm disappears. Yet hookworm itself is one of the factors in the persistence of poverty and misery, for the disease saps the will and vitality of whole communities; likewise, the removal of hookworm may start a community on the road to self-improvement and eventual health. Culture, the ability of man to plan and execute his own fate, has become an evolutionary factor with which the hookworm cannot cope. Social evolution thus destroys the dynamic balance which once kept man and hookworm on equal terms throughout much of the world. In short, hookworm can be seen as an illuminating example of host-parasite relationships in human welfare. In order to perceive the repeated patterns of parasitism described in this and the preceding chapters, the reader may think of other examples which he may wish to investigate on his own.

SUGGESTED READINGS

BRECKENRIDGE, R. L., "Infestation of the Skin with *Demodex folliculorum,*" *Am. J. Clin. Path.*, 23:348–352, 1953.

DALMAT, H. T., "Cutaneous Myiasis of the Scalp due to *Dermatobia hominis* (Linnaeus, Jr.) (Diptera, Cuterebridae),*" Am. J. Trop. Med. Hyg.*, 4:334–335, 1955.

VON BONSDORFF, B., "*Diphyllobothrium latum* as a Cause of Pernicious Anemia," *Parasitological Reviews, Exptl. Parasit.*, 5:207–230, 1956. '

MUELLER, J. F., E. P. HART, and W. P. WALSH, "Human Sparganosis in the United States," *J. Parasit.*, 49:249–296, 1963.

HALLBERG, C. W., "*Dioctophyma renale* (Goeze, 1782), A Study of the Migration Routes to the Kidneys of Mammals and Resultant Pathology," *Trans. Amer. Micros. Soc.*, 67:257–261, 1953.

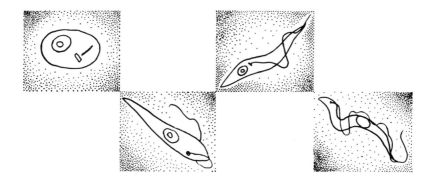

CHAPTER 29

EVOLUTION

Two ideas are quite important in understanding the evolution of parasites. The first is the concept of the host as environment. Such an environment encourages both dependency (lack of competence) and cooperation (lack of virulence) on the part of the parasite. Also, the host is itself a living organism, a member of a living population of a living species; it changes in response to its own evolutionary stimuli, which it transmits, however indirectly, to its parasites. The second idea is the concept of the adaptability of parasites, an adaptability which is probably high compared with that of free-living animals in general. A review of some of the facts found in the first part of this text will serve to illustrate these ideas and to prepare the reader for a theoretical treatment of the evolution of parasites.

PLASMODIUM FALCIPARUM AND THE SICKLING TRAIT

Plasmodium falciparum through virulence eliminates many Africans who lack the sickling hemoglobin, and it thereby preserves this semilethal gene in the populations exposed to malignant tertian malaria. This conspicuous host-parasite interaction shows one specific mechanism by which mutual tolerance between host and parasite populations is maintained. Briefly, there is a gene which, when present in both members of the pair of matching chromosomes which can carry it, causes a deficiency in the hemoglobin molecule. This deficiency is expressed in two ways:

(1) by a severe anemia which is often fatal, and (2) by a peculiar change in the red blood cells when these are placed on a slide. The cells change shape to become crescentic or "sickle-shaped." When the gene for sickling is accompanied by a normal hemoglobin determiner (another way of saying it is present in a single dose only), anemia does not develop, but the blood can still be induced to "sickle" in some degree. Therefore the gene can be recognized not only in the victims of the hereditary anemia but also in carriers of the trait. When geneticists tested large numbers of people for the presence of the sickling gene, they found that only in equatorial Africa was there a high incidence of this harmful mutation. It was discovered that this high frequency coincided with the highest infection rate of *falciparum* malaria, to which natives of equatorial Africa are notably resistant. Further studies showed that only the individuals who carry the sickling gene are resistant to malaria. It thus became apparent that a harmful gene, which would have no value in most parts of the world, had increased in frequency through selection in central Africa. Although persons homozygous for sickling die of anemia, those who are heterozygous (having one instead of two such genes) resist malaria. The death of susceptible members of the population by *falciparum* malaria leaves the sickling-trait carriers to survive and reproduce, and it perpetuates the protective gene, in spite of its lethal effect when homozygous.

By another specific mechanism—premunition—low grade infections maintain resistance to severe attacks of disease. Premunition is an adaptation of individuals to ever-present malaria, an effective adjustment which depends only indirectly on heredity.

Thus it appears that *Plasmodium* is capable of variation in its biology, including its effect on its hosts. Also, the relationships between parasite and host populations are dependent on a variety of adaptive mechanisms on the part of each population.

VARIATION IN THE HEMOFLAGELLATES

The hemoflagellates, like other asexual organisms, can be expected to vary from host to host and from area to area. There is no chance for recombination of mutant traits to occur in such "vegetative" forms, and while this fact implies lower variability in asexual than in sexual organisms, it also implies eventual formation of many distinct strains. Such strains have been recognized in the genus *Leishmania*, where the kala-azar organism causes at least three kinds of visceral disease and the organisms producing tropical sores cause several different types of dermal and mucocutaneous leishmaniasis. Probably all these parasites are morphologically indistinguishable from each other, but serological tests confirm the differences so dramatically shown by the pathologies which these protozoa cause. A possibly rapid variation of *Trypanosoma* was shown by the drug resistance which developed during the attempt to mass-treat African cattle with "Antrycide" and by the well-known experiments of Taliaferro upon *Trypanosoma lewisi* in rats. The latter host animal produces an antibody, "ablastin," which prevents reproduction of the trypanosomes and thus effectively stabilizes the infection without permitting "relapse." Presumably hosts without ablastin, such as guinea pigs, succumb to repeated relapses, that is, to repeated population surges of antibody resistant parasites; ordinary lysins are effective against each relapse population, but they do not prevent nonlysable varieties of the parasite from initiating new strains. Although they are asexual organisms, the hemoflagellates are versatile and opportunistic both in the invention of geographical and drug-resistant strains and in the repeated production of antibody-resistant "varieties" within a single host.

VARIATION IN HOST-PARASITE RELATIONSHIP

Entamoeba histolytica illustrates the variability of host-parasite interaction. The two strains—the small cyst (benign) and the large cyst (virulent)—may cause an almost limitless variety of pathological conditions in the host. This parasite is harmless (certainly symptomless) in the majority of hosts, while causing an acute and fatal disease in some. The presence of two major strains suggests the probability that many minor strains exist, but it is hard to explain the extreme variations in host reactions as a result of strain differences alone. Probably these parasites find the human intestine and the other human tissues (lungs, liver) quite varied habitats. It is not surprising that the variable human species shows many different reactions to an invasive but originally intestinal parasite. The situation is probably ecological; it is a complex of factors—intestinal flora, amoebic strain, host condition, and host genetic endowment—which so far have defied analysis. This problem rivals in difficulty those other complexities, i.e., the communities of free-living animals and plants in nature.

OPPORTUNISM OF PARASITIC SPECIES

Schistosomiasis illustrates only too well some opportunistic features of parasitism. With the development of new irrigation systems, the schistosomes are engaged in a population explosion of their own and are expanding to fill new niches in a classical manner. (The increase in size of any area suitable for a species' occupation inevitably leads to the expansion of that species into the area.) The blood flukes of man also demonstrate speciation in process through the Formosan "animal" strain of *S. japonicum*. This strain probably represents a type which is ancestral to other *S. japonicum*, the "human" strains. The latter, from their severe effect on man, seem to be recent acquisitions, since the well-adapted blood flukes, *S. mansoni* and *S. haematobium*, are better tolerated by their human hosts and better resisted than *S. japonicum*.

In clonorchiasis, the role of reservoir hosts is important both in maintaining this disease of piscivorous mammals and in showing how a parasite which avoids host specificity multiplies its chance for survival. This group of parasites, like all digenetic trematodes, is limited in the number of molluscan hosts it can use. But the handicap of this specificity is somewhat overcome by *Clonorchis* and *Opisthorchis* through the "dormancy" of the eggs. Hatching occurs only when the egg is ingested by a mollusc. Miracidial energy is conserved, the infective period of eggs is prolonged over that of most trematodes, and the life cycle is closely linked to sewage-eating snails. The second phase of the cycle (metacercarial encystment in fish) and the last stage (in the livers of fish-eating mammals) are both relatively nonspecific. Thus the whole cycle shows economy and ingenuity through avoidance of extreme host-specificity and the use of effective ecological factors. Of course, man may remove himself from danger by a change in food habits; but it seems that, thanks to cycles which have several alternate pathways from snail to snail, *Clonorchis* and its relatives will survive most attempts at elimination.

VARIABILITY OF FILARIOIDEA

The filarial parasites also show variation. Although they belong to one species, the African and American forms of *Onchocerca volvulus* occupy different sites in the human body and cause dissimilar lesions. These differences are conceivably adaptive, for the spread of microfilariae through the skin begins at the nodules where adult worms live, and the concentration of these larvae in areas accessible to black flies favors transmission. Only the face, head, and neck are exposed by the clothing customs of Central America, whereas the whole body is exposed in Central Africa. In the respective geographic areas, localization of nodules in the exposed parts of the body would favor in each area the efficient transmission of onchocerciasis. Since it is believed that the American variety is a relatively recent introduction from Africa (because it has not yet occupied its total available range in the New World), the variation between the two forms may be of recent origin, and evolutionary flexibility on the part of *Onchocerca* is evident. A similar, and perhaps also recent, variation of *Wuchereria bancrofti* has been noted between the form which exists in most of its range and the variety "pacifica" found in the Polynesian islands. The once puzzling difference between these strains—a difference in larval periodicity —is now linked to two factors. One of these is a physiological or genetic response of the larvae to host activity such that the microfilariae remain in the lung capillaries except during the night, when vector mosquitoes are biting. The other is an ecological factor—the biting habits of the vectors. In the Pacific strain, larval periodicity has been lost; vectors of this strain bite at all hours. The conclusion that periodicity is evolutionary has seemed obvious to many students of the cosmopolitan strain; the conclusion that the Pacific strain has lost its periodicity seems reasonable from the known history of the Polynesian peoples, who reached Hawaii not more than 1000 years ago and may have started to spread across the great ocean not many centuries earlier. Thus two examples of perhaps recent evolution of filarial worms can be cited.

HOST-PARASITE ADJUSTMENT OF HOOKWORMS

Hookworms, like the above parasites, give evidence of the intricate nature of host-parasite adjustment. The entry of hookworms seems to involve an evolutionary (racial) pattern. Thus members of the Negro race are much less susceptible to hookworm infection than members of the Caucasian race; this is presumably because of differences between the various barriers—skin, blood, lungs—in members of the two races. The effect of the worms once established within the intestine, however, is dependent on individual host conditions. In regions where hookworms are common and where nearly everyone harbors an infection, the symptoms depend on both worm burden and the nutritional well-being of the victims. In well-fed individuals hundreds of worms may be present before detectable injury occurs; while in poorly nourished hosts, less than 100 worms may cause anemia, emaciation, and other typical effects. The interaction is a self-generating effect, a positive feedback, whereby reduced protein intake weakens immune responses; thus heavier infections are permitted. They deplete further by blood-letting the already low protein reserve. High-protein diet reverses this trend and increases antibody production until "self-cure" is effected. It is impossible to measure exactly the part which the host, or the parasite, plays in these complex interrelationships.

REVIEW OF PRINCIPLES

The foregoing examples support the principles defined in the three preceding chapters. Host-parasite relationships, as seen in Chapter 26, were said to be adaptive. It is now clear that even in disease both parasite and host interact in a mutually beneficial way; the adapted parasite is not lethal (as an adventitious parasite may be), and the normal host provides suitable housing for its parasites even while it limits parasite numbers and invasiveness by immune mechanisms. In Chapter 27 the details of host reaction were described, and it was seen that in various ways (classified as resistance, tolerance, and immunity) the host adjusts to its parasites and at least survives even in sickness. In Chapter 28 the parasite's specific harm was recognized; its mechanisms of invasion and tissue destruction were analyzed and the struggle for tissue repair and physiological balance on the part of the host was described. Usually parasitisms were seen to be neither acute nor fatal. The dynamic nature of health and disease appeared in that discussion as a necessary part of parasitism; the latter was portrayed as a conflict between parasitic ingenuity and host response. It remains for these principles—adaptive connection between host and parasite, mutual responses, invasiveness vs. resistance, etc.—to be brought into a general theory of evolution of parasitism, which in turn must be related to the broader evolution which molds and changes all living things.

ORIGIN OF PARASITISM

Speculations on the origin of parasitism are interesting; they are mentioned here not because of their present value but because they may, like much speculation, eventually lead somewhere. Three hypotheses on the origin of parasitism will be discussed—ingestion, injection, and invasion.

Ingestion of a protoparasite by the host seems to be a reasonable explanation of the origin of endoparasitism. Every higher animal accidentally eats a great number of other animals; among the latter are a few which could conceivably benefit by being eaten. For example, if their oxygen requirement were low enough, fly maggots could find ample food and protection in the mammalian gut. Various soil nematodes, especially coprophagous species or saprophytes (feces-eaters or decay organisms), should be able to survive well in various host intestines, and they must often be ingested. Indeed, it is generally agreed that ingestion offers an excellent opportunity for certain arthropods (bots, see Chapter 24) and nematodes to become parasites. For ingestion of various maggots and free-living or plant-parasitic nematodes has occasionally resulted in alarming cases of spurious parasitism, and in at least one case, that of *Aloionema*, a free-living nematode may become established temporarily as a parasite when ingested by an appropriate invertebrate host.

A second theory, injection, seems quite useful to explain the origin of certain kinds of tissue parasitisms. The hemoflagellates, for instance, include plant parasites transmitted by sap-sucking insects. A shift of insect habit toward blood-sucking would theoretically make possible the injection of plant flagellates into vertebrate blood. Other injected parasites—malaria organisms and filarial worms—might have been first acquired by accidental injection.

A third theory, invasion by the parasites themselves, seems well suited to explain the origin of such nematode parasitisms as strongyloidiasis and hookworm disease. Nematode larvae capable of active movement through soil and equipped with stylets or other organs of penetration might frequently enter the skin of an exposed vertebrate.

Entry into a host is perhaps explained for certain parasites by the above speculations concerning ingestion, injection, and invasion. Such entry is, of course, only a first step in parasitic adaptation. Any theory of the evolution of parasitism must account for migration within the host, growth, mating, egg production, and transmission to a new host. Since these adaptations are extremely varied and since practically nothing is known about most of them, it seems wise to speculate no further on the origin of parasitism and to discuss instead some rather definite observations.

HISTORICAL DATES OF ORIGIN

The Branchiobdellidae are a family of annelid worms (Oligochaeta) which are found in the branchial chambers of the crayfishes of the Northern Hemisphere. Crayfishes of the Southern Hemisphere (especially South America) are hosts of another kind of gill-chamber parasite, the turbellarian flatworms of the suborder Temnocephalida (see Chapter 7). Since crayfishes of these two regions have not had a common range of distribution since the late Cretaceous or early Tertiary period (about 50 million years ago), it is clear that neither branchiobdellids nor temnocephalids could have assumed their parasitic habit earlier than that ancient time; if they had, then these parasites would now be found in both hemispheres. Crayfishes and their ancestors existed, of course, for millions of years before the Cretaceous period; so did both dallyellioid rhabdocoeles (the presumed ancestors of Temnocephalida) and various nonparasitic annelids (the prebranchiobdellids).

The avian cestode genera *Idiogenes* and *Chapmania* occur in the bustards of Eurasia and the cariama of South America. Ornithologists have had some difficulty placing the latter bird in its proper group, for they have been uncertain whether it should be classified as a member of the order Gruiformes, an order which includes the cranes, coots, and bustards, or whether it should be placed in an order of its own. Since it shares the above genera of cestodes with the bustards and since these cestodes are found in no other host, it seems very reasonable to conclude that the cariama belongs with the bustards, i.e., that the cariama descended with the bustards from common ancestors which once shared the same range and the same parasites. Thus the age of the tapeworm genera *Idiogenes* and *Chapmania* must be at least as great as the period of time which has elapsed since their hosts occupied a common range. That range was split more than 50 million years ago when South America was cut off from the Holarctic continents.

While the above observations tell nothing of the mode of origin of particular forms of parasitisms, they make it possible to estimate the ages of certain host-parasite relationships. They are in a sense data concerning the time and place of historical events. Any speculation about the origin of parasitism must be compatible with these data.

EVOLUTIONARY PLASTICITY OF PARASITES

An actual observation of the origin of parasitism has never been made. However, the creation of drug-resistant strains and the existence of populations of parasites which parasitize animals other than their "normal" hosts are both evidence of the plasticity which may be considered as being characteristic of ancestors of parasites.

Drug resistance, for example, has been an important element in both malaria and the diseases caused by the hemoflagellates. The excellent antimalarial drug, Daraprim, is ineffective against the Daraprim-resistant strains of *Plasmodium* which occasionally develop during therapy. The medical significance of drug resistance is obvious, and the phenomenon can be countered by the use of drugs other than the one against which resistance has developed. Drug resistance, in addition to insecticide resistance occurring in mosquitoes, was a major factor in the world shift in emphasis from malaria control to malaria eradication (see Chapter 2); the genetic plasticity of parasites and vectors will be a continuing challenge to pharmaceutical research.

In the attempt to control trypanosomiasis in African cattle, a promising drug, Antrycide, was useful at first but became markedly less effective within the short time of a year or two; presumably populations of *Trypanosoma*, like those of *Plasmodium*, are variable enough so that drugs act as selective mechanisms and allow the survival of specifically resistant strains. Such strains can invade drug-treated hosts and may be considered as newly evolved.

Although the phenomenon of relapse in malaria is not fully understood, it seems reasonable to suppose that while the original episode of disease may stimulate the production of antibodies against the blood forms (perhaps the merozoites) then present, nevertheless these antibodies are ineffective against much later merozoites coming from the exoerythrocytic centers of residual infection. The relapse invasion of the blood may thus represent a new strain of parasite, which may differ from the old in its ability to live in the presence of certain antibodies. Thus relapse may be an evolutionary phenomenon, in which old antigenic strains become extinct, and newly selected strains (the relapse organisms) survive—at least until they stimulate new antibody-production.

The few studies which have been made of existing populations and strains of parasitic animals give further evidence of the origin of parasitisms; such evidence shows that parasites of the present are highly capable of such opportunistic variability which protoparasites must have exhibited. *Leishmania donovani* causes visceral leishmaniasis (see Chapter 4) of five types, each of which is both geographically and clinically (that is, biologically) distinct. There are Chinese, Indian, Mediterranean, Sudanese, and South American forms which differ among themselves with respect to reservoir hosts, pathological lesions, and response to drugs, for instance. It is not known whether this parasite has become geographically differentiated as a consequence of man's parallel subspeciation, but it must be assumed either that such divergent evolution occurred or that by parallel evolution each strain arose independently from strains indigenous to the "reservoir" host in each locality. Whichever hypothesis is true, there must be considerable genetic flexibility in the species *Leishmania donovani*. A similar situation exists in the case of *Leishmania tropica* (which causes oriental sore) ; there are numerous kinds of sores, and each is characteristic of a certain region of the world. Moreover, the evidence of local "emergence" of human infection with *L. tropica* from a source native to the region is much clearer than the doubtful hypothesis of a parallel evolution in man of *L. tropica*, which is similar to that just proposed for *L. donovani*. The South and Central American sores are acquired by man through transmission from wild animals. Thus these parasites must be indigenous to various lower forms, a number of which have been identified; yet still the protozoa are capable of establishment and survival when injected into man. Distinctively human strains of *L. tropica* could develop in the New World, just as they undoubtedly have in the Old.

The evolutionary pattern mentioned above may be characteristic chiefly of asexual species. In this pattern it can be seen that parasites have formed varieties in rather short lengths of time (the duration of man's migration to the continental regions), and some (the *L. tropica* group) are still in the process of acquiring new hosts. The impossibility of the exchange of genes assures that such species, if at all capable of variation, will form a number of strains or varieties. The population structure of such organisms is non-Mendelian, and each individual is potentially the parent of a new species. It will be explained later (see discussion of stability of parasitism) how such a genetic system may either speed or delay the evolution of parasites.

Sexually reproductive parasites also form strains, varieties, and incipient species. The schistosomes, especially *S. japonicum* (see Chapter 9), exhibit variation. The cestode *Hymenolepis nana* has been widely studied because of its effect on man, its interesting life cycle (it is one of two tapeworms known to be directly infective through its eggs to a vertebrate host), and its ability to utilize either rodents or man as final host. This cosmopolitan worm varies a great deal in size, infectivity, and even the number of hooks it bears on its scolex. Much of this variation is, of course, environmentally produced, but some of it (the scolex hooks, for instance) can be considered taxonomic, that is, genetic. Probably any numerous well-known parasite that can be studied minutely and in population-sized samples will show considerable variation of this nature. It is a curious fact that the literature of the classification and naming of cestodes, like other such taxonomies, contains a number of examples of uncertain species. Several papers have been written disagreeing on some particulars of measurement, host, or life cycle of such groups. It has been pointed out that some of this uncertainty is derived from the instability of the species concerned and is not merely an expression of the human frailty of taxonomists. While it is probably impossible to compare variability in large categories of animals, one can say at least that parasitic animals are capable of variation and that this ability helps explain parasitic adaptation, including the origin of parasites.

TAXONOMIC DATA ON ORIGINS

Still another source of information on the probable origin and progress of parasitism is the taxonomic data, i.e., information analogous to the facts of comparative morphology upon which the evolutionary patterns of vertebrates and other animals can be reconstructed.

An example of this kind of data is the comparative morphology of the genera in the family Trypanosomidae (the hemoflagellates and their close relatives). Descriptions of these genera (see Chapter 4) are based on the occurrence in each genus of two or more of four distinct forms of organism (Fig. 29–1). These forms are (1) the leptomonas, an elongated organism with central nucleus and anterior kinetoplast and flagellum; (2) the crithidia, in which the kinetoplast is near the nucleus and a short undulating membrane attaches part of the flagellum to the body; (3) the trypanosoma, with posterior kinetoplast and complete undulating membrane; and (4) the leishmania, with rounded or oval body, a kinetoplast, but no flagellum or membrane. At least seven genera are recognized; some have names identical with those used (above) for the various body types. Thus the genus *Leptomonas* has only leptomonas and leishmania forms; it is parasitic in various insects. The genus *Leishmania* is morphologically the same as *Leptomonas,* but its species parasitize vertebrates and are transmitted by various insects. *Phytomonas* is morphologically the same as both *Leptomonas* and *Leishmania,* yet its species inhabit plants with milky sap and are transmitted by plant-feeding insects. (The three above-named genera are distinguished, therefore, by their host relationships; we will discuss this important point later.) The genus *Crithidia* has leptomonad, crithidial, and leishmania stages, and its species parasitize insects. *Herpetomonas* has the above three forms in its life cycle, and in addition it has a form very nearly trypanosomal; as in trypanosomes, the flagellum arises posteriorly but is attached closely to the body, and a true undulating membrane is not present. The species of this genus parasitize insects only. *Schizotrypanum* includes all four forms: leptomanas, crithidia, leishmania, and trypanosoma. It utilizes vertebrates and insects to complete its life cycle; in vertebrates it multiplies intracellularly by leishmania bodies. *Trypanosoma* also may include all forms, and it parasitizes insect and vertebrate hosts. In the latter host, only the trypanosome form occurs.

The above description can be interpreted in a phylogenetic way. The most primitive (least complex) members of the family are in the genus *Leptomonas* and the similar genera *Phytomonas* and *Leishmania*. This fact suggests that at some early stage in their evolution, insect parasites (*Leptomonas*) became transmitted by injection into plants (*Phytomonas*) and vertebrates (*Leishmania*). The above-mentioned plasticity of *Leishmania* (kala-azar, oriental sore) shows that without change of form these parasites can become adapted to a variety of hosts and habitats. Increased complexity, illustrated by the forms crithidia and trypanosoma, developed in insect hosts. Transmission by injection or ingestion may have followed, so that vertebrates became hosts also. The paucity of developmental forms in vertebrates, compared with the insect hosts, supports the hypothesis that vertebrates are the latest acquisition among the hosts of hemoflagellates. According to the above speculation, the genera of the Trypanosomidae recapitulate the phylogeny of this group; the retention of primitive forms in the life cycles of the hemoflagellates further strengthens this view.

One would not expect to find many examples as clear as the above to illustrate the history of a group. Larval adaptations in the life cycles of metazoan parasites must obscure all or most evidence of phylogeny. The parasite's habitat must usually be so restrictive that only evolutionary fitness, not origin, is apparent. Therefore one cannot say whether or not the Trypanosomidae illustrate anything general; yet they do provide persuasive arguments, based upon their present-day taxonomy, for a theory of their own evolution.

FIG. 29–1. Evolution in the family Trypanosomidae. At left, hypothetical pathways stem from a supposed leptomonaslike ancestor of the parasitic trypanosomidae. The present genera which most resemble such an ancestor, both in habitat (plants, for instance) and in morphology (only two forms, a flagellate and a nonflagellate, being present), are at the bottom of the chart. These genera (*Leptomonas*, *Phytomonas*, and *Leishmania*) are differentiated in terms of life cycle only, a fact which suggests that small physiological mutations could have permitted plant parasites to become insect- and vertebrate-dwellers. Progressive complexity of form is seen in the other genera. *Crithidia* adds the short undulating membrane; *Herpetomonas* has a longer but very narrow membrane; and *Trypanosoma* and *Schizotrypanum* possess the full length membrane characterizing the trypanosome body form. These genera are further distinguishable by their hosts, as illustrated, and by their growth patterns (especially *Schizotrypanum* of Chagas' disease). It appears that the higher forms are modifications upon the primitive ones; moreover, in development the higher forms "recapitulate" their ancestry, since *Crithidia*, *Herpetomonas*, *Trypanosoma*, and *Schizotrypanum* possess leishmania and leptomonas stages in their typical life cycles.

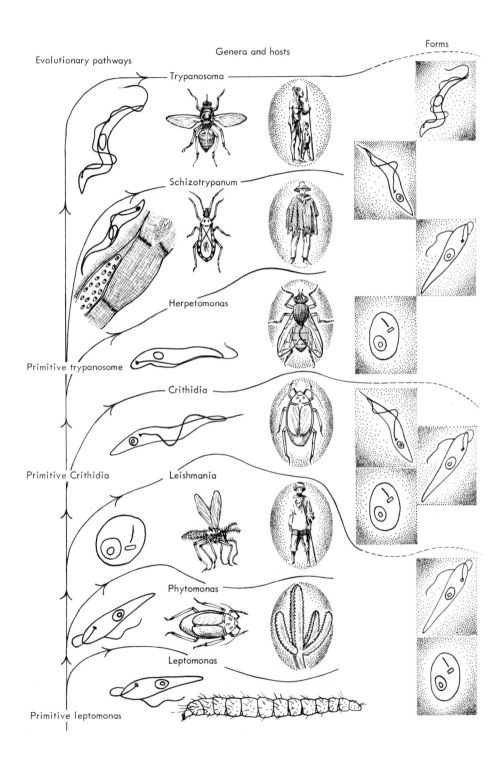

Evolutionary pathways

Genera and hosts

Forms

Trypanosoma

Schizotrypanum

Herpetomonas

Primitive trypanosome

Crithidia

Primitive Crithidia

Leishmania

Phytomonas

Leptomonas

Primitive leptomonas

Another sort of taxonomic data on the origin of parasitism is the overall charac-teristics of the major groups of parasites.

These groups may be considered analogous to the animal-plant inhabitants of various geographic regions; that is, the parasitic protozoa, nematodes, etc., repre-sent isolates from their nonparasitic relatives, just as the inhabitants of a large island are isolates which were cut off at some remote time from the stream of life in other lands. Therefore we would expect some parasitic groups to reveal by their present-day characteristics a history similar to that of the Australian mammalian fauna, which are the diversified but closely related descendants of a rather limited marsupial stock; the modern Australian marsupials are derived from a single origin (single, at least, compared with the origins of the animals of other continents). Other groups of parasites resemble the inhabitants of an intermittently isolated land mass such as South America. In that continent the old native forms (the edentates and generalized marsupial opossums) coexist with later arrivals (the neotropical simians, the tapirs, certain rodents) and very recent immigrants (man, for instance). The South American fauna is thus seen as polyphyletic when com-pared with the monophyletic fauna of Australia. Geologists and geographers know that these facts are related to the history of the two continents concerned; Australia has been isolated since the Cretaceous period (about 50 million years ago) and South America has been joined intermittently by land bridge to the holarctic conti-nental mass from which in successive waves its invading fauna came.

The parasitic protozoa, nematoda, and arthropoda are large and diversified groups. Comparable to the fauna of South America, the groups are the result of frequent waves of immigration from the free-living to the parasitic way of life. The relationships of the members of these groups are best seen, at least theoretically, in terms of their free-living relatives, for a system of classification of parasitic protozoa or arthropoda would be highly artificial, indeed impossible, without refer-ence to many categories of free-living animals. This view is supported by the present difficulty of classifying nematodes, where so little is known about free-living forms that the extremely disjunctive groupings of the parasitic members satisfies no taxonomist at present; the unknown free-living nematodes must provide con-nections between the known groups of parasitic forms.

On the other hand, the Acanthocephala and the parasitic flatworms are like the fauna of long isolated continents. Because of the obviously close relationship among the groups of Acanthocephala and because of the fact that no clearly visible relationship exists between them and any other animals, the Acanthocephala may be taken as an analogue of the Australian marsupials. Like the marsupials of Australia, the parasitic thorny-headed worms must be the descendants of a single stock and are still not completely diversified. The same principle applies to the Digenea, whose relationship to the Turbellaria is clear but whose unique life cycle (obligatory molluscan host with polyembryony) indicates a close relationship within the subclass; the Digenea are descended from a single stock. So, too, are the true tapeworms, the subclass Cestoda; however, there are such differences among these, especially in their specificities for phylogenetically distinct groups of hosts, that one must conclude that they became isolated very early from the flatworm stock. Yet the tapeworms must have come from a single source.

Thus it is possible to speak of some groups (the parasitic insects, nematodes, or protozoa, for example) as being polyphyletic in origin and to recognize that members of other groups of parasites must have had a common origin. Taxonomy shows that parasitism arose many times in some animal phyla and only once or twice in others.

FACTORS AFFECTING THE EVOLUTION OF PARASITISM

Parasites, of course, must follow in their evolution the same basic pathways and must be influenced by the same fundamental factors as other organisms. These pathways are the divergent ones of speciation, and lead from mere difference in the genetic materials of populations through geographic or otherwise isolated subspecies to true noninterbreeding species and finally into the higher categories such as genera, families, etc. The guiding factors determining which organisms will be represented at a given moment of time are the well-known principles of neo-Darwinism: mutation, isolation, recombination, competition, selection, and genetic drift. Let us now examine the ways in which parasites are evolutionarily unique and the pathways and guiding factors which may be altered or exaggerated in the evolution of the parasitic animals.

One feature of parasite evolution is the way in which hosts and parasites influence the evolutionary changes in each other. Parasitic harm, or virulence of a disease, becomes limited in various ways. In theory a successful parasite does not harm the host, since the parasite's survival depends upon the host's survival. Therefore on strictly logical grounds one might suppose that only the most recently emerging parasitisms are harmful or virulent and that the age of each relationship is inversely proportional to the harm which it causes. Unfortunately, not enough is known about a sufficient number of parasites to support or contradict this theory. There is a theoretical objection to it (the age-virulence inverse relationship). The objection comes from the need of successful parasites to produce large numbers of offspring because of the transmission factor, which is a weak link in nearly every parasite's life cycle. High reproductive potential of the parasite implies considerable drain upon the resources of the host. Thus one survival factor (fecundity) opposes another (harmlessness to the host). But before we examine the consequences of these opposing factors, we should look at some of the mechanisms by which virulence is limited, i.e., mechanisms which seem chiefly host-operated.

Immunity is a direct response of the host to various parasitisms. Tissue parasites may be surrounded, encapsulated, and killed, or their growth and migration may be slowed by immune response. A successful invasion by parasites may render the host insusceptible to subsequent invasions by the same kind of parasite so long as the original parasites are present; this response is known as premunition. Sometimes immunity is not protective, but it may be actually harmful. In onchocerciasis most of the symptoms of disease stem from allergy or sensitization of the skin, eyes, etc., to larval worms. One may speculate that such hyperimmunity is self-limiting, since an increase in immune response to *Onchocerca volvulus* might be fatal to the host. It is possible that some such limiting factor exists in various intermediate hosts; certainly the limited number of microfilariae which black flies or mosquitoes can safely ingest has a limiting effect on the level of human infections with cutaneous and systemic filariasis.

Other host responses (adaptations) which tend to mitigate the harm caused by parasites are covered by the rather vague concepts of "tolerance," "self-cure," and "age resistance." Tolerance may be considered as the complex of factors which prevent a parasitized host from becoming sick. Diet is said to be such a factor. In hookworm disease it is known that the loss of blood protein causes a weakening of the immune mechanisms, and that this effect is proportional to the number of worms present but may be corrected by increased dietary protein. Tapeworm anemia may be corrected by vitamin B_{12} supplement. It is probable that most parasites upset the nutritional balance of their hosts, and restoring that balance by supplement may be an effective way to cure the symptoms of parasitism. The kind of "depraved appetite" noted in those hookworm sufferers who crave earthy or gritty substances may in fact be an imperfect adaptation to hookworm disease, i.e., a response to loss of mineral and protein. Self-cure is a phenomenon by which the number of parasites in a host becomes greatly reduced with time, although neither treatment nor control measures have been instituted. It is believed that both immune responses and local changes (alterations in the intestinal mucosa, for instance, in self-cure of tapeworms) may be involved. At present it is impossible to differentiate self-cure from the effects of tolerance (nutritional, especially), on the one hand, and age resistance, on the other. Age resistance is a general phenomenon. Young animals are more susceptible to harm by parasites than are old animals. Perhaps this difference is due to the fact that young animals are physiologically less specialized than old; hence they are susceptible to a greater variety of foreign organisms. Such organisms, while able to invade the unspecialized younger host, may not be adapted to life in the mature host. But even where well-adapted parasites are concerned, a degree of age resistance has been observed. In the latter cases, immunity may be the explanation. Probably each example of age resistance should be analyzed separately in order that dietary factors, physiological aging effects, and developing immunity may be given their proper weights.

Control or limitation of the number of parasites affecting a host is determined not only by the factors just mentioned but also by dispersion factors. By these we mean such things as population density of the host, infection rates in intermediate hosts, and survival times of eggs, larvae, spores, cysts, or other transmission stages.

The populations of parasites in coccidiosis—a severe epidemic disease of poultry—become enormous only because poultry are usually crowded together, so that innumerable coccidial oocysts are ingested directly with droppings of infected birds. Such an ecological situation would not develop in nature. Presumably the virulence of coccidiosis is due to crowding (a recent man-caused condition) rather than to failure of the coccidia to reach a tolerable level of adaptation to their hosts. Cattle and horse bots cause serious diseases in these animals. The fully grown larvae drop to the soil and pupate. If animals are removed from the areas of pupation, the emerging flies cannot readily find hosts on which to lay their eggs, and they must die without perpetuating the life cycle. However, restriction of the host animals to the same range throughout a period of months permits reinfection and the build-up of a large population of bots. Man's restriction of the range of domesticated animals thus subjects the latter to heavier parasitisms than they would suffer in the wild. Since the effect of crowding on human health is so well known, it will not be discussed here. It should be obvious that parasite population levels are usually related to host population density and that man has altered the importance of many parasitic diseases by his cultural and agricultural practices.

Other population effects are less well known than the above crowding factors. The infection rate in vectors of malaria is an important element in the infection rate of man, but other factors (zoophilic vs. anthropophilic species of mosquitoes, biting frequency, population size of vectors, flight range) complicate attempts to assign a value to infectivity of *Plasmodium* species for the intermediate host. Survival of eggs, larvae, etc., also affects population size and hence harmfulness of many parasites. Very high individual worm burdens of *Ascaris* occur where eggs can accumulate in the environment; the extraordinary durability of such eggs makes accumulation possible. In England, the spreading of human taenia eggs in sewage—a sort of dilution—has made the dissemination of infective eggs by sewage-feeding birds a common method of spread, but the incidence of bovine and human infection remains small. In desert lands, where contamination of relatively small corrals, pastures, or oases by the feces of infected human hosts is usual, cattle acquire the eggs of cestodes in rather large numbers, and bovine as well as human infection rates remain high. Many examples of such relationships between dispersion mechanisms and parasite populations can be found; they result from interacting factors in parasite evolution. Dispersion is a necessary feature of any parasite life cycle; it also automatically reduces the population size in individual hosts.

Such ambiguous tendencies—toward and away from overwhelming the host—lead to a concept mentioned earlier, i.e., the notion of host-parasite balance. This notion was stated first in terms of evolution: old parasitisms are mild, new ones are virulent. Perhaps the idea should be restated in terms of population dynamics: parasitism tends to balance host damage by host tolerance. Is it possible that this is not a balance but a trend? It has been postulated that nature is essentially cooperative and that the community of life is a system of interdependencies. A trend of parasitism toward commensalism or even mutualism would conform to that worldwide view. But there is no historic evidence of such a trend. The balance between host health and parasite well-being is usually achieved at the expense of the host's health, and in innumerable parasitisms it seems to be stable at some point far short of benefit to the host. Therefore it is fruitless to try to find in the existing data evidence of a trend toward mutualism. We may, how-

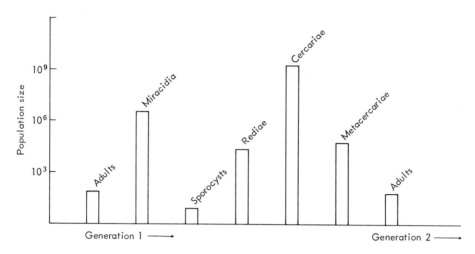

FIG. 29–2. Population fluctuation in Digenea. For populations of adults to remain constant, each adult must lay thousands of eggs, some of which produce miracidia. A few of the latter succeed in penetrating molluscs, where many rediae are produced from each sporocyst. Each redia produces many cercariae, a few of which successfully become adults in the final host. The sporocyst and adult stages represent the population at its smallest, the point where genetic drift is most effective.

ever, look for evidence of stability in parasitic adaptations, since such evidence may establish the possibility, or the impossibility, of tendencies in one direction or another.

But before considering the difficult question of the stability of parasitism, we should consider some of the mechanisms of evolution which apply particularly to parasites.

As the Sewall Wright hypothesis (genetic drift) emphasizes, population size affects evolution. (See Fig. 29–2.) All animal populations vary with seasons, food supply, predator activity, and many other factors. Some animals (and for most of his history man was among these) exist in very small groups and have limited ranges; in the evolution of such animals, the probability factors (mutation, accidental death, etc.) play a leading role. Genetic change in an individual member of a small population has greater effect upon the genetic constitution (gene pool) of the group than does a similar change in a member of a large population. Therefore, population size affects the probability of genetic change, and the extent of probable change (drift) is dependent on the smallness of the population.

How small or large are populations of parasites? First, a population must be defined as all the members of a group which are able to interbreed with other members of the group but not with the individuals beyond the group's limit. The population, so defined, is the unit upon which evolutionary forces operate; it is potentially a species. Some populations of parasites are extremely small. A single strobila of a rare cestode, like *Taenia solium* (the pork tapeworm), may actually be such a population. Its members would be the proglottids, of which each would be capable perhaps of "mating" with others in the chain. A mutation in the germinal (neck) region of such a "population" would become a part of all the progeny; drift would be a one-generation phenomenon. Perhaps most populations of parasites undergo fantastic fluctuations in size, since the transmitting stages, while produced in large numbers, must largely die out as the necessary few become established in new hosts. Trematode cycles (Fig. 29–2) are of this nature, and nematode cycles must be also. Malaria parasites and other protozoa pass between hosts through similar evolutionary bottlenecks. Probably most parasites reach population peaks and descend into population valleys; the latter would give genetic drift a significant role in parasite evolution. The effect of drift is, of course, to speed speciation. Genetic drift must be considered a factor in the variability—instability—of parasitic adaptations.

Small populations also show another behavior—inbreeding. It is not known whether digenetic trematodes must cross-fertilize or whether they can be self-fertile. Nematodes, being dioecious, obviously do exchange genetic material; the breeding populations of these worms may be quite large. Cestodes are perhaps largely self-fertile, and consequently they may be ordinarily homozygous by means of the same genetic mechanism which makes self-fertile plants, such as Mendel's original lines of peas, homozygous. Inbreeding increases the uniformity of a population not always to the extent of homozygosity but usually to some degree. Thus this factor may oppose the factor of drift, which tends to change the genetic balance of a population. Inbreeding is a stabilizing influence in parasite evolution.

Isolation, which evolutionists consider an essential first step in speciation, has peculiar importance in parasitism. Each host is in a sense both the habitat and the limited range of its populations of parasites. New species of parasites have been described usually from rather small samples of such limited populations. Yet the modern concept of species, applicable to many animals and plants, requires some description of a biological group; such a description should include not only typical but also atypical forms and the mean as well as the extremes of a measured variable.

This variable, the species, was once (at the time of Linnaeus) considered an invariant created entity. Thus it was the duty of taxonomists to define this entity for the convenience of biologists and to reveal more completely a universe which was believed to be orderly. Type specimens were enshrined in museums, and their measurements were usually recorded in the dead, hence rigorous (*rigor mortis*), language of the ancient Romans. Expert taxonomists became the more or less priestly authorities who guarded these type specimens and attempted to preserve against modifications and emendations the original descriptions which inevitably began to record variation in the organic world. The more conservative taxonomists, called irreverently "lumpers," defended established categories from attack by the radical "splitters." Even today, the discourse shows few signs of quieting. Thus in 1959, the highly respected authority, Stunkard, stated his view that parasitologists cannot yet use a genetic definition of species (that is, one based on population norms, reproductive isolation, etc.) but must adhere strictly to the classical system of types, original descriptions, and internationally established nomenclatural rules. His argument in favor of old methods is enthusiastically presented, but it was probably not intended to defend absolutely a *status quo*, for he has himself introduced revolutionary elements, such as the study of larval forms and life cycles, into the taxonomy of parasites. Stunkard sensibly decries disorder in classification; he believes that a shift in emphasis away from types would increase disorder. Nevertheless, as parasites become better known biologically, studies which are cytological, genetic, and, above all, ecological can be expected to alter taxonomic concepts. This alteration has already occurred in such important groups as the hemoflagellates and the schistosomes, where the species concept now appears extremely fluid. Yet for the great majority of described species of parasites, most taxonomic parasitologists still depend on museum specimens and published descriptions for a stable and useful classification.

EFFECT OF THE HOST

The above evolutionary factors—population structure, genetic drift, inbreeding, and isolation (including the species concept)—have to be considered in the light of a fifth factor, natural selection. This factor is identical with the parasite's environment, that is, the living host. The structure and physiology of the host, its food habits, its habitat, and its behavior all determine to a large extent the kinds of parasites that can successfully find shelter and food in such a host. And evolutionary (hence very slow) or cultural (hence rapid) changes in the life of the host will have profound effects upon the established parasites. The host is the chief agent of natural selection in the survival or extinction of the parasite.

There can be no doubt that within the host the parasite causes stress and that the host adapts itself accordingly. The rule suggested earlier, that old parasitisms are mild and recent ones may be virulent, is derived from the examples (seen throughout this book) of the various kinds of host resistance, tolerance, and accommodation to parasitism. Since we think of parasites most of the time as a hostile part of the environment of man and his animals, we find it difficult to assume another point of view and to consider the various hosts, including man, as sometimes perilous and stressful environments to which the parasite must adjust for survival. But from the latter point of view we can best see how environment plays a part in the shaping of parasitic ways of life. The host is that environment.

Environments change. The physical and biological world has undergone tremendous changes throughout its history. These changes have, through selection, brought about great alterations in the fauna and flora of the earth. The parasites of animals and plants, although protected in a trivial sense from environmental change, were actually living in an organic world which was being altered by geologic and climatic upheavals; the parasites were affected by the evolutionary changes in their hosts.

Some of these changes must have caused ecological stress in the parasite life cycle. An interesting analogy suggests itself here. Continental drift, the majestic movement of granitic masses upon the surface of a turbulent and viscous sphere, has split asunder regions which were once contiguous. Most of the animals inhabiting such masses show by their evolution evidence of the antiquity of these events; each geographic realm has its rather distinct fauna of related forms. But the migratory birds have often retained a habitat in both parts of their historically sundered world, and the heroic distances they annually traverse measure the long memory of these species. The shift of continents has stretched what must have once been seasonal but short migrations into hazardous journeys of some thousands of miles. Change in the environment has stressed the established patterns of behavior, and has forced evolutionary experimentation to give rise to that still not understood complexity of bird migration—navigation by the stars over featureless wastes of ocean.

Some of the paradoxes in the present-day life cycles of parasites may be the results of similar historic stress. Thus the digenetic trematodes (predominantly today parasites of fishes, and probably very ancient parasites at that) have managed to invade many of the completely terrestrial animals and even the birds. Important in this invasion is the role of a second intermediate host. (See Fig. 29–3.) Perhaps because of the success of polyembryony as a probability factor, the Digenea have been unable to modify the requirement for a mollusc as first host. However, by means of a fish, the aquatic larva of a flying insect, or the growing blades of grass in a wet pasture, certain Digenea can move into positions where their infective cysts are likely to be ingested by a piscivorous mammal, an insect-eating bird or frog, or a grazing ruminant, respectively. The second host makes possible these extensions of an essentially aquatic life cycle or the preservation of a cycle which began ages ago in a simpler form. Interestingly, the loss of a second intermediate host (assuming that one may have existed) by the schistosomes is another device by which a cycle may be maintained; the blood flukes enter the final host directly without an intervening cyst. In still other Digenea the second intermediate host seems to be suitable, or nearly so, for completing the life cycle. For example, certain fish trematodes, which utilize crayfish as intermediate hosts, can reach sexual maturity and actually produce eggs in the intermediate host, so that the final host becomes superfluous. This type of evolutionary change, by which a preadult form achieves reproduction, is called neoteny, a general term meaning persistence of youthful characteristics into maturity or achievement of sexual maturity of an immature form. Neoteny may be illustrated by those strange tapeworms, the gyrocotylids and amphilinids. These resemble sexually precocious larvae, but no adults to match them are known. The notion that these larvae are fragments of life cycles which once included extinct Mesozoic reptiles may seem fantastic, but it would illustrate, if true, survival of a parasite with the loss of its adult stages. Neoteny certainly contains that potentiality. Such anomalous changes may be considered as responses to stress, that is, as the results of selection by a changing environment. Of course, we cannot distinguish such adaptations from such new adaptations as the invasions of new hosts. But then, stress and opportunity are often the same.

FIG. 29–3. Divergence in transmission mechanisms in the Digenea. (A) The "conservative" parts of a digenetic trematode's life cycle: the molluscan host, the polyembryonic larvae within the mollusc, and the adult, which resembles most adult trematodes relatively closely. The remaining parts of the figure show various intermediate hosts of infective larvae: (B) The branched, brightly colored, and pulsating sporocyst of *Leucochloridium*, a trematode which causes the tentacles of infected snails to become highly visible to the final host, snail-eating birds. (C) The fruit of a water ling, which is usually peeled with the teeth and eaten raw in the Far East, with cysts of *Fasciolopsis buski*, the human intestinal fluke, upon its surface. (D) Slime balls containing cercariae of *Dicrocoelium dendriticum*, the lancet fluke or small liver fluke of sheep. (E) Ants eat these slime balls and are then eaten by the final host. (F) In the foot of a frog are many encysted metacercariae of *Leptophallus nigrovenosus*. (G) A dipterous larva (*Culicoides*) is penetrated by many cercariae, which lose their tails after entering. (H) A schistosome cercaria is capable of penetrating its final host directly. The variety among second intermediate hosts illustrates the opportunism which overcame ecological stress, that is, the difficulty of transmission between the ancestral host (the mollusc) and various vertebrates.

The tapeworms of carnivores have responded to ecological or biological stress in several ways. It is well known that carnivorous predators feed irregularly over a relatively wide range. Therefore the opportunities for transmission of their tapeworms must depend on infection of large numbers of the intermediate (herbivorous) host. Most taenid worms are fantastically prolific; they literally seed the environment with eggs over a period of years. But one, *Echinococcus granulosus*, lays relatively few eggs as an adult; instead it produces larval cysts which may contain thousands of "heads" capable of growing into egg-laying adults. Either by vast numbers of eggs or by multiplying cysts, the tapeworms of carnivores achieve reinfection by superfecundity.

Cultural changes (i.e., changes in man's behavior) or changes in other animals influenced by man have been known to affect parasites greatly. Chandler remarked that modern plumbing threatens to eradicate *Ascaris* and that the human body louse fights a losing battle against the Saturday night bath. But it is noteworthy that the body louse is itself a product of culture; its habit of laying eggs in cloth fibers implies that its evolution is historically as recent as the invention of weaving. The fact that body lice can be transformed (in several generations) into head lice, and *vice versa*, illustrates the evolutionary plasticity of these insects. Many of man's parasites have been affected by cultural change; it remains to be seen whether their tolerance will change and whether the extreme selection pressure exerted by civilization can bring about extinction of some of the ancient parasites of man. It would be unwise to predict; *Enterobius*, the ubiquitous pinworm, is most numerous among the highly civilized.

STABILITY OF PARASITISM

The foregoing speculations about parasite responses to environmental change certainly suggest that the plasticity of parasites is considerable and that parasites are capable of a degree of evolutionary change matching that of their hosts. Yet much of the data of parasitism supports the widely held view that parasitism is a conservative way of life. As Noble and Noble (1963) have stated, many of the parasitic adaptations are blind alleys, end products of an "inexorable march of evolution"; such animals are doomed to extinction by their extreme specialization. Other

FIG. 29–4. Evolutionary divergence of hosts; mutualism in flagellates of the common ancestors of termites and wood roaches. At lower left is a protoblattid, an ancient insect common to the Carboniferous period. Surrounding it are hypothetical flagellates of four families which probably lived in the gut of such insects. At upper left are specimens of *Cryptocercus*, the modern wood roach. This primitive, semisocial insect is clearly a member of the blattid branch of the order Orthoptera. At right are some termites, members of the order Isoptera. Flagellates parasitizing these separate groups of modern insects belong to the same families and sometimes to the same genera. This fact is evidence of a host-parasite relationship (mutualism, in this case) which extends back to the time when the host insects occupied a single taxon, that is, to the Carboniferous period, some 300,000,000 years ago. The parasites have changed very little, while the hosts have separated into different orders of Insecta.

adjustments, like those of the cestodes, are so old that quite similar worms now occupy quite dissimilar hosts, i.e., hosts whose common ancestors can be found not later than many millions of years in the past. In fact, it is well known that the conservatism of parasites may sometimes be used by taxonomists who wish to discover the relationship between hosts. The cariama-bustard relationship was revealed by the common genera of parasites which these geographically separated birds possess. Thus we raise a question of whether parasites are radical opportunists in their evolution, or whether they are evolutionary conservatives lagging behind their hosts in the tempo of their change. This question is, of course, more than theoretically important; answering it should illuminate some problems in taxonomy and may well clarify the meanings of host-parasite adjustment, mutual tolerance, and the possibility of "trends" toward or away from mutualism; these matters have been discussed elsewhere in this text.

The generally held opinion is that parasites are more conservative than their hosts. Examples supporting this opinion certainly exist. The fact that much less change has occurred in the parasites than in their hosts is indicated by the presence of opalinid protozoa in the gut of tailless amphibia living in all the major regions of the world and by the existence of the same genera of these protozoa in places as far apart as Africa and Australia and in frogs of separate taxonomic groups. It is well known that termites, members of the insect order Isoptera, utilize symbiotic protozoa which belong to many of the same genera of symbionts found in the wood-eating roach, *Cryptocercus;* this roach is itself a member of another order of insects. (See Fig. 29–4.) These symbionts have maintained their taxonomic identity at the genus level, while their hosts diverged to an extreme degree. The above-mentioned similarities between the parasites of Eurasian bustards and those of the neotropical cariama are another example.

Theoretical support for the opinion that parasites are conservative comes from several ideas. First, parasites inhabit a relatively uniform and secure environment, the "steady-state" environment of the host tissue or gut; environmental stress that would speed up evolution is not present, and the successful parasite, well-protected, settles down to an evolutionarily stagnant way of life. Second, the host provides a limited environment which is quite unlike the outside world; in the latter, expansion into all sorts of niches is made necessary by competition and is made possible by mutation. If mutation occurs in the parasite, it cannot "improve" a well-adapted organism; in the parasite's environment, there are no niches into which the parasite can expand or speciate. These two arguments are, of course, almost identical. The first emphasizes stress; the second, opportunity. A third argument states that parasitic animals are evolutionarily conservative because of certain unusual features of their structure, life cycles, or populations. Many parasites are essentially asexual, either because they lack sexual forms (as in the hemoflagellates) or because they are hermaphroditic (monoecious), like the flatworms. "Selfing," which may occur in the latter, results in genetically uniform populations; genetic uniformity must necessarily be very high in effectively asexual animals. The small populations which are formed by many parasites lead, through genetic drift, to fixation of genes and uniformity of the gene pool. Also, inbreeding, in small populations, amounts genetically to the same thing as selfing; homozygosity results. Thus it seems that the opinion in favor of the evolutionary conservatism of parasites is supported by both evidence and theory.

However, the contrary opinion is not worthless. There is evidence, as well as theory, to support it.

First, instead of lagging behind their host's evolutionary advance, many parasites seem to have overtaken their hosts. While it is true that the parasites belong to primitive phyla in comparison with host phyla, nevertheless the acquisition by ancient and primitive organisms (such as protozoa or flatworms) of a recent host (such as bird or mammal) is evidence of evolutionary opportunism. And this opportunism, or plasticity, as stated in Chapter 26, is characteristic of parasites in general.

Second, parasite life cycles are extremely ingenious. The conservatism of parasites is usually discussed in terms of the definitive form—the inhabitant of the gut or tissues of a vertebrate host. The evolutionary plasticity of parasites in their transmission phase can be illustrated by the utilization of a great variety of intermediate and transport hosts, the matching of infective forms with the behavior of the transmitting or final host (as in the larval periodicity of *Wuchereria,* for instance, and the utilization of desert water sources by *Dracunculus*), or the loss of intermediate hosts (by *Hymenolepis nana* and *Trypanosoma equiperdum*) as well as the possible loss of final hosts (by neotenous Digenea or Cestodaria). Therefore it appears that only part of the parasite's life is spent in a nonstressful environment and that much of the life cycle involves highly precarious transitions, which require great adaptability of the successful parasite.

Third, theoretical considerations of form, population structure, and environment of parasites do not necessarily support the view that parasites are conservative.

Asexual reproduction does result in homogeneous populations; but any mutation in an asexual line immediately increases in number, without mixing or dilution, as the population grows. It is theoretically possible for an asexual organism to give rise to a new species in one generation. As was pointed out earlier in this chapter, variant lines and stocks are actually well known in the asexual hemoflagellates. Isolation of small populations, a characteristic of parasitism in solitary animals, allows genetic drift to have an exaggerated effect. While it is true that each small population should, through drift, lose genes completely or become homozygous for other genes (hence become genetically invariable), still the several isolates in a species would, by drift, be expected to become different from each other; thus genetic drift is seen as a factor in intraspecific variation.

Fourth, the effect of a stable, rich environment upon parasites in their definitive host may not be so conservative as the earlier discussion had suggested. The rich nutritional milieu—host gut, blood, or tissue—may be analogous to the enriched waters of some ocean currents or to the energy-filled canopy of the tropical rain forest. In these environments life is abundant both in numbers and in kinds; perhaps the environmental factors which affect variation are neither positive nor negative in the enriched milieu, and the effect of such an environment upon evolution is actually not significant.

Thus it appears that the supposed conservatism of parasites may not be real and that parasites may be, not less, but more evolutionarily progressive than their hosts.

The above arguments have been presented as if only one of them could be correct. But, as is well known in science, disagreements on the meaning of data cannot be settled by argument, for argument serves properly only to illuminate a problem and to suggest further means of study. Therefore in conclusion it should be stated that while the question of the rate of the evolution of parasites is an intriguing one, there is at present no satisfactory answer to it. Continued research upon the morphology, taxonomy, life cycles, ecology, distribution, and evolution of parasites (and their hosts) may eventually make clear the way in which these animals evolved and the rate at which they are evolving.

SUMMARY

From the many descriptive chapters of this text, several evolutionary facts were gleaned for restatement in this chapter.

The existence of immunologically or biologically distinct strains and varieties was noted in the malaria organism, in the hemoflagellates, in filarial nematodes (both systemic and cutaneous), in *Entamoeba histolytica,* and in the schistosomes. It was shown that some of these different strains may have arisen through selection, and the implication was drawn that such differences are examples of continuing evolution. Other evolutionary facts noted were the host-parasite adjustment in *falciparum* malaria whereby the harmful genetic trait—sickling of erythrocytes—protects against the effects of malaria and is maintained at a high level in the exposed human population despite the lethality of the trait when homozygous. There was mention of other host reactions, some of them imperfect, by which levels of mutual tolerance are maintained between host and parasite. These reactions include premunition (as in malaria and perhaps amoebiasis), generalized or localized immunities (as in the worm infections), and allergic responses (as in onchocerciasis); the utility of such responses is unknown and may represent adaptations which are in an evolving state. The relative importance of the lack of specialization was outlined with respect to *Clonorchis;* alternative or reservoir hosts are valuable safeguards against a parasite's extinction. Levels of host response and the effect of diet upon pathology were illustrated by reference to the hookworms.

The generalization that parasites are usually not fatal to their host and that both parasite and host through evolutionary processes support some kind of mutual tolerance was stated both as a description of parasitism and as a derivative from more general evolutionary principles.

The origin of parasitism was freely speculated upon; the acquisition of parasites by ingestion, injection, or invasion was considered. Then certain fragments of evidence were introduced. The earliest probable date of acquisition of the sibling parasites of crayfishes (branchiobdellids in the Northern Hemisphere, temnocephalids in the Southern) was set at about 50 million years ago, and, to give some idea of the stability of particular tapeworms, the latest date for the emergence of the cestode genus *Idiogenes* of some South American and Eurasian birds was set at about the same time. Several groups of parasites (the Acanthocephala, Cestoda, and Digenea) were compared with other groups (parasitic Protozoa, Nematoda and Insecta) in order to show that the former are monophyletic and the latter are polyphyletic in origin. This comparison was illustrated by analogy, and the taxonomic groups were considered as geographic realms, continents, or islands, which had been invaded either only once by foreign animals (the Australian—monophyletic example) or repeatedly (the South American—polyphyletic example).

We discussed some of the features which affect, more or less uniquely, the evolution of parasites. In successful parasitism, population sizes of host and parasite are balanced by mutual interaction. This is similar to but simpler than the balance of nature, and each example involves only two species of organism. The question of whether this kind of balance represents a trend toward mutualism was raised but not answered. The extreme isolation of some parasite populations, together with their occasionally very small size, was mentioned as a factor which increased the importance of genetic drift in parasite evolution. Also, because of the above peculiarities of parasite population, the nature of the parasite species was questioned, and it appeared advisable to adopt a newer systematics to describe well-known groups but to cling to the old taxonomic methods for the sake of stability and convenience in most cases.

We concluded with a discussion of the relative stability of parasitism and a question of whether parasites in general are more conservative animals than their hosts or whether they are perhaps more plastic or radical in their evolution. The inconclusiveness of this discussion indicated that much needs to be known about the nature of many host-parasite relationships.

SUGGESTED READINGS

ALLISON, A. C., "Malaria in Carriers of the Sickle-Cell Trait and in Newborn Children," *Exptl. Parasit.*, 6:418–448, 1957.

BAER, J. G., *Ecology of Animal Parasites.* Univ. of Illinois Press, Urbana, Ill., 1951, pp. 155–172.

CLAY, T., "The Mallophaga of Birds," *Premier symp. sur la spec. parasit. des paras. de Vertebres. Neuchatel,* 1957, pp. 120–158.

COX, D. D., H. CIORDIA, and A. W. JONES, "Variations in *Hymenolepis serrula* Oswald, 1951 (Cestoda: Hymenolepididae), a Cestode from the Smoky Shrew, *Sorex fumeus* Miller, 1895, with Special Reference to Three Geographic Areas," *J. Tennessee Acad. Sci.*, 31:287–299, 1956.

FAUST, E. C., "The Multiple Facets of *Entamoeba histolytica* Infection," *Internat. Rev. Trop. Med.*, 1:76, 1961.

HSU, H. F., and S. Y. LI HSU, "*Schistosoma japonicum* in Formosa: a Critical Review," *Exptl. Parasitol.*, 12:459–462, 1962.

JOB, P. S., A. W. JONES, J. A. DVORAK, and R. L. KISNER, "Decreased Variability in the Cestode *Hymenolepis diminuta* following Irradiation of Successive Generations," *Evolution*, 17:163–169, 1963.

NOBLE, E. R., and G. A. NOBLE, *Parasitology, the Biology of Animal Parasites.* Lea and Febiger, Philadelphia, 1961, pp. 684–785.

ROGERS, W. P., *The Nature of Parasitism.* Academic Press, New York, 1962, pp. 242–259.

RUIZ-REYES, F., "Datos históricos sobre el origen de la oncocercosis en América," *Rev. Med.* (México), 32:49–56, 1952.

STUNKARD, H. W., "Intraspecific Variation in Parasitic Flatworms," *Syst. Zool.*, 6:7–18, 1957.

CHAPTER 30

HEALTH

INTRODUCTION

The famous writer and sage, Gertrude Stein, muttered as she lay dying "What is the answer? What is the answer?" Finally, after seeming to listen for a while, she said with startling clarity "In that case, what is the question?" Then she died.

Miss Stein was far from being a scientist, but if the story of her last words is true, she became aware, before she died, of what every scientist knows—that knowledge and progress come not so much from answers as from questions.

In this final chapter of an introductory textbook, some questions will be raised. It will become obvious that these questions, if answerable at all, have quite imperfect answers. Parasitologists are therefore in as healthy a condition as most scientists; they know enough about their science to see the directions in which they must go to achieve more knowledge and control, and they are in no danger of completing their work in the foreseeable future.

IMPORTANCE OF PARASITES

Parasites are important because they cause disease in man and his domestic animals. But this obvious fact tells nothing about the importance of parasites in

430

relation to other agents of disease and misery. The question of importance may be tentatively answered by comparing some diseases caused by parasites with other diseases in terms of numbers of victims, total human misery involved, and, where data are available, economic cost.

In the United States, the estimated number of sufferers from heart disease is 12,000,000. This figure extended to the world comes to about 180,000,000. The number is somewhat larger than recent estimated figures for the incidence of malaria and schistosomiasis. Other major diseases of similar incidence are arthritis and mental illness. The incidence of venereal disease is unknown, but this as well as yaws (tropical treponematosis), cholera, tuberculosis, and, above all, malnutrition, are important human diseases. In terms of numbers, some of the most dreaded diseases are not important; for instance, poliomyelitis, cancer, and leprosy are less than one tenth as prevalent as any of the diseases mentioned earlier. If numbers alone can be considered, undoubtedly the hosts of hookworm, pinworm, *Entamoeba histolytica,* or *Ascaris* are more numerous than the hosts of any other parasite or the victims of any other pathogen. Certain infectious diseases—epidemic typhus, cholera, plague, and kala-azar, for example—may become tremendously widespread during epidemics (see typhus in Chapter 23); but during times of retreat and quiescence, these diseases have few victims. The above examples of incidence and comparisons of several different kinds of disease in terms of numbers affected suggest that both numbers of host and severity of the disease are important. In parasitism, the severity of disease is usually, but not always, related to the number of parasites.

As was pointed out earlier, the prevalence of hookworm in southeastern India (80%) has little public health significance, because the average worm burden is very light. The pinworm, certainly the commonest parasite of civilized and urban man, is perhaps the most common of all parasites; but the damage it causes is so slight that it can scarcely be considered important. *Entamoeba histolytica* is certainly (if we include the species *E. hartmanni*) or probably (if we exclude that species) harbored by vast numbers of people, perhaps 10 to 20% of the world's inhabitants. It cannot be said, however, that such a tremendous number suffer clinical symptoms; of those who do have symptoms, many are not seriously ill. Similarly, *Ascaris,* while almost universal in many parts of the world and once believed to infect half the world's population, does not in most cases cause serious harm. Malaria, on the other hand, while infecting fewer by far than hookworms, pinworms, amoebas, or *Ascaris,* is a harmful disease in all who have it, and, as we have seen, it can be held in check by natural forces only where it is stable, that is, where continuing infection provides continuing immunity. Likewise, schistosomiasis is a very harmful condition, which is sure to take lives and cause great pain and misery to a substantial number of people in an area where the disease exists. In fact, each of the major health problems discussed in this book deserves to be ranked not far below the great organic, bacterial, treponemal, and virus diseases and should be ranked thus not only on grounds of the number of victims alone but also on the basis of severity of the disease itself.

Therefore it seems that two criteria—relative number of sufferers and relative severity of effects—can be employed in an attempt to answer the question of the importance of parasites.

A third criterion of importance, point of view or bias of the evaluators, cannot be neglected. It was suggested in Chapter 3 that the parasites of burros had received less attention than the parasites of horses, although the former are essential parts of the labor force in many countries, while the latter have the status of being luxuries and not necessities in much of the modern world. In reference to human disease, citizens of the developed countries may be much concerned with rare diseases, like poliomyelitis or muscular dystrophy, while the rest of the world suffers from ancient, common ills, with little aid from the more fortunate few. In the United States, for example, parasite-caused diseases are of minor importance, while in neighboring countries malaria may still exist, hookworm and schistosomiasis persist, and *Ascaris* and *Trichuris* may still infect millions of children and adults. The very fact that disease is apt to be worse and more prevalent among the poor and powerless than among the wealthy tends to obscure from the latter the real importance of disease. Elaborate and expensive equipment has been invented, for example, to prolong the lives of sufferers from rare diseases in the United States, with per patient costs of tens of thousands of dollars; while in other countries schistosomes, malaria, and hookworm, to say nothing of malnutrition, infectious bacterial disease, etc., injure millions of people yet could be controlled or treated at relatively small expense. The people in the underdeveloped countries, moreover, may themselves be indifferent to human misery; their long experience with disease may have reinforced a fatalism that is actually part of many cultural-religious structures throughout the world. If sympathy for others could be generated among the fortunate, and rebelliousness against the "human condition" could be stimulated among the wretched, a major obstacle to progress toward world health would be removed.

COST

Early in this text, malaria was used to introduce some important ideas in parasitology; one of these ideas is the concept of cost of disease. This concept should be discussed again to remind the reader that it is possible to evaluate disease in realistic terms, and to suggest a basis for investment of time and energy in attempting to solve health problems. Malaria, affecting some 150 million persons, threatens a vastly larger number living in lands where anopheline mosquitoes occur and where temperatures are at least seasonally optimal for malaria transmission. The malaria-caused losses due to inefficiency and absenteeism of labor, as well as the costs of caring for the victims, have been estimated to be of the order of magnitude of several billions of dollars annually. The total cost of eradicating malaria from most of its range (the regions of "stable" malaria perhaps excepted) is a fraction (perhaps one sixth) of the annual loss due to malaria. Actually, programs of eradication are far advanced, and malaria has experienced a spectacular decline during the last decade. Many nations are already in a position to realize a return on their investment in public health. Their freedom from malaria has given them both increased national product and increased potential for economic growth at a cost not equal to one year's former losses due to malaria. Yet it cannot be said that the success of one program, such as malaria eradication, proves that all public health problems can be solved. This example, like other medical and scientific achievements, emphasizes the importance of knowledge and the necessity of formulating significant questions. What are some of the questions?

NUTRITION

A basic problem in the control or suppression of many parasite-caused diseases is nutrition. Malaria, until its eradication is complete, will subject many millions to anemia caused by the destruction of erythrocytes. Resulting iron deficiencies affect the oxygen-carrying capacity of the blood. This function is essential to growth, to mental and physical development, and to the utilization of energy in performing work. In regions of stable malaria, dietary supplement will be an important, and perhaps permanent, adjunct to malaria suppression or eradication programs. Hookworm disease causes an even more severe kind of anemia, since the loss of whole blood depletes the immunity reserves of the body and creates protein deficiency in addition to hemoglobin loss. Worm burdens which would not affect well-nourished individuals can severely injure the poorly nourished; indeed, a vicious cycle is set up. Probably one of the most difficult aspects of hookworm control is the weakness and listlessness of the victims and their inability to help themselves. Dietary supplement may actually be an important prerequisite to a successful antihookworm campaign. What seems so clear for malaria and hookworm can be discerned in other parasitic diseases. Almost all of these are more severe in the poorly nourished. Thus the problem of parasitism is often accompanied by the problem of starvation. The great question of how to feed the world's expanding population which is so badly crowded in the poorest lands can merely be asked, not answered, by public health parasitologists.

IGNORANCE

A second great problem, or question, is ignorance. Everyone believes that ignorance is a chief foe to progress. In parasitology, this is certainly true, and it is true in two somewhat different respects.

First, there are many diseases allied with the ignorance of the victims themselves. If the people of Thailand, Laos, and many other eastern nations were aware that liver flukes are acquired from raw fish, perhaps their liking for raw fish might decrease, but perhaps not. Until such ignorance is dispelled, however, these people have no choice but to continue to harbor *Clonorchis* and *Opisthorchis* in their livers. In the many irrigated farmlands where schistosomes are common, the victims seldom know the source of their disease. In parts of Africa where urinary schistosomiasis is prevalent, natives bind up the urethra when working in water, because they believe that the parasites affecting this organ enter it directly. The relationship between pollution and snails is unknown to these people; no measures to control schistosomiasis can be successful so long as the people who must carry out these measures are ignorant of the reason for use of molluscicides and treatment of sewage. The complex cycles of hookworms and *Strongyloides* are unknown to most people. Amoebiasis infects millions to whom the value of frequent hand-washing is not at all apparent. Perhaps more people know about the transmission of malaria than about most diseases, because of the many attempts in the past to control malaria through mosquito control; but other insect-borne parasitisms, like leishmaniasis, African and South American trypanosomiasis, and the filariases, are not so thoroughly understood. While knowledge by the endangered population might do little to enable them to protect themselves from the above diseases, such knowledge would make the work of public health officers and physicians easier. Local resistance to mass treatment for trypanosomiasis in Africa, to crack-filling programs in Brazilian houses to discourage triatomids, or to surgery upon the tumorous scalps of Guatemalans infested with *Onchocerca* has often seriously obstructed health programs and projects, and such resistance would disappear with good educational preparation of the people. Ascariasis and trichuriasis, worm infestations almost universal in moist and crowded tropical regions, are the result of grossly unsanitary disposal of human feces. Some knowledge that such customs greatly increase the above and other enteric diseases might bring acceptance of new sanitary practices. But the hope that knowledge in lieu of ignorance will be a panacea is certainly vain. Although very many people in the United States either are aware of the facts about *Trichinella,* or can easily learn them, the U.S. still has the highest rate of trichinosis in the world. Thus the problem of ignorance—folk ignorance, that is—is really part of the great problems of education, cultural modification, and the possibility of changing "human nature."

Second, there is another kind of ignorance with which disease is allied, but it seems to yield readily to educational advance. This is not the ignorance discussed above. Instead it might be called the ignorance of experts, and it includes those deficiencies in scientific knowledge which concern the parasites and their epidemiologies, treatments, and cures. Probably no parasitism of man or animal is so well known that further research would be fruitless, but there are several diseases which cannot yet be efficiently attacked without more expert knowledge than we now possess.

Trypanosomiasis in Africa, a major disease of cattle and man, is still mysterious in many ways. The identity of the parasites—especially the "species" *brucei, gambiense,* and *rhodesiense*—is in doubt. Establishment of "standard laboratory strains," defined by growth rates on minimal media or by the responses of laboratory animals to infection with standard strains, which can be compared with new strains recovered from tsetse flies or vertebrate hosts, will eventually provide a catalogue of reference for the epidemiologist, but at present the question of species in *Trypano-*

soma must remain open. Another open question in African trypanosomiasis is the actual cause of disease. Physiology of host and parasite must be further studied; the pathological effects cannot now be rationally counteracted, since there is no convincing theory to explain them. Also, treatment (especially mass treatment of men as well as cattle) needs to be improved. The problem of drug resistance has arisen several times already when promising trypanosomicidal drugs were tested. Drug-fast strains may be expected to arise in the future, and this fact suggests the need for an increasingly varied arsenal of pharmaceuticals.

Chagas' disease is a still more difficult problem for research. Until a drug suitable for large-scale treatment is developed, this insidious chronic infection will continue to afflict the poor in what may become, pessimistically, a more suitable environment in the future for triatomids and *Schizotrypanum*. Unless population growth is checked, South America may greatly expand its "rural slums" with more mud-walled, thatch-roofed dwellings and thus permit a population explosion by the insect vectors of Chagas' disease.

Amebiasis, a condition affecting the health of hundreds of millions of people, cannot be easily diagnosed or cheaply and effectively treated. Moreover, unless reinfection can be prevented, treatment of individual cases in regions of high incidence does not have much lasting value. Thus there are three unanswered questions in amebiasis; two are in the field of epidemiology. The latter problem involves sanitary engineers and perhaps educators as well as parasitologists.

Recent rapid improvement in the scientific approaches to filariasis should not permit complete optimism here. Although filaricidal drugs, first in systemic filariasis and later in onchocerciasis, have made mass treatment and eradication theoretically possible, there are formidable obstacles in the way to success. For instance, many existing cases of systemic filariasis have resulted in elephantiasis; the cases are difficult to treat and essentially impossible to cure. The etiology of this condition is not actually known, for there are conflicting theories. One theory proposes a bacterial adjunct to *Wuchereria* or *Brugia* adults as a necessary cause of lymph gland enlargement with its sequel. A major problem in onchocerciasis is reaching the victims of the disease over difficult trails in Africa and Central America. Perhaps the desired economic improvement of these regions will make onchocerciasis accessible. Thus in filariasis, as in amebiasis, both laboratory research and epidemiological investigations are needed.

Schistosomiasis is a major health problem which has been aggravated and extended by the very processes—improved agriculture and heightened prosperity—which may assist in solving other problems. The chief obstructions to controlling schistosomiasis are technical. Mollusc control requires special knowledge of each area and watershed. Mass treatment to stop transmission is not yet practical because of the kind of treatment required—injection of schistosomacidal drugs. Moreover, mass treatment of *S. japonicum*, and perhaps of other human schistosomes, permits the survival of a reservoir of parasites in other hosts. Work on immunization is proceeding; hopefully, this solution may provide protection for exposed persons and at the same time prevent the expansion of the disease to new areas. It is no accident that a great many of the current published papers in parasitology deal with schistosomiasis.

Two relatively unimportant conditions will conclude this list of examples which show the need for scientific knowledge.

While relatively rare (not being in the same class in numbers or importance with any of the above examples), human infections with the larvae of two taeniid cestodes, *Echinococcus granulosos* and *Taenia solium,* are always medically important to the persons concerned, for the first can cause large and harmful hydatid cysts, while the second produces generalized cysticercosis with bladder worms in brain, muscles, or other organs. Treatment of hydatid cyst must be surgical. Destruction of small cysts of *Echinococcus* or of the cysticerci of *Taenia* by drugs would be very desirable, but such treatment has not been devised. Thus there exists a clearly medical problem, which can be solved only by research.

Enterobius vermicularis, the pinworm, presents an almost unique combination of problems for the parasitologist. First, the question of pathogenicity is unanswered. Since very heavy infestations of groups of people, such as workers living under crowded conditions, have been shown to be responsible for nervousness, loss of working efficiency, and insomnia, there is no question but that pinworms are harmful. But there is little evidence of damage to tissues or other specific pathology. Second, while diagnostic procedures have been refined to a state of high accuracy and efficiency and treatment is a matter of two doses of a well-tolerated drug taken one week apart, still there is not much evidence of general decline in incidence. Of course, the reason for this is obvious; neither the doctors nor the public take pinworms seriously. Therefore the pinworm problem awaits solution not by parasitological experts, who have done all that is necessary, but by experts in public relations, who must find ways to persuade a literate and prosperous public, the people of the highly developed nations, that they ought to control their most prevalent helminthic disease.

Fortunately, the above list is not a permanent one. Even at the moment that this book is published—too late for last-minute revision—perhaps more than one of the problems involving research and expert knowledge will have been solved, and part of a chapter will have been rendered obsolete. However, other questions will continue to be asked. Perhaps the most difficult one involves diseases allied with poverty.

POVERTY

Parasites which thrive where living conditions are poor include hookworms, *Ascaris* and *Trichuris, Entamoeba histolytica,* and *Schizotrypanum cruzi.* Although each of these has already been discussed, the peculiar connection with poverty needs to be pointed out.

Hookworm, historically in the United States, was allied with poverty in a region unfortunate in other ways. The American Southeast lost a long and bitter war in the 1860's. Parts of the country recovered, for cotton and tobacco were very profitable enterprises. But the poorer farms and their owners, in the hill country of the Cumberland Plateau, were already exhausted by soil-wrecking farming practices and by the war. Housing, food, clothing, and education were all inadequate. These people made no recovery from the war; they suffered from a nutritional disease, pellagra, and from a still worse disease, grinding poverty. When Stiles called attention to the prevalence of hookworm in these blighted and "underdeveloped" areas, the people themselves were powerless to do anything about the problem. But a government program sparked by private philanthropy came to their aid and helped establish among them the bases of sanitation and medical care upon which at last they could build a future. Now the American Southeast is a thriving part of a rich nation. Its hookworm problem is all but forgotten. But history repeats itself in other impoverished lands, where hookworms bleed the young and old of their ambition and even of the physical materials with which the bodies of better nourished individuals can fight and mitigate the effects of hookworm. Part of the worldwide effort to improve the human condition is wisely directed toward the control of hookworm. Efforts at economic improvement will have a special effect here.

Ascariasis has only recently been revealed as a serious condition. Like trichuriasis, it afflicts only the poor, i.e., those whose surroundings are by necessity dirty and who cannot protect themselves from neighborhood contamination of food and drink. Although wealthy tourists or landed aristocrats in the tropics run some risk of acquiring these worms, they will acquire, at most, only a few and not the hundreds which the average child of a slum will harbor. And the prosperous can, of course, be treated for what to them would be a harmless but embarrassing infestation. On the other hand, unless their environment changes, the poor must live with their parasites. *Ascaris* and *Trichuris* have all but disappeared from the "developed" countries of the world. When poverty disappears, or when at least a minimum base for human dignity is built in every land (a base of public sanitation and adequate nutrition), *Ascaris* and *Trichuris* will become mere medical curiosities.

We will consider only one other example of the relationship between poverty and parasites—Chagas' disease. Until recently it was believed that this disease could neither be relieved nor controlled under conditions prevailing in the parts of South America where *Schizotrypanum cruzi* and its hosts reside. The transmitting hosts (triatomid bugs) abound in the cracked walls of ordinary dwellings. Reservoirs of infection, the armadillos (possibly the "original" hosts of this unique trypanosome), range throughout the affected area; of course, other reservoirs also occur. Long-term control of insects by insecticides is apt to encounter insecticide resistance and to become both more expensive and less effective. Control by residual spraying of insects which live in the nests of rodents, in cracks in the walls, or within roof thatch is particularly difficult. A few years ago, most epidemiologists believed that Chagas' disease, like malnutrition, was a disease of poverty and that it would persist so long as poverty existed unless a specific remedy were discovered in the laboratory. In the late 1950's, a surprising discovery was made (but not in the laboratory). Fresh cow dung mixed with sand makes a cheap but temporarily odoriferous plaster. Sealing the cracks of mud walls with this cement closes off most of the breeding and hiding places of *Triatoma* and *Panstrongylus*, greatly reduces the population of these vectors, and gives hope that the incidence of Chagas' disease can be drastically curbed. It appears from the success of the cow-dung program in Brazil that not all diseases linked with poverty should be viewed fatalistically, nor must the control of these diseases await the end of poverty. Solutions can be sought under existing conditions, and sometimes they will be found close at hand within the resources and will of the people most concerned.

CONCLUSION

The above references to unanswered questions in human parasitology can be summarized in three words—food, knowledge, wealth. The lack of these three things makes harmful parasitism necessary. This lack necessitates other harmful elements —misery, starvation, cruelty between individuals and among nations, colonialism, revolution, and war. If mankind had food enough, if education were universal, and if comfort and leisure were everywhere the fruit of labor, these age-old enemies of humanity would be defeated. Almost incidental would be the defeat of disease, including parasitism. The parallel between the question of human misery through parasitism and the question of human misery in general is clear.

On the brighter side, the parallel between these problems shows that even very difficult problems can be solved. Hookworm in the United States was largely eliminated through education, public health efforts, and a rise in general prosperity. This was accomplished under a mixed political and economic system, where good will was the common ingredient. Malaria is being eradicated throughout the world by the efforts of a similar combination of individuals and agencies with private and public funds, but now on an international scale, under the leadership of the United Nations. *Ascaris* and *Trichuris*, already disappearing from some countries because of increased prosperity and the improvement in sanitation that accompanies industrialization and educational advance, can be expected to dwindle elsewhere if humanity moves forward to a rational destiny. It is encouraging to think that the aim of a better world is shared by many and that the efforts of statesmen who work for their country's welfare and for the world's peace, the efforts of educators who strive to remove the darkness from men's minds, the efforts of businessmen and government economic planners who are designing a world without want, the efforts of scientists and manufacturers to make such a world technically possible, and the efforts of uncounted individuals who help each other in uncounted ways are matched by the efforts of medical workers, including parasitologists, who are attacking directly some of the obstacles to health. In the larger context, all these efforts must interact for the unlimited enrichment of mankind.

INDEX

45515

Jones

Introduction to parasitology